T4-AIP-687

Rational Drug Design

ACS SYMPOSIUM SERIES **719**

Rational Drug Design

Novel Methodology and Practical Applications

Abby L. Parrill, EDITOR
University of Memphis

M. Rami Reddy, EDITOR
Metabasis Therapeutics, Inc.

American Chemical Society

Library of Congress Cataloging-in-Publication Data

Rational drug design : novel methodology and practical applications /
Abby L. Parrill, M. Rami Reddy, editors.

p. cm. -- (ACS symposium series ; 719)

"Developed from a symposium sponsored by the Division of Computers in Chemistry at the 214th National Meeting of the American Chemical Society, Las Vegas, Nevada, September 7-12, 1997."

Includes bibliographical references and index.

ISBN 0-8412-3603-8

1. Drugs--Design--Congresses. 2. Drugs--Structure-activity relationships--Congresses. 3. Combinatorial chemistry--Congresses.

I. Parrill, Abby L. II. Rami Reddy, M., 1934-. III. American Chemical Society. Division of Computers in Chemistry. IV. American Chemical Society. Meeting (214th : 1997 : Las Vegas, Nev.) V. Series.

RS420.R37 1999 98-32087
615'.19--dc21 CIP

The paper used in this publication meets the minimum reqirements of the American National Standard for Information Sciences—Permanence of Paper for Printed Library Materials, ANSI Z39.48–1984.

Distributed by Oxford University Press

PRINTED IN THE UNITED STATES OF AMERICA

Advisory Board

ACS Symposium Series

Foreword

THE ACS SYMPOSIUM SERIES was first published in 1974 to provide a mechanism for publishing symposia quickly in book form. The purpose of the series is to publish timely, comprehensive books developed from ACS sponsored symposia based on current scientific research. Occasionally, books are developed from symposia sponsored by other organizations when the topic is of keen interest to the chemistry audience.

Before agreeing to publish a book, the proposed table of contents is reviewed for appropriate and comprehensive coverage and for interest to the audience. Some papers may be excluded in order to better focus the book; others may be added to provide comprehensiveness. When appropriate, overview or introductory chapters are added. Drafts of chapters are peer-reviewed prior to final acceptance or rejection, and manuscripts are prepared in camera-ready format.

As a rule, only original research papers and original review papers are included in the volumes. Verbatim reproductions of previously published papers are not accepted.

ACS BOOKS DEPARTMENT

Contents

Preface

Now that you've picked up this book, don't set it down! We sincerely believe you will find value in this publication. This book is based on a symposium held at the 214th American Chemical Society National Meeting and Exposition during the Fall of 1997 in Las Vegas. The symposium was co-organized by an academic chemist with strong interests in novel methodology for rational drug design (Parrill) and an industrial chemist with strong interests in applications of rational drug design (Reddy). The mix of papers in this book therefore reflects a broad spectrum of current topics in rational drug design. The authors also come from diverse work environments where research in rational drug design takes place, including the pharmaceutical industry, software development indus-try, and academia. The broad range of topics, perspectives, and approaches makes this book appealing to an equally distributed group of readers.

Rational drug design is a multidisciplinary field. Specific sections in the book are likely to be useful to a variety of chemists not specifically involved in drug design. In particular, the papers in the second section should have appeal to many computational chemists, offering new methods for modeling solvent effects and dynamic processes in chemistry.

Although the value of this book to computational chemists and researchers in-volved in rational drug or materials design is evident from a quick glance at the Table of Contents, there are several features of value to those wanting an introduction to the field. The first chapter offers an overview of rational drug design and provides explanation for many of the topics that occur in later chapters. Chapters 2 through 8 then discuss more detailed aspects of energetics and solvation in drug design. Chapters 9 through 12 describe applications of drug design methods. Chapters 13 through 16 detail new methods for QSAR, combinatorial chemistry, and docking. In Chapter 17, David Clark provides a review of evolutionary algorithms in drug design that provides an excellent introduction to an optimization method that is rapidly finding application in almost all areas of rational drug design as well as in more general areas of chemistry. The following four chapters, 18 through 21 cover specific uses of evolutionary algorithms in drug design. The final chapter, "Is Rational Design Good for Anything?," by Donald Boyd, offers a perspective on successes and failures of rational drug design.

If you are interested in drug design, molecular interactions, free energy computation, solvation, optimization problems, combinatorial chemistry, or QSAR, don't put this book down! Whether you are looking only for an introduction or in-depth coverage you should find much to interest you in this volume.

Finally, we express our gratitude for generous support from the Computers in Chemistry Division of American Chemical Society, the Petroleum Research Fund, Metabasis Therapeutics, Inc., Molecular Simulations, Inc., Merck, Amgen, Inc., and Lederle.

Abby Parrill
Department of Chemistry
The University of Memphis
Memphis, TN 38152

M. Rami Reddy
Metabasis Therapeutics, Inc.
9390 Towne Centre Drive
San Diego, CA 92121

Chapter 1

Overview of Rational Drug Design

M. Rami Reddy[1] and Abby L. Parrill[2]

[1] Metabasis Therapeutics, 9360 Towne Centre Drive, San Diego, CA 92121
[2] Department of Chemistry, University of Memphis, Memphis, TN 38152

Traditional Drug Design.

Drug discovery programs in the pharmaceutical industry prior to the 1960's were based entirely on screening thousands of natural and synthetic compounds for activity. Once a potential drug compound was selected by this process, medicinal chemists would then synthesize hundreds of related compounds to develop the safest, most effective drug for patients use. However, the costs and risks associated with this process have become enormous; the cost of completing the research and development process for a single new drug has more than doubled in the last decade. Various sources estimate this cost to to be anywhere from $200-$500 million. Each year researchers test hundreds of thousands of chemical compounds, yet in the United States only about 25 new drugs are introduced. Even worldwide the introduction of new drugs only reaches 40-45 per year. Many of these new drugs are only “me-too compounds", as the various companies attempt to apply their patented “molecular manipulations” to other companies' top selling drugs.

A major limitation of the drug screening strategy is that it does not reveal why a compound is active or inactive, or how it might be improved. It also provides no assurances that an active compound is specific for a given human target protein. The lack of such specificity can be a major source of undesirable side effects which can halt the clinical development of a drug. Drug screening is essentially a blind process, indicating the reason for the need to test approximately 20,000 compounds in order to find one that becomes a marketable drug.

Drug screening is often followed by structural optimization of lead compounds in order to improve potency and other properties, but deciding when to move from screening to synthesis is a problem.[1] Although screening has produced the vast majority of existing drugs, it has not proven to be a wholly satisfactory strategy. There are many important therapeutic needs for which screening has failed.

Notable work began to appear in 1962 which led to drastic changes in the process used to optimize chemical structures for medicinal purposes.[2,3] This work established the foundation for the multi-parameter QSAR methods in common use today. Subsequent publications outlined how the Hansch approach could be applied to drug design.[4-6]

Challenges of Drug Design

Researchers have worked for a long time to overcome the limitations of screening by designing molecules to perform specific therapeutic tasks. This vision of rational drug design was made more plausible by improvements in our understanding of the similarities between the actions of different biologically active compounds. Almost all drug molecules achieve a biological response through interaction with a target or receptor biomolecule. Descriptions of our earliest understanding of this interaction compared the drug molecule entry into a crevice of the target protein to a key in a lock, thus inhibiting the protein's normal biological function. Current descriptions liken drug interactions with biomolecules to a handshake, where both the ligand[7] and the protein adjust somewhat to accommodate the other.

The general drug - target scheme suggests that structure-based rational drug design can be accomplished by three basic tasks. First, the appropriate protein target for a given therapeutic need must be identified. Second, the distinguishing structure of the target protein must be determined. Finally, the structure of a drug must be designed to interact with the target protein. However, a number of technical barriers have hindered work in the area of structure-based drug design. First, many important human diseases are not sufficiently well-understood at the molecular level to permit scientists to identify an appropriate drug target. Second, even when an appropriate target has been known, its molecular structure has generally not been known in adequate detail for drug design. Finally, the design of structures complementary to the target requires consideration of both the three-dimensional as well as the functional aspects of chemical structures.

In cases where an appropriate biological target cannot be identified or characterized, rational drug design requires a different strategy. This alternate strategy makes use of structural information about drugs that produce the same biological response at different doses. It is often reasonable to assume that such drugs interact with the same, albeit unknown, biological target. They must, therefore, have some common set of structural features that are required in order to evoke the aforementioned biological response. This common set of structural features is the pharmacophore. This assumed similarity of drugs with similar effect suggests an alternative set of tasks that can accomplish rational drug design. First, the structural features important for biological activity must be determined. It is important that these features provide three-dimensional information either implicitly or explicitly. Second, optimal combinations for these features must be determined. Finally, the structure of a drug must be designed which exhibits the optimal combination of these features. Drug design efforts that seek to accomplish this alternative set of tasks are classified as pharmacophore based approaches.[8] Drug design efforts using pharmacophore based approaches have their own set of challenges. First, determining important structural features when a variety of chemical structures demonstrate the same biological activity requires an understanding of the structural correspondance. An additional complication arises due to the fact that some drugs which elicit the same biological activity display multiple modes of interaction with the target. Challenges for the structure design portion of the pharmacophore based approach are the same as the challenges during the corresponding activity when used in structure-based drug design approaches.

Rational Drug Design

Since the early 1980's, advances in molecular biology, protein crystallography, and computational chemistry have greatly aided Rational Drug Design (RDD) paradigms and the accuracy of their binding affinity predictions.[9-11] Figure 1 shows a flowchart that describes the different approaches that may be employed by drug discovery groups during RDD or ligand design. Further discussion of RDD will be organized into four main areas. Two of these areas, pharmacophore based approaches and structure-based approaches depend on whether the three-dimensional structure of the biological target is available. The other two areas, new lead generation and structure evaluation, will be performed regardless of whether the biological target structure is known.

Pharmacophore-Based Approaches. The path at the first decision point is determined by the availability of the 3-dimensional structure of the enzyme or complex. If the structure of the biological target is unknown, various methods that utilize active (and inactive) analogs can be used to develop a working model of the requirements for biological activity, in other words, the pharmacophore. There are several evolving quantitative methods that utilize active compounds such as 2D-QSAR,[12-15] 3D-QSAR[16] and neural networks.[17] Comparative Molecular Field Analysis (CoMFA) is a very widely used 3D-QSAR technique.[16] CoMFA represents a significant achievement due to its ability to develop a three-dimensional quantitative model that relates steric and electrostatic fields to biological activity. An initial problem with the method was the need to select both conformations and alignments of the molecules to be modeled. Due to this problem, many initial uses of the CoMFA method involved molecules with rigid ring systems. For example, Allen *et. al.* predicted the binding affinities of six analogs of beta-carbolines for the benzodiazepine receptor (BzR) prior to synthesis[18] using a previously published CoMFA model.[19] The standard error of prediction for these six analogs is significantly lower than the standard error estimate of the cross-validation runs on the training set, hence the predictions made using this model are much better than expected. Even now, nine years after the first description of the CoMFA method, papers are appearing in the literature that offer new solutions to the alignment[20] and conformer selection[21] problems. In addition to such three-dimensional models, pharmacophore hypotheses may also be developed by more qualitative methods.[22] Using any of these methods one could propose new analogs of a lead compound based on the pharmacophore model.

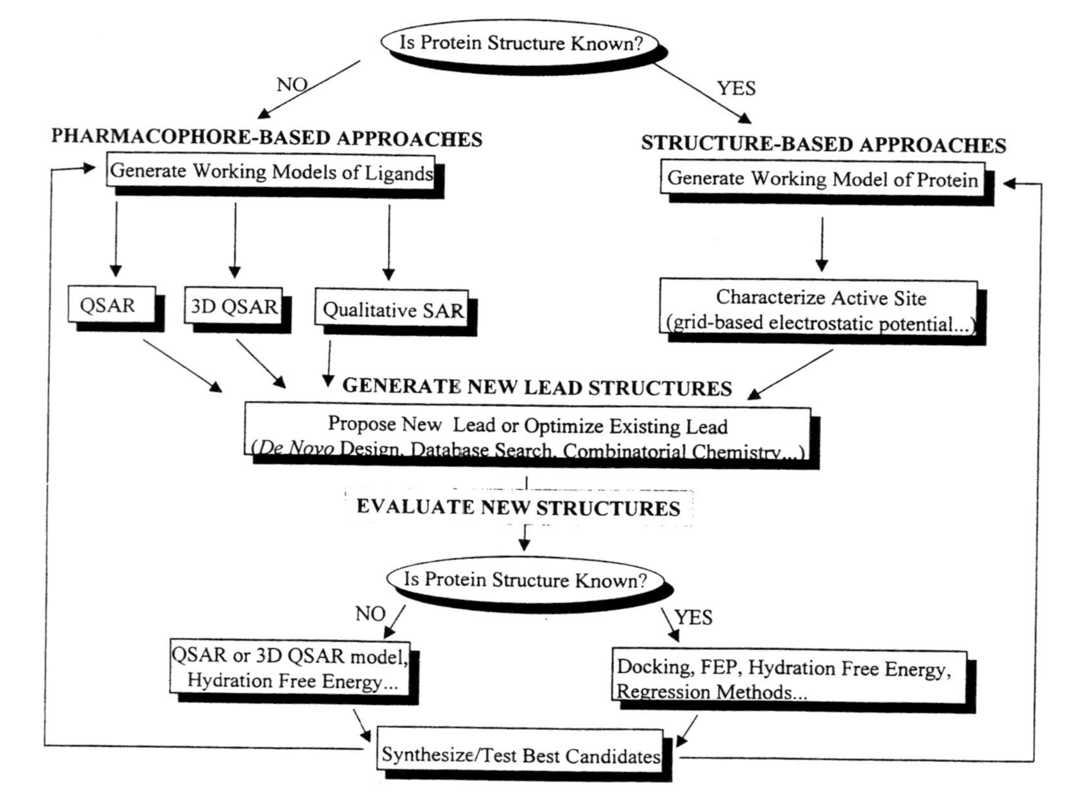

Figure 1. An overview of the many types of methods that provide an understanding of drug action and their integration in the process of rational drug design.

Structure-Based Approaches. The other branch at the first decision point is used when the 3-dimensional structure of the enzyme or complex is known. The process typically begins by generating a working computational model from crystallographic data, but methods to develop models of the binding site from active ligands are becoming more prevalent.[23-26] Development of the working model may include developing molecular mechanics force field parameters for non-standard residues consistent with the force field for standard residues, modeling any missing segments, assigning the protonation states of histidines, and orienting carbonyl and amide groups of asparagine and glutamine residues based upon neighboring donor and acceptor groups. Characterization of the active site is then aided by a variety of visualization tools. For example, hydrophobic and hydrophilic regions of the active site are readily identified by calculating the electrostatic potential at different surface grid points, and hydrogen bond donor and acceptor groups can be highlighted in the active site. The information gained by the characterization of the active-site is very important for proposing new lead compounds or analogs of a known leads.

New Lead Generation. Generation of new lead compounds can be accomplished using *de novo* design methods to design new structures[27,28] or by searching databases[22,29-35] of known chemicals for particular structural features. *De novo* molecular design methods may design structures by sequentially adding or joining molecular fragments to a growing structure,[36-38] by adding functionality to an appropriately-sized molecular scaffold, or by evolving complete structures[39-41]. Some *de novo* design methods have concentrated on the design of diverse molecular scaffolds,[42] or on the development of diverse substituents to place on a single scaffold.[35] Database search methods have been developed that search based on separation of molecular functionality by a particular number of bonds or distance ranges. More chemically intuitive database search methods seek for chemicals with particular steric and electrostatic fields.[33]

A growing number of drug leads are being generated by combinatorial methods in combination with high-throughput screening. Computational chemistry is currently being used to assist efforts in this area by ensuring that the library of structures generated for use in high-throughput screening assays incorporates a great deal of molecular diversity.[34,35,43-48] This ensures that a diverse set of lead compounds can be found and optimized at much lower cost than if the entire library of possible structures were synthesized and tested. The diverse set of leads that can be found by combinatorial chemistry can give important insight into the requirements for biological activity. This is particularly valuable for relatively new drug targets for which insufficient information is available for application of the structure-based or pharmacophore based approaches.

Structure Evaluation. With new drug leads proposed, rapid and accurate prediction of *in vivo* activities are needed in order to evaluate and thereby prioritize these structures prior to chemical synthesis. In reality, evaluation methodologies are limited to *in vitro* measurables such as binding affinity although an *in vivo* property such as clinical effect would be ideal. The challenge within this limitation is to develop evaluation methods that rapidly and accurately predict absolute binding affinities of the aforementioned large, diverse set of potential ligands. Currently available evaluation methods can either provide qualitative rank ordering of a large number of molecules in a relatively short period of time[49] or generate quantitatively accurate predictions of relative binding affinities for structurally related molecules using substantial computing power.[50,51] Consequently, biological activity evaluation techniques that increase speed without greatly compromising accuracy (or vice versa) are of value to drug discovery programs. Methods of ligand evaluation include graphical visualization of the ligand in the binding site,[52] substitution of parameters from the new ligand into SAR models, and calculation of relative binding affinities.[53,54] Usually about 50% of proposed new leads or optimized analogs can be eliminated by evaluating their expected binding affinities based on docking, visualization, conformational analysis and desolvation costs. The remaining analogs will be ranked for synthesis using one or all of the following methods, depending on computational power, time and resources, namely; 1) Free Energy Perturbation (FEP) calculations, which give very accurate quantitative predictions, but are computationally very expensive,[50,51] 2) molecular mechanics calculations, which will give only qualitative predictions, but these calculations are very fast,[10,49] 3) regression methods[55] that incorporate interaction variables (intra and intermolecular interaction energies, hydrophobic interactions) and ligand properties (desolvation, log P etc.), which will give semi-quantitative predictions, and are much faster than FEP calculations, and 4) relative hydration free energies.[56] Calculation of relative hydration free energies is important in the design and optimization of molecules that act as enzyme inhibitors only after undergoing covalent hydration. For example, a class of adenosine deaminase and cytidine deaminase inhibitors are known by X-ray crystallography to bind in the hydrated form.[57,58] Calculation of both the relative hydration free energy and the relative binding free energy for the hydrated species provides an accuarte method for calculating relative inhibitor potencies since it accounts for differences in both hydration equilibrium and binding. Then, top scoring compounds are synthesized and tested for activity. Depending on the convergence criteria of the biological activity, the flow chart is repeated.

Free Energy Perturbation Methods. Since FEP methods provide very accurate quantitative predictions, we discuss its use in the comparison of similar ligands binding to an enzyme. This task is of particular value during the lead optimization phase of drug design. We considered two examples where FEP calculations were used successfully to predict binding affinities of ligands to enzymes prior to synthesis. The first example considered was one of the earliest successes of FEP calculations and involves transition state ligands bound to thermolysin, carried out by Merz and Kollman.[51] In this work these authors predicted that the replacement of an -NH group with a methylene group would not be detrimental to binding affinity despite a loss in a hydrogen bond between the NH and an amide carbonyl. The principal reason was related to ligand desolvation. This prediction, which was made ahead of biochemical measurements, was later confirmed experimentally. In the second example, several research groups[59-62] used the FEP method for calculating relative binding affinities for HIV-1 protease inhibitors and obtained good agreement with experimental results. More recently, Reddy *et. al.*[50] used a computer-assisted drug design method that combines molecular mechanics, dynamics, FEP calculations, inhibitor design, synthesis, and biochemical testing of peptidomimetic inhibitors and crystallographic structure determination of the protein-inhibitor complexes to successfully design novel inhibitors of HIV-1 protease. This study involved a large set of molecules whose relative binding affinities were predicted using FEP methods prior to synthesis, and were later confirmed by experimental measurements.

Fast Methods for Qualitative Binding Prediction of Binding Affinities of Ligands. Though the FEP method is theoretically more accurate and provides quantitative predictions between two similar ligands, it suffers from some practical limitations as applied to ligand design. Therefore, faster methods that can accommodate structural diversity are being developed. In some cases predictions of ligand binding has been based on soley on a visual analysis of structures without any force field calculations.[52] These methods relied on graphical analysis of features such as steric and electronic complementary of the docked inhibitor to the target protein, the extent of buried hydrophobic surface and the number of rotatable bonds in the ligand. Quantitative descriptors based on molecular shape[63] and grid-based energetics[64] have also proved to be useful. More advanced methods have used an empirical scoring function[65] derived from crystal structure data and experimental binding affinities. Though molecular mechanics methods appear to be more useful in this regard, these methods met with only limited success initially,[66] due to the large approximations involved in the analysis (e.g., binding conformations, solvent model used, lack of entropic terms etc.). Recently, Montgomery *et. al.* adopted some improvements to the molecular mechanics methods by using Monte Carlo techniques to derive the binding conformations of inhibitors followed by energy minimizations.[10,67] This method allowed the prediction of binding affinities for proposed purine nucleoside phosphorylase inhibitors prior to synthesis. The calculated results suggested that differences in solvation and entropy would contribute minimally to binding affinity. Although the binding conformations were accurately predicted in this study, analysis of interaction energies across the inhibitors was less informative, presumably because of unaccounted factors such as desolvation and entropy.

Future Directions

Rational drug design methods are continually improving, and a wider variety of drug targets are being approached by these methods. A wide variety of additional improvements can be anticipated in the future as well. Improved computer hardware will allow the use of more rigorous methods to be applied to large molecular systems. It will not be surprising to see fully quantum mechanical docking studies appearing in the future. A second trend in computational methods that should continue in the future is the development of both hybrid methods (currently known examples include genetic neural networks,[15] k-nearest neighbor genetic algorithms,[68] among others) and integrated tools for drug design. Advances in the modeling of protein structures will promote more widespread use of structure-based drug design for drug targets that do not crystallize.

New experimental methods will also lead rational drug design in new directions. Combinatorial chemistry and high-throughput screening would not be as highly useful as they are today without solid-phase synthesis methods. Improvements in areas such as catalyst design to allow rapid access to an ever-increasing range of chemical structures, biological activity assays to allow the use of a wider variety of biological targets, and experimental structure determination methods to provide a wider selection of structural information for structure-based approaches will have significant impact on how rational drug design is performed in the future.

For lead optimization, the quantitative FEP methods provide an accurate prediction of relative binding affinities between inhibitors only for structurally similar molecules, whereas the qualitative methods provide qualitative trends for relative binding affinities across a more structurally diverse set of compounds. Ideally, methods that combine both of those features will greatly enhance the utility of computational methods to drug design. Increased structural diversity, however, requires accurate calculation of additional factors that significantly impact the compounds binding affinity. For example, the larger the difference in structure, the greater the chance that solvation, entropy, inter and intramolecular interaction energies of ligand both in solvent and in the complex, hydrophobic effects, conformational flexibility etc., will influence relative binding affinities. Understanding the magnitude of each contribution is key to an accurate prediction. Since an equation that incorporates each factor accurately has not been derived, we cannot expect accurate predictions using any of above mentioned methods for the diverse set of molecules. Therefore, regression equations which incorporates many of the properties discussed above would greatly strengthen the rational drug desion methods for fast screening (prior to synthesis) of diverse set of inhibitors to an enzyme semi-quantitatively.

In conclusion, rational drug design is an exciting and constantly growing field of research. Its impact on quality of life and health ensure the vitality of the field.

References

1. Young, S. S.; Sheffield, C. F.; Farmen, M. *J. Chem. Inf. Comput. Sci.* **1997**, *37*, 892-899.
2. Hansch, C.; Maloney, P. P.; Fujita, T.; Muir, R. M. *Nature* **1962**, *194*, 178.
3. Hansch, C. *Acc. Chem. Res.* **1969**, *2*, 232-239.
4. Topliss, J. G. *J. Med. Chem.* **1972**, *15*, 1006-1011.
5. Topliss, J. G. *J. Med. Chem.* **1977**, *20*, 463-469.
6. Topliss, J. G. *Perspect. Drug Disc. Design* **1993**, *1*, 253-268.
7. Nicklaus, M. C.; Wang, S.; Driscoll, J. S.; Milne, G. W. A. *Bioorg. Med. Chem.* **1995**, *3*, 411-428.

8. Marshall, G. R.; Barry, C. D.; Bosshard, H. E.; Dammkoehler, R. A.; Dunn, D. A. *The Conformational Parameter in Drug Design: The Active Analog Approach*; Marshall, G. R.; Barry, C. D.; Bosshard, H. E.; Dammkoehler, R. A.; Dunn, D. A., Ed.; ACS: Washington, D.C., 1979; Vol. 112, pp 205-226.
9. Appelt, K.; Bacquet, R. J.; Bartlett, C. A.; Booth, C. L.; Freer, S. T.; Fuhry, M. A. M.; Gehring, M. R.; Hermann, S. M.; Howland, E. F.; Janson, C. A.; Jones, T. R.; Kan, C.; Kathardeker, V.; Lewis, K. K.; Marzoni, G. P.; Matthews, D. A.; Mohr, D. A.; Morse, C. A.; Oatley, S. J.; Ogden, R. O.; Reddy, M. R.; Reich, S. H.; Schoettlin, W. S.; Webber, S. E.; Welsch, K. M.; White, J. *J. Med. Chem.* **1991**, *34*, 1925-1934.
10. Montgomery, J. A.; Niwas, S.; Rose, J. D.; Secrist, J. A.; Babu, S. Y.; Bugg, C. E.; Erion, M. E.; Guida, W. C.; Ealick, S. E. *J. Med. Chem.* **1993**, *36*, 55-69.
11. Whittle, P. J.; Blundell, T. L. *Annu. Rev. Biophys. Biomol. Struct.* **1994**, *23*, 349-375.
12. King, R. D.; Hirst, J.; Sternberg, M. J. E. *Perspect. Drug Disc. Design* **1993**, *1*, 279-290.
13. Luke, B. T. *Journal of Chemical Information and Computer Sciences* **1994**, *34*, 1279-1287.
14. Rogers, D.; Hopfinger, A. J. *Journal of Chemical Information and Computer Sciences* **1994**, *34*, 854-866.
15. So, S.-S.; Karplus, M. *J. Med. Chem.* **1996**, *39*, 1521-1530.
16. Cramer, R. D. I.; Patterson, D. E.; Bunce, J. D. *J. Am. Chem. Soc.* **1988**, *110*, 5959-5967.
17. Baskin, I. I. *J. Chem. Inf. Comput. Sci.* **1997**, *37*, 715-721.
18. Allen, M. S.; LaLoggia, A. J.; Dorn, L. J.; Martin, M. J.; Costantino, G.; Hagen, T. J.; Koehler, K. F.; Skolnick, P.; Cook, J. M. *J. Med. Chem.* **1993**, *35*, 4001-4010.
19. Allen, M. S.; Tan, Y.-T.; Trudell, M. L.; Narayanan, K.; Schindler, L. R.; Martin, M. J.; Schultz, C.; Hagen, T. J.; Koehler, K. F.; Codding, P. W.; Skolnick, P. *J. Med. Chem.* **1990**, *33*, 2343-2357.
20. Robinson, D. D.; Barlow, T. W.; Richards, G. W. *J. Chem. Inf. Comput. Sci.* **1997**, *37*, 943-950.
21. Jiang, H.; Chen, K.; Tang, Y.; Chen, J.; Li, Q.; Wang, Q.; Ji, R. *J. Med. Chem.* **1997**, *40*, 3085-3090.
22. Van Drie, J. H. *J. Comput.-Aid. Mol. Design* **1997**, *11*, 39-52.
23. Hahn, M. *J. Med. Chem.* **1995**, *38*, 2080-2090.
24. Crippen, G. M. *J. Comput. Chem.* **1995**, *16*, 486-500.
25. Walters, D. E.; Hinds, R. M. *J. Med. Chem.* **1994**, *37*, 2527-2536.
26. Crippen, G. M. *J. Med. Chem.* **1997**, *40*, 3161-3172.
27. Murcko, M. A. *Recent Advances in Ligand Design Methods*; Murcko, M. A., Ed.; VCH: New York, 1997; Vol. 11, pp 1-66.
28. Clark, D. E.; Murray, C. W.; Li, J. *Current Issues in De Novo Molecular Design*; Clark, D. E.; Murray, C. W.; Li, J., Ed.; VCH: New York, 1997; Vol. 11, pp 67-125.
29. Ewing, T. J. A.; Kuntz, I. D. *J. Comput. Chem.* **1997**, *18*, 1175-1189.
30. Hahn, M. *J. Chem. Inf. Comput. Sci.* **1997**, *37*, 80-86.

31. Neamati, N.; Hong, H.; Mazumder, A.; Wang, S.; Sunder, S.; Nicklaus, M. C.; Milne, G. W. A.; Proksa, B.; Pommier, Y. *J. Med. Chem.* **1997**, *40*, 942-951.
32. Sadowski, J. *J. Comput.-Aid. Mol. Design* **1997**, *11*, 53-60.
33. Thorner, D. A.; Willett, P.; Wright, P. M.; Taylor, R. *J. Comput.-Aid. Mol. Design* **1997**, *11*, 163-174.
34. Higgs, R. E.; Bemis, K. G.; Watson, I. A.; Wikel, J. H. *J. Chem. Inf. Comput. Sci.* **1997**, *37*, 861-870.
35. Murray, C. W.; Clark, D. E.; Auton, T. R.; Firth, M. A.; Li, J.; Sykes, R. A.; Waszkowycz, B.; Westhead, D. R.; Young, S. C. *J. Comput.-Aid. Mol. Design* **1997**, *11*, 193-207.
36. Böhm, H.-J. *J. Comput.-Aided Mol. Design* **1992**, *6*, 61-78.
37. Böhm, H.-J. *Perspect. Drug Disc. Des.* **1995**, *3*, 21-33.
38. Luo, Z.; Wang, R.; Lai, L. *J. Chem. Inf. Comput. Sci.* **1996**, *36*, 1187-1194.
39. Gehlhaar, D. K.; Moerder, K. E.; Zichi, D.; Sherman, C. J.; Ogden, R. C.; Freer, S. T. *J. Med. Chem.* **1995**, *38*, 466-472.
40. Westhead, D. R.; Clark, D. E.; Frenkel, D.; Li, J.; Murray, C. W.; Robson, B.; Waszkowycz, B. *J. Comput.-Aided Mol. Design* **1995**, *9*, 139-148.
41. Glen, R. C.; Payne, A. W. R. *J. Comput.-Aided Mol. Design* **1995**, *9*, 181-202.
42. Todorov, N. P.; Dean, P. M. *J. Comput.-Aid. Mol. Design* **1997**, *11*, 175-192.
43. Agrafiotis, D. *J. Chem. Inf. Comput. Sci.* **1997**, *37*, 841-851.
44. Agrafiotis, D. *J. Chem. Inf. Comput. Sci.* **1997**, *37*, 576-580.
45. Brown, R. D.; Martin, Y. C. *J. Med. Chem.* **1997**, *40*, 2304-2313.
46. Gillet, V. J.; Willett, P.; Bradshaw, J. *J. Chem. Inf. Comput. Sci.* **1997**, *37*, 731-740.
47. Lewis, R. A.; Mason, J. S.; McLay, I. M. *J. Chem. Inf. Comput. Sci.* **1997**, *37*, 599-614.
48. Warr, W. A. *J. Chem. Inf. Comput. Sci.* **1997**, *37*, 134-140.
49. Holloway, K.; Wai, J. M.; Halgren, T. A.; Fitzgerald, P. M.; Vacca, J. P.; Dorsey, B. D.; Levin, R. B.; Thompson, W. J.; Chen, J. L.; deSolms, J. S.; Gaffin, N.; Ghosh, A. K.; Guiliani, E. A.; Graham, S. L.; Guare, J. P.; Hungate, R. W.; Lyle, T. A.; Sanders, W. M.; Tucker, T. J.; Wiggins, M.; Wiscount, C. M.; Woltersdorf, O. W.; Young, S. D.; Darke, P. L.; Zugay, J. A. *J. Med. Chem.* **1995**, *38*, 305-317.
50. Reddy, M. R.; Varney, M. D.; Kalish, V.; Viswanadhan, V. N.; Appelt, K. *J. Med. Chem.* **1994**, *114*, 10117-10122.
51. Merz, K. M.; Kollman, P. A. *J. Am. Chem.Soc.* **1989**, *111*, 5649-5658.
52. Bohacek, R. S.; McMartin, C. *J. Med. Chem.* **1992**, *35*, 1671-1684.
53. McCammon, J. A. *Current Opinion in Structural Biology* **1991**, *1*, 196-200.
54. Beveridge, D. L.; DiCapua, F. M. *Annu. Rev. Biophys. Chem.* **1989**, *18*, 431-492.
55. Head, R. D.; Smythe, M. L.; Oprea, T. I.; Waller, C. L.; Green, S. M.; Marshall, G. R. *J. Am. Chem. Soc.* **1996**, *118*, 3959-3969.
56. Erion, M. D.; Reddy, M. R. *J. Am. Chem. Soc.* **1997 (accepted)**.
57. Wilson, D. K.; Rudolph, F. B.; Quiocho, F. A. *Science* **1991**, *252*, 1278-1284.
58. Sharff, A. J.; Wilson, D. K.; Chang, Z.; Quiocho, F. A. *J. Mol. Biol.* **1992**, *226*, 917-921.

59. Reddy, M. R.; Viswanadhan, V. N.; Weinstein, J. N. *Proc. Natl. Acad. Sci. USA* **1991**, *88*, 10297-10291.
60. Tropshaw, A. J.; Hermans, J. *Prot. Eng.* **1992**, *5*, 29-33.
61. Ferguson, D. M.; Radmer, R. J.; Kollman, P. A. *J. Med. Chem.* **1991**, *34*, 2654-2659.
62. Rao, B. G.; Tilton, R. F.; Singh, U. C. *J. Am. Chem. Soc.* **1992**, *114*, 4447-4452.
63. Kuntz, I. D.; Meng, E. C.; Shoichet, B. K. *Acc. Chem. Res.* **1994**, *27*, 117-123.
64. Goodford, P. A. *J. Med. Chem.* **1985**, *28*, 849-857.
65. Böhm, H.-J. *J. Comput. Aid. Mol. Des.* **1994**, *8*, 243-256.
66. Sansom, C. E.; Wu, J.; Weber, I. T. *Protein Eng.* **1992**, *5*, 659-667.
67. Erion, M. D.; Montgomery, J. A.; Niwas, S.; Rose, J. D.; Ananthan, S.; Allen, M.; Secrist, J. A.; Babu, S. Y.; Bugg, C. E.; Guida, W. C.; Ealick, S. E. *J. Med. Chem.* **1993**, *36*, 3771-3783.
68. Raymer, M. L.; Sanschagrin, P. C.; Punch, W. F.; Venkataraman, S.; Goodman, E. D.; Kuhn, L. A. *J. Mol. Biol.* **1997**, *265*, 445-464.

Chapter 2

Conformational and Energetic Aspects of Receptor–Ligand Recognition

J. D. Hirst, B. Dominy, Z. Guo, M. Vieth, and C. L. Brooks

Department of Molecular Biology, TPC-6, The Scripps Research Institute, 10550 North Torrey Pines Road, La Jolla, CA 92037

In this chapter, we describe ongoing work in our laboratory on the development of methodologies to enhance the process of inhibitor discovery and optimization. We review developments of energy (scoring) functions for flexible ligand, all atom docking and a scoring function to assess the efficiency of docking protocols. Using approximate free energy methods, we explore the similarity between bound and unbound conformations of ligands. When an appropriate anchor point descriptor is utilized, the lowest conformations in solution correspond well with the ligand-bound conformation in most cases. We also use approximate free energy methods to explore binding specificity of a ligand, LP149, binding to a model for resistance-evolved HIV protease, by comparing the characteristics of binding of this compound to proteases from HIV 1 and FIV. Finally, we provide an outline and applications of the λ-dynamics methodology to a rigorous free-energy based screening calculation of multiple ligands in a common receptor.

1. Overview and Introduction

This chapter presents an overview of ongoing work in our group on the development of techniques to study the broad spectrum of questions that arise in research in the area of drug discovery and optimization. The studies described in the section below comprise the presentations of the authors in the symposia titled "New Methods in Computational Chemistry" and "Rational Drug Design" at the *214th National Meeting of the American Chemical Society* Meeting that took place in Las Vegas, Nevada during September 7 and September 11, 1997.

The theoretical study of ligand-receptor interactions, as practiced in the discovery and design of new therapeutics, requires the fusion of methodologies and techniques from many different areas of theoretical and computational chemistry. The process of binding, the identification of binding sites and the assessment of ligand "viability" involve the full description of the ligand and protein (receptor) in solution, the movement of the possibly flexible ligand into the receptor binding site and the energetic components of ligand-receptor interactions (*1*). The search for a comprehensive solution to the drug design problem leads to several questions that may be addressed and answered using today's tools of structural and energetic modeling. Key questions that drive much work in this area include: What is the relationship between receptor bound conformations of the ligand and those it populates in solution prior to binding? Does flexibility of the ligand or receptor influence the efficiency of ligand binding? What determines optimal binding affinity in a series of ligands binding to the same receptor? What determines binding affinity to similar receptors, as occurs in cases where drug resistance develops due to evolutionary pressure in a pathogen? Can molecular origins of drug resistance be identified and quantified?

Accompanying these fundamental questions are the technical issues that are embodied in the development of theoretical methods for their treatment. Issues like: How can one assess the efficacy of an approximate energy function used for describing ligand-receptor binding? Can one find an optimal scoring function for such an assessment? Are there preferred methods for docking ligands to known receptors? Can approximate free energy "estimators" effectively be used to explore or screen ligands binding to their receptors? Aspects of many of these questions are addressed in the sections of the chapter that follow.

In the initial section, Section 2 below, we consider the question of ligand-receptor docking. Our primary objective is the development of robust techniques that employ a complete molecular mechanics description of the energy for the docking of ligands to different receptors. Critical in this treatment is consideration of the flexibility of the ligand, the efficiency with which one can search for and find optimal sites of binding in ligand-receptor complexes, and the establishment of an appropriate scoring function to distinguish correctly docked and incorrectly docked structures from each other. The study that we outline below considers the binding of five different ligands to their respective receptors. The ligands have different degrees of flexibility and are representative of the range of ligands often encountered in drug discovery efforts. The receptors too, though all proteins, are believed to be representative. We note in this section that one may consider the problem of docking as composed of at least two major components. The first is whether the energy (scoring) function has the ability to correctly discriminate between docked and mis-docked ligands. This is clearly a necessary condition for the successful *de novo* prediction of new ligand receptor pairs, yet it seems to be a requirement often ignored, or at least glossed over, in many current approaches to ligand docking. We describe studies that explore 144 possible energy functions and devise a general scoring function for ranking these energy functions for their ability to discriminate between correctly docked and mis-docked conformations. Also critical are those features of the "docking landscape" that influence the efficiency of the search strategy in finding optimal docking solutions. We develop and describe a docking efficiency assessor by considering a range of energy functions and their influence on docking efficiency. We conclude the first section with a summary of the simple molecular mechanics energy functions found to provide optimal docking efficiency and selectivity.

In Section 3 of this chapter, we explore the question of the influence of the receptor on the conformation of the bound ligand. We first note that the concept of a single static ligand conformation in the receptor pocket does not adequately account for the dynamic character of the ligand-receptor interaction. This consideration also leads us to the idea of "anchor points" within a ligand, atoms or groups that show the least motion and have the largest interactions with the receptor in the complex. We then go on to show that when appropriate descriptors are chosen to represent the ligand, i.e., the anchor points, that its lowest free energy solution conformation is in good agreement with that found in the ligand-receptor complex. In this study we examine ten different ligands in solution and their associated receptors. Our findings presented in this section suggest that in the development of pharmacaphore maps from families of ligands in the absence of information about their receptor structure it is essential to use a "blurred" description of atomic details and focus instead on functional anchor points.

Drug resistance is a growing problem in the treatment of many diseases that affect us in modern times. Thus an understanding of the atomic basis of resistance is a first step toward the development of compounds that are resilient to resistance. One particular area of resistance that is of significant social concern is that associated with the HIV I virus. In this system many of the compounds that have been used to reduce virus levels in infected individuals are of limited use when used as the sole antiviral agent. Thus the development of methods and the study of drug resistant compounds is a critical element in fighting diseases such as AIDS. In Section 4 of this chapter we describe a modest first step toward our search for the atomic origins of resistance. We present a study that considers the binding of a single common ligand to both the HIV I protease and the analogous protease from the feline version of the virus, FIV protease. The FIV protease differs from HIV in a number of ways that are common to resistance-evolved proteases of HIV I, and thus serves as a model system for our examination of differential ligand-receptor interactions. Our efforts in this section focus on (i) the exploration of efficient and accurate ways to develop ensembles of ligand-protease complexes to permit a more statistically based study of ligand-receptor affinity and (ii) the development of a "free energy approximator" built up from computationally undemanding protocols, e.g., assessment of protein and ligand energies with conventional molecular mechanics potentials and inclusion of solvent effects with continuum-based Poisson-Boltzmann (PB) techniques. In this study we find that the major determinant of the difference in binding affinity between a peptide mimetic inhibitor of HIV/FIV binding to these two proteases is the ligand's internal strain. We suggest that differences in protein-ligand interactions can be "redistributed" throughout the protein-ligand interface by small adjustments of protein-protein interactions, whereas different degrees of "crowding" of the ligand will be manifest in the energy of the ligand. We also examine the role of individual residues, that differ between proteases from the two species, in differentiating the ligand and note one Asp to Ile difference that appears not to be fully compensated for.

The free energy difference between a ligand and receptor in their unbound conformations and in their bound conformations is the ultimate "discriminator" in relative binding affinity assays. Thus, the screening of ligands against a common receptor should be based as closely as possible on this free energy difference. In the final section of the chapter we describe our developments of the new λ-dynamics approach to free energy calculations and ligand-receptor screening. We provide an overview of the formalism and show how it may be exploited as a free energy based technique for the screening of ligands for a common receptor in a time that is faster than conventional free energy calculations. We demonstrate the λ-dynamics approach by considering a series of para-substituted benzamidine derivatives bound to the protein trypsin. The application of this methodology to the simultaneous calculation of free energy changes of ligands binding to a common receptor illustrates how λ-dynamics provides an efficient means of both free energy-based screening and more conventional free energy calculations.

2. Optimizing an energy function for ligand-receptor docking

We begin by discussing one of the more familiar computational approaches to rational drug design, that of ligand-receptor docking. The interactions of ligands with their biological receptors determine the process of molecular recognition. In computer assisted drug design, molecular recognition is modeled using docking algorithms, which predict the structure of the ligand-receptor complex from the structures of free ligand and free receptor (*2*). There are many approaches to the docking problem (*2-10*). In rigid docking (*10*), both receptor and ligand are treated as rigid bodies, following the lock and key mechanism of binding (*11*). In the majority of flexible docking approaches, ligands only are allowed conformational flexibility (*9*). Of course, in reality both protein and ligand may be flexible, as embodied in the induced fit model of binding (*12*).

A successful docking algorithm requires a good scoring/energy function and a good search/optimization algorithm. Both are essential components. A good scoring function should be selective, i.e. able to distinguish the correct solution (the crystal structure of the ligand-receptor complex) from all others. It should also be efficient. By this we mean that the potential energy surface should not have too many large barriers and thus should be amenable to rapid searching, so that a given search algorithm can quickly find the correct solution. In this section we present a systematic way to design a molecular mechanics scoring function for a flexible docking algorithm. We allow ligand flexibility and treat the receptor as a rigid body. The idea behind this choice is to illustrate the scoring function design process for the simpler case where the sampling problems are almost nonexistent. Simulated annealing molecular dynamics (MD) is used as the search strategy.

2.1 Optimization. To optimize the selectivity and efficiency of a scoring function we use a test system comprising five protein-ligand complexes.

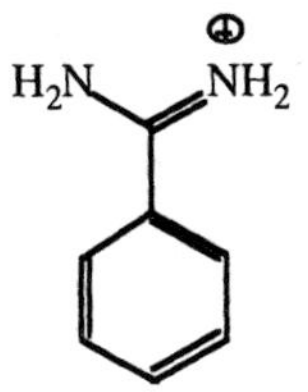

Benzamidine

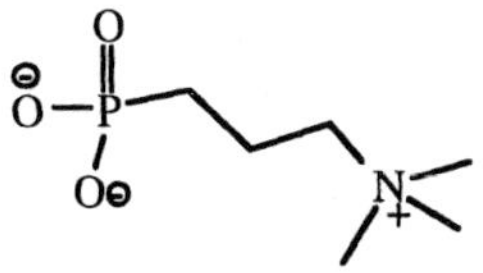

Phosphocholine

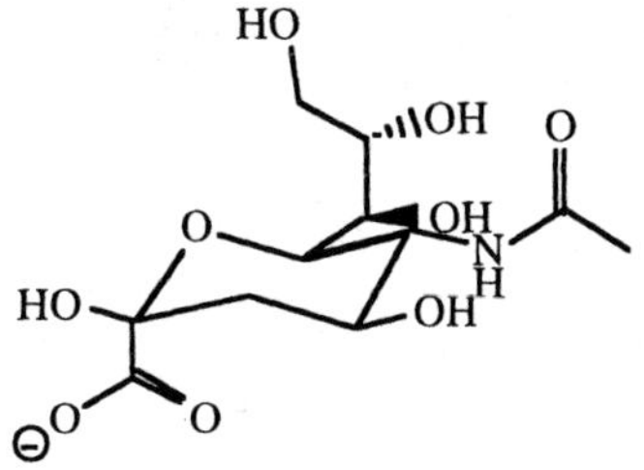

Sialic acid

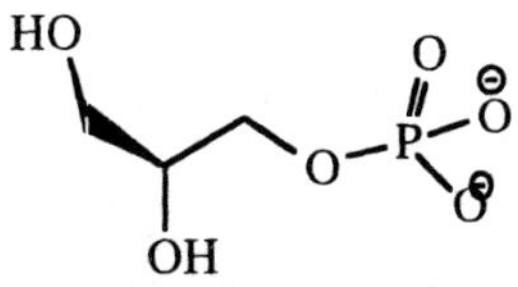

Glycerol 3-phosphate

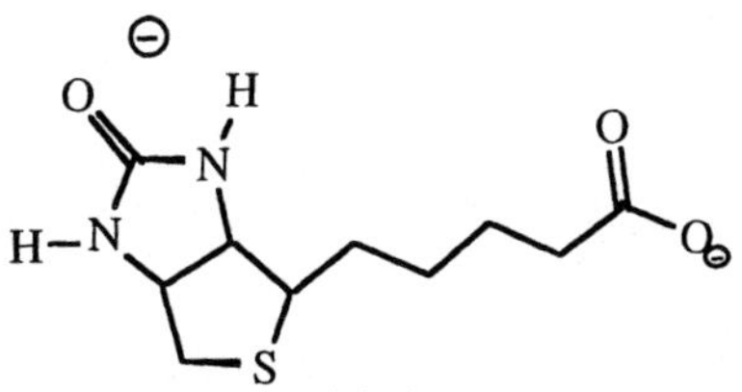

Biotin

The complexes include a small rigid ligand in an open active site, benzamidine/trypsin (PDB code 3ptb) (*13*), flexible ligands in open active sites, phosphocholine/FAB McPC-603 (PDB code 2mcp) (*14*) and sialic acid/hemagglutinin (PDB code 4hmg) (*15*) and flexible ligands in relatively inaccessible active sites, glycerol 3-phosphate/triose phosphate isomerase (PDB code 6tim) (*16*) and biotin/streptavidin (PDB code 1stp) (*17*). We believe that these ligand-receptor complexes are sufficiently diverse to allow us to draw general conclusions about the selectivity and efficiency of an energy function.

For protein receptors, the parameter set was based on CHARMM param19/toph19 topology and parameter libraries (*18*). For the ligands, the charges were generated by the template method in Quanta with smoothing of charges over all atoms. For the nonbonded interactions, a switching function for the van der Waals (vdW) and electrostatic interactions (*18*) was used. In order to restrain the ligand to the neighborhood of the active site, a harmonic restraining potential was applied. It acted only when the center of geometry of the ligand was further than 11Å from the center of the active site. The restraining potential was implemented through the NOE module in CHARMM.

Our scoring function for docking comprises the CHARMM receptor-ligand interaction energy and the ligand self energy. In our optimization we modify the nonbonded interaction parameters. In particular, we examine different nonbonded truncation parameters (with three values of 8Å, 9Å and 10Å), different models for electrostatics (distance dependent dielectric, constant dielectric and Poisson-Boltzmann (PB) solvation based on the electrostatic field generated by the protein), different dielectric constants (ε = 1, 2, 3, 4), the reduction of surface side chain charges and changing the hard core vdW potential to a soft core.

The reduction of the charges of the surface side chains is implemented as follows:

$$C_{new} = \left(1 - \frac{S_{protein}}{S_{tripeptide}}\right) C \tag{1}$$

where C_{new} is the reduced charge of the side chain atom, $S_{protein}$ is the surface exposure of side chain X in the protein, $S_{tripeptide}$ is the surface exposure of this side chain in a GLY-X-GLY trans tripeptide and C is the original charge of the side chain atom. The modification of the vdW interaction involved the reduction of vdW interactions at r_{ij} = 0.8σ for the initial annealing stage and was implemented through:

$$E_{VDW} = \begin{cases} \sum_{i \neq j} \left(\frac{12A_{ij}}{(f\sigma_{ij})^{13}} - \frac{6B_{ij}}{(f\sigma_{ij})^{7}} \right) (f\sigma_{ij} - r_{ij}) + E_{VDW}(f\sigma_{ij}) \; ; \; r_{ij} \leq f\sigma_{ij} \\ E_{VDW}(r_{ij}) = \sum_{i \neq j} \left(\frac{A_{ij}}{r_{ij}^{12}} - \frac{B_{ij}}{r_{ij}^{6}} \right) sw(r_{ij}^2, r_{on}^2, r_{off}^2) \quad ; \; r_{ij} > f\sigma_{ij} \end{cases} \tag{2}$$

where E_{VDW} is the vdW energy, A_{ij} and B_{ij} are non-bonded parameter, *sw* is a switching function, σ_{ij} is the mean of the vdW radii of atoms i and j and r_{ij} is the distance between atoms i and j. This equation is discussed in greater detail in Brooks et al. (*18*). The essence of this modification is that for distances shorter than $f\sigma_{ij}$ the force acting on atoms i and j is the same as at $f\sigma_{ij}$. In other words, the vdW interaction from $f\sigma_{ij}$ is a linear function of r_{ij}. f is the fraction of σ_{ij} that the soft core potential starts at (0.8 in most applications). The modification makes the vdW repulsion very small (the maximum value is on the order of hundreds of kcal/mol) and the resulting conformational transition barriers are much smaller than with the original vdW potential. A similar modification is made to the electrostatic potential and forces. All combinations of the modifications were examined, giving a total of 3*4*2*2*3 = 144 tested potentials.

2.2 Optimizing selectivity. The first stage of designing a good scoring function for docking is the optimization of selectivity, that is, the ability to distinguish the docked structures from mis-docked ones. Docked structures are defined as structures consistent with the crystallographic conformation, within a root mean square deviation (RMSD) of 2Å. Mis-docked structures have a RMSD greater than 4Å from the Brookhaven Protein Data Bank (PDB) (*19*) structure of the complex. The selectivity of the scoring function is evaluated based on the separation of the energy distributions for docked and mis-docked structures for five protein-ligand complexes. The scoring function is analogous to those used in protein folding parameter design (*20*), sequence design studies (*21, 22*) and inverse folding approaches (*23-25*), in that energies are converted to Z scores (*26*) to account for the nature of the energy distribution:

$$Z(E) = \frac{\left(E - \overline{E}\right)}{\sigma} \tag{3}$$

where $\overline{E}$ is the average energy of the distribution and σ is the standard deviation. Selectivity is measured using the function,

$$SG_j = \begin{cases} \frac{1}{N}\sum_{i=1}^{N}\left(Z_{D,m}^{ij} - Z_{M,m}^{ij}\right)f_{ij}(Z_{D,m}^{ij} < Z_{M,m}^{ij}) & ; \prod_{i=1}^{N} f_{ij} \neq 0 \\ 0 & ; \prod_{i=1}^{N} f_{ij} = 0 \end{cases} \tag{4}$$

where j refers to the parameter set (j = 1, 144), i refers to the complex (i = 1, 5) $Z_{D,m}^{ij}$ and $Z_{M,m}^{ij}$ are the minimum Z scores for the docked and mis-docked conformations respectively. f_{ij} is the fraction of the docked structures whose Z scores are lower than those of the mis-docked structures. N is the number of complexes - in our case study N = 5. The score is zero for a parameter set if, for any complex, the lowest Z score of mis-docked structures is less than the lowest Z score of the docked structures. A lower score corresponds to a better selectivity.

2.3 Optimizing efficiency. The efficiency of a scoring function is defined as the mean number of docked structures per unit of time and is averaged over five protein-ligand complexes,

$$SE_j = \begin{cases} \frac{1}{time}\sum_{i=1}^{N}\left(f_{ij,<2} + 0.5(f_{ij,<3} - f_{ij,<2})\right) & ; \prod_{i=1}^{N} \mathrm{f}_{ij,<3} \neq 0 \\ 0 & ; \prod_{i=1}^{N} \mathrm{f}_{ij,<3} = 0 \end{cases} \tag{5}$$

where $f_{ij<a}$ indicates the fraction of structures for parameter set j with RMSD less than a Å and *time* is the total computer time used for docking of all five ligands to their respective receptors. All timings and fractions of docked structures utilized the same annealing schedules with 108,000 energy evaluations. The score is zero for a given parameter set if for any complex $f_{ij,<3}$ is zero.

2.4 Selectivity. The selectivity of the scoring functions was computed based on a library of docked and mis-docked structures generated by MD simulated annealing with ten randomly chosen scoring functions. For each receptor-ligand complex, there were 400 orientations of the ligands, with roughly a 50:50 mix of docked and mis-docked structures. For each ligand position, the score was computed as the ligand-receptor interaction energy plus the ligand self-energy. Table I presents the 22 potentials out of the 144 possibilities that had non-zero selectivity. The most selective potentials have a distance dependent dielectric, short (8Å) nonbonded truncation cutoffs and hard core vdW interactions. The advantage of the regular vdW interactions over soft core vdW arises from the fact that tight packing in the active site is favored over other binding sites much more with a hard core potential than with a soft core potential. In false binding sites there is often looser packing, which a soft core potential tends to favor, as it imposes fewer constraints on shape complementarity.

Table I. Ranking of parameter sets, based on selectivity

Rank	Efficiency *SG*	Energy gap[a]	Solvation model[b], ε	Reduce surface side chain charges ?	vdW soft core ?	Non-bonded cutoff (Å)
1	-0.78	-10.2	rdie, 3	YES	NO	8
2	-0.77	-9.6	rdie, 4	YES	NO	8
3	-0.72	-9.4	rdie, 4	NO	NO	8
4	-0.70	-11.4	rdie, 2	YES	NO	8
5	-0.67	-9.8	rdie, 3	NO	NO	8
6	-0.45	-10.0	rdie, 2	NO	NO	8
7	-0.36	-5.9	rdie, 3	NO	YES	8
8	-0.35	-5.4	rdie, 4	NO	YES	8
9	-0.35	-5.9	rdie, 3	YES	YES	8
10	-0.33	-5.1	rdie, 4	YES	YES	8
11	-0.31	-6.4	rdie, 2	YES	YES	8
12	-0.26	-5.9	rdie, 2	NO	YES	8
13	-0.23	-4.7	rdie, 2	NO	NO	10
14	-0.22	-3.5	rdie, 4	YES	YES	10
15	-0.22	-4.5	rdie, 2	YES	NO	10
16	-0.21	-3.6	rdie, 4	NO	NO	9
17	-0.21	-3.5	rdie, 4	YES	NO	9
18	-0.17	-3.4	solv, 4	YES	NO	10
19	-0.16	-4.0	solv, 4	YES	YES	8
20	-0.13	-4.8	rdie, 1	YES	NO	10
21	-0.09	-5.6	rdie, 1	YES	YES	8
22	-0.04	-3.5	rdie, 1	NO	NO	10

[a]Energy gap between the correctly docked structure of lowest energy and the minimum energy mis-docked structure. [b]Solvation model: rdie - distance dependent dielectric constant, cdie - gas phase, solv - continuum solvation contribution with constant dielectric; ε is the value of dielectric constant (the distant dependent dielectric is given by ε*r*).

Short nonbonded truncation favors the specific close interactions in the binding pocket over the more delocalized interactions in other nonspecific binding sites. Most highly selective potentials benefit from reducing the side chain surface charges, because most false binding sites are located on the receptor surface. Thus, reducing surface interactions favors active site structures. In addition, a distance dependent dielectric is almost always present in the highly selective potentials. Neither a constant dielectric nor an approximate continuum solvation model are desirable for the discrimination of docked conformations from mis-docked ones. The poorer selectivity of the set with approximate solvation may be understood in terms of the energy gap between the docked and mis-docked structures, which is reduced due to the smaller solvation penalty at the surface sites. On the other hand, a constant dielectric without any account of solvation effects exaggerates the electrostatic contribution and diminishes the influence of tight packing in the active site.

2.5 Efficiency. To test the efficiency of the docking potentials, we selected nine parameter sets from Table I (ranks 1-2, 7-11, 14 and 19). For each parameters set, 20 ligand replicas were subjected to simulated annealing MD, utilizing the multiple copy simultaneous search method (*27*). Initial conditions were the same for each parameter set, as were the cooling schedules, which involved an annealing from 700K to 300K in 100 picoseconds (ps) followed by a second annealing from 500K to 300K in 100 ps and subsequent quenching from 300K to 50K in 16 ps. Table II shows the ranking of the tested nine parameter sets based on efficiency.

Table II. Comparison of docking for different parameter sets

SE Rank	*SE* score	$f_{<3,a}$ [a]	*SG* score	f_G [b]	Solvation model	Reduced side chain charges ?	vdW soft core ?	Non-bonded cutoff (Å)
1	0.039	0.55	-0.35	0.45	rdie, 3	YES	YES	8
2	0.038	0.53	-0.33	0.44	rdie, 4	YES	YES	8
3	0.038	0.55	-0.35	0.45	rdie, 4	NO	YES	8
4	0.038	0.53	-0.36	0.46	rdie, 3	NO	YES	8
5	0.036	0.52	-0.31	0.44	rdie, 2	YES	YES	8
6	0.025	0.36	0.0	0.25	solv, 4	YES	YES	8
7	0.022	0.56	-0.23	0.39	rdie, 4	YES	YES	10
8	0.00	0.32	-0.77	0.76	rdie, 4	YES	NO	8
9	0.00	0.22	-0.78	0.77	rdie, 3	YES	NO	8

[a]Average fraction of structures with RMSD from crystallographic complex less than 3 Å on all heavy atoms. [b]Fraction of docked conformations with lower energies than the best mis-docked conformations.

It is apparent that a soft core vdW potential is necessary for the efficient docking for all five complexes. This is because two receptors, streptavidin and triose phosphate isomerase, have rather inaccessible active sites. For the other complexes, where the active sites are more accessible, the efficiency does not strongly depend on the form of the vdW potential. We find that the most selective potentials are not the most efficient. The most efficient potential seems to be a distance dependent dielectric with $\varepsilon = 3$, reduced surface charges, soft core vdW and an 8Å nonbonded truncation. Most potentials with soft core vdW have similar docking efficiencies, however shorter nonbonded cutoffs lead to lower computational times and thus better efficiency. The most efficient docking potential is reasonably selective, with a fair separation of docked from mis-docked structures. This is shown in Figure 1, where the energy histograms for the most selective potential and the most efficient potential are shown. We select the best docking potential as the most efficient one, as it has an acceptable separation of energy distributions.

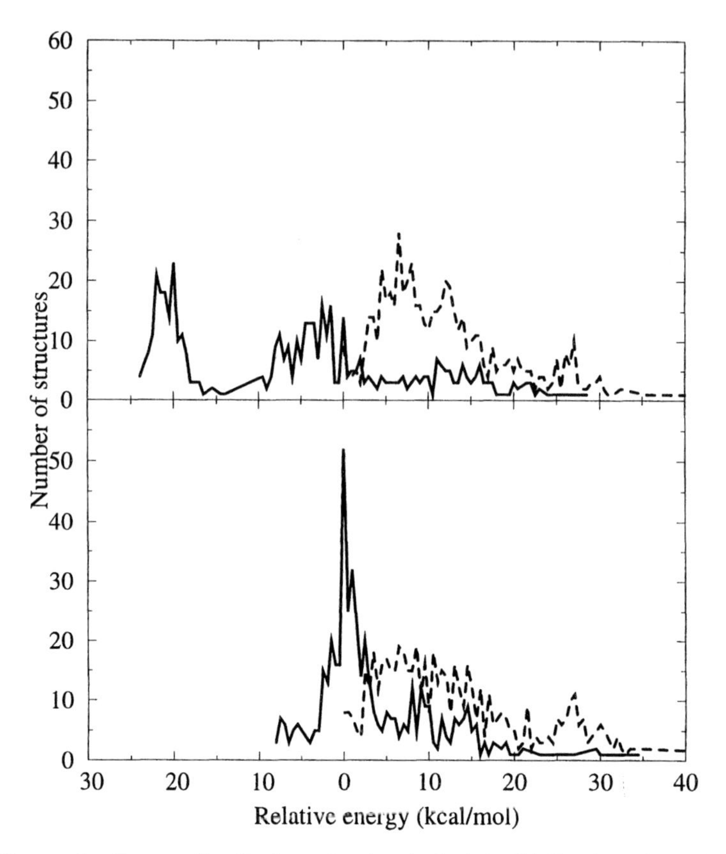

Figure 1. Energy distributions for the docked (solid lines) and mis-docked (dashed lines) conformations averaged over five complexes. Energies are relative to the minimum energy of the mis-docked structures. Upper panel: the most selective parameter set (Rank 1, Table I). Lower panel: the most efficient parameter set (Rank 1, Table II).

2.6 Optimal energy functions for docking. We have examined a number of potentials that discriminate correctly docked structures from mis-docked structures. The most discriminating potentials have short nonbonded cutoffs, a distance dependent dielectric (with $\varepsilon = 3$ or 4), reduced surface charges and regular (hard core) vdW potential. Potentials with soft core vdW are less discriminating. We found that soft core repulsion was critical for the kinetic accessibility of the binding site. In receptors with relatively small entrances to the active site, the use of the regular vdW potential precludes ligands from entering in a reasonable amount of time. For receptors with closed active sites, such as HIV protease, soft core potentials will be essential for successful efficient docking. In general, the most selective scoring functions may not be the best scoring functions for docking. Thus, we see a separation of kinetic and thermodynamic effects. In contrast to a commonly held belief, we show that MD can be successfully used when paired with a smooth energy surface that has a clear global minimum. While our approach remains to be extended to allow receptor flexibility, we have demonstrated the general guidelines and ideas that can be used to develop a scoring function for any docking problem.

3. Ligand conformation in the active site and in free solution

In this section we step back and re-visit one of the fundamental assumptions underlying much of rational drug design by addressing the question: How similar is the conformation of an unbound ligand to its bound conformation? The construction of pharmacophore models (a pharmacophore is a spatial arrangement of atoms that may give rise to a biological activity) is a key process in computer assisted drug design (*28-30*). In pharmacophore mapping techniques (*2, 31-33*) the assumptions are often made that (i) there exists a commonality between isolated structures of ligands (2) and (ii) their low energy solution structures are similar to the conformations they adopt in the receptor binding site. The importance of this similarity extends beyond the scope of computer assisted drug design to questions related to the physics of binding. For example, how does a receptor deform the solution structure of a ligand? Is the lock and key mechanism (*11*) a good approximation? There have been a number of attempts to relate the structures of isolated ligands to their receptor bound conformations (*34, 35*). It has been concluded from these studies that "... any local minima identified for the isolated state are of little or no relevance for the situation in a protein" (*34*).

Here, we present another view of similarity between the conformations of small ligands free in solution and in receptor complexes. We investigate a principal assumption underlying pharmacophore mapping, namely that conformations of the isolated ligands are relevant to binding, but we do not address the topic of pharmacophore identification without knowledge of the receptor. The concept of similarity should take into account the flexibility exhibited by some ligands in active sites, which may be quantified by simulation. With this in mind, we examine two metrics of similarity. First, we compare the low energy solution structures of a ligand to the family of conformations observed in the active site, with similarity identified by values of torsion angles. Secondly, we examine similarity defined in terms of the spatial orientation of anchor points (key atoms responsible for binding) in the low free energy solution structures and in the active site conformations.

3.1 Assessing similarity metrics. We tested the two similarity metrics on ten protein-ligand complexes, the four flexible complexes used in the previous section (i.e. not the benzamidine-trypsin complex) and an additional six complexes: tricarballylic acid (bound to aconitase), Ile-Val (bound to pancreatic trypsin inhibitor), Gly-Tyr (bound to carboxypeptidase A), Val-Trp (bound to thermolysin), chloramphenicol (bound to chloramphenicol acetyltransferase), and N-phosphoryl-L-leucinamid (bound to thermolysin). The ligands vary from dipeptides and peptide-like molecules to small flexible organic molecules. The number of important rotatable bonds in these ligands varies from two to five, where rotation about an "important" rotatable bond is defined as involving the rotation of at least one other torsion. The anchor points (shown as gray circles) of a ligand were identified as those atoms with the largest contribution to the interaction energy with the receptor, based on simulation results. Each ligand was divided into functional groups. Within each functional group, anchor points are the heavy atom (or atoms) with higher than average interaction energies per atom. Anchor points are also characterized by low mobility. Some functional groups, in which none of the atoms interacted strongly with the receptor, have no anchor points assigned.

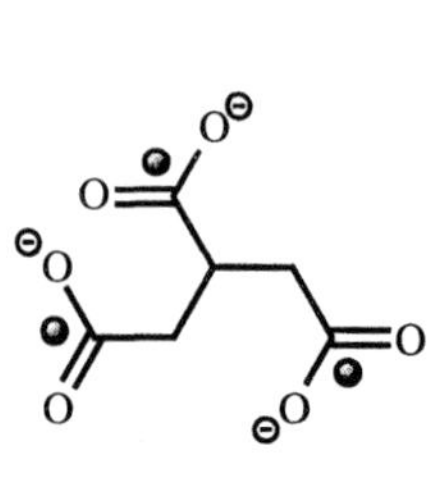

Tricarballylic acid 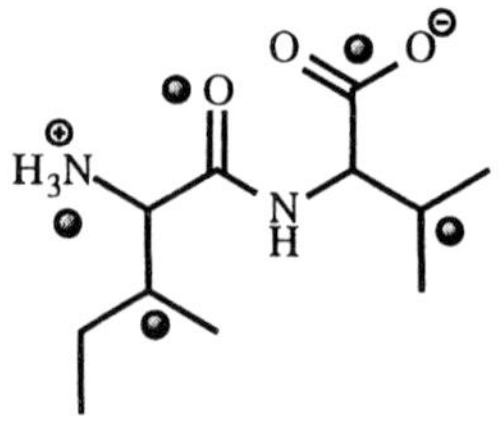 Ile-Val 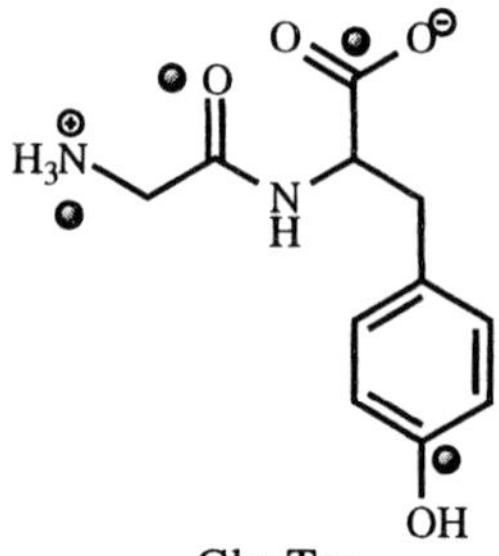 Gly-Tyr

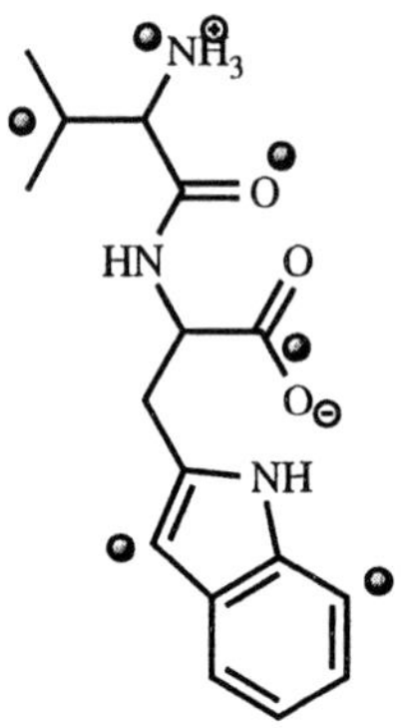

Val-Trp 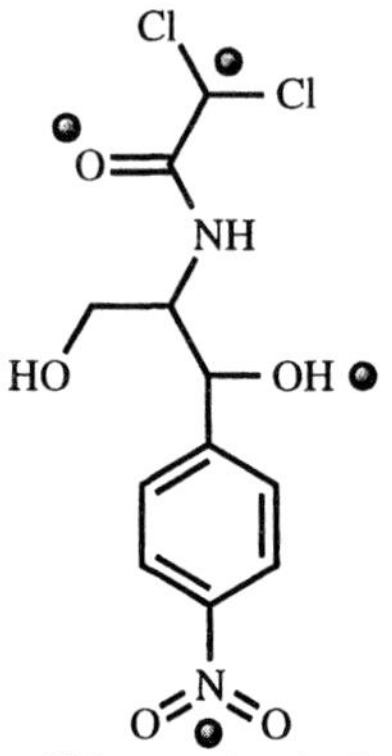 Chloramphenicol 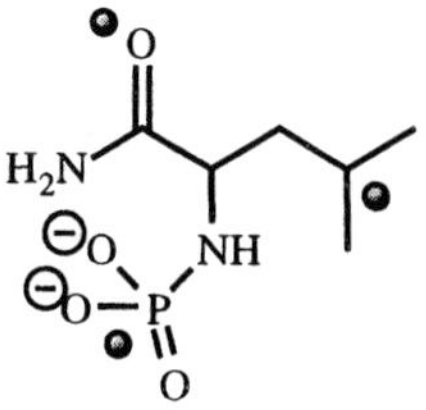 N-phosphoryl-L-leucinamid

For all peptidic ligands we use the polar hydrogen CHARMM param19 parameter set (*18*). The force field parameters for nonpeptidic ligands are obtained by the template method in Quanta. The active site families of ligands were generated by one hundred iterations of room temperature MD simulation of 75 ps with a rigid receptor. A distance dependent dielectric with $\varepsilon = 2$ was used. The resulting 100 structures were minimized by 1000 steps conjugent gradient and subsequently clustered by a hierarchical agglomerative method (*36*) based on the torsion angle metric and the anchor point distance metric. Representative structures from solution were generated by a systematic search (*33*), sampling important rotatable bonds in 15° increments. The resulting structures were minimized using 1000 steps of conjugent gradient minimization and then clustered based on the two metrics. Clustering reduces the number of structures by roughly 500 fold. The largest number of individual structures was 316,954 for biotin, which was reduced to 270 representatives. The free energy was calculated based on the population of a cluster (the entropic contribution) and the solvation and internal energy of that structure that was closest to the geometric cluster center,

$$A = E_{solv,PB} + E_{\text{int}} - k_B T \ln N \qquad (6)$$

where A is the free energy of a cluster, $E_{solv,PB}$ is the solvation free energy of a cluster center calculated using the finite difference PB equation (*37*), E_{int} is the internal free energy of a ligand, $k_B T = 0.6$ kcal/mol and N is the population of a cluster. For continuum solvation calculations a grid size of 0.4Å was used with an interior dielectric of 2.

The next step is to rank the clusters based on their free energies. For each ligand, cluster representatives from solution are compared to cluster representatives from the active site. For all solution cluster representatives within the clustering radius (15° for the torsion angle metric, 0.5Å for the anchor point distance metric) from an active site representative the one of lowest free energy is taken as the solution structure. As cluster representatives are selected based on energies calculated with a distance dependent dielectric and not on the PB solvation energies (*38*), all structures with free energies within 3 kcal/mol of the minimum free energy cluster are considered to be of equal free energy.

3.2 Comparison of similarity metrics. Table III presents results of a comparison of the lowest free energy structures in solution similar to (i.e., belonging to the same cluster as) the active site conformations as judged by the torsion metric.

Table III. Results for the torsion angle metric

Ligand	ΔA[a]	Population[b]
chloramphenicol	0.0	0.451
N-phosphoryl-L-leucinamid	3.0	0.005
phosphocholine	0.3	0.238
tricarballylic acid	5.9	0.000
biotin	4.5	0.000
glycerol 3-phosphate	2.8	0.009
Ile-Val	4.4	0.001
Gly-Tyr	7.7	0.000
Val-Trp	9.0	0.000
sialic acid	5.7	0.000

[a]Free energy difference (in kcal/mol) between the lowest free energy solution cluster and the lowest free energy solution cluster most similar to any of the active site clusters. [b]Population (based on Boltzmann probability at 300K) of the lowest free energy solution cluster similar to any active site cluster.

In only four out of ten cases do the ligands in free solution populate the torsions that are the same as those of the ligand in the active site conformation. Thus, in agreement with Bohm et al (*34, 35*), we conclude that low energy solution structures of small, polar ligands do not have similar torsions to their active site conformations. In general, solution search strategies aiming to provide exact values of torsions for the active site conformations will not be successful.

Table IV shows the comparison of low energy structures for ten complexes having the same position of anchor points as the active site ligand conformations. It is apparent that for nine out of ten complexes the positions of the anchor points in the lowest free energy solution structures are similar to their positions in the crystal structure of the complexes. This observation may well also hold for larger ligands, but this remains to be verified. Based on the results from Table IV we find support for the hypothesis that "there is a strong similarity between the positions of anchor points (atoms responsible for tight binding with the receptor) of the ligand in solution and in the active site".

Table IV. Results for anchor point distance metric (notation as in Table III)

Ligand	ΔA	Population
chloramphenicol	0.0	0.761
N-phosphoryl-L-leucinamid	1.1	0.129
phosphocholine	0.0	1.000
tricarballylic acid	0.0	0.998
biotin	2.7	0.006
glycerol 3-phosphate	0.0	0.956
Ile-Val	0.0	0.998
Gly-Tyr	0.0	0.752
Val-Trp	0.6	0.198
sialic acid	6.5	0.000

3.3 Bound and Unbound Conformations of Ligands are Similar. This similarity has a number of implications. Firstly, receptors may change the solution structure of ligands that bind to their active sites, however the majority of changes occur in regions that contribute less to binding interactions. The positions of atoms responsible for tight binding seem to be very similar in low energy solution structures and in the active site of the receptor. Thus, receptors may bind ligands for which the positions of these points match up with key active site atoms. The similarity between low energy solution structures and active site ligand conformations strongly supports current efforts to build pharmacophore models based on solution structures and finding commonalities in position of key atoms in a series of ligands binding to a given receptor.

4. Binding Specificity of HIV and FIV Proteases

In the following two sections we focus on more detailed analyses of binding specificity. Drug resistance in bacterial pathogens has many concerned about a "post-antimicrobial age" (*39*). Many potent drugs against viral pathogens, such as HIV, have been rendered ineffective (*40-42*). Drug resistance may be acquired through a variety of mechanisms, one of which is the evolutionary modification of a drug target that significantly reduces the efficacy of an inhibitor while maintaining sufficient activity to survive (*40, 43, 44*). Understanding the mechanism of this process is an important goal. The problem may be posed as: What is the basis of ligand binding specificity among closely related protein targets?

We are working to understand the energetic details of this problem by studying proteases from Human Immunodeficiency Virus (HIV) and Feline Immunodeficiency Virus (FIV). This model system embodies the properties of strong homology and distinct specificity observed in drug resistant systems. The feline has also been demonstrated to exhibit symptoms similar to those seen in human subjects with the analogous infection. Because of this, it is hoped that a more thorough understanding of the FIV infection and viral components may lead to a viable animal model for AIDS. The FIV model system is useful both in understanding the mechanisms of binding specificity, as observed in drug resistant HIV strains, as well as developing an animal model for AIDS that may enhance the search for novel and effective therapies.

HIV protease and FIV protease are closely related enzymes that demonstrate altered specificity (*45, 46*). We have studied the energetic and molecular origin of this specificity. HIV protease is a primary target for AIDS therapy, the other important target being reverse transcriptase. The function of HIV protease within the virus is to cleave a poly-protein synthesized by the infected cell's machinery using the reverse transcribed RNA genome from HIV (*47*). During or following cleavage of this poly-protein, the new infectious viral particle is formed. Inhibition of the HIV protease prevents the formation of infectious virions (*48*) and halts the progression of the disease. HIV protease is a homodimer, with each monomer containing 99 amino acids. The active site, which has C_2 symmetry, is at the interface of the two monomers (*49*). Two aspartate groups, one donated from each monomer, form the fundamental catalytic unit (*50*).

4.1 Homology and Specificity. FIV protease is homologous to HIV protease, as reflected in its similar function, structure, and sequence. Structurally, the backbone positions overlap very well, particularly in the active site where the RMSD of the α-carbon atoms is approximately 0.5Å (*46*). The amino acid sequence within the active site of FIV protease is 42% identical to the active site in HIV protease. However, despite the strong homology between FIV and HIV protease, FIV protease is highly specific for its own natural substrates and designed inhibitors (*45, 46*). HIV protease inhibitors and substrates are barely functional within the FIV protease environment, while HIV protease can cleave FIV protease substrates although at slower rates (*45*). FIV protease also requires a longer substrate than HIV protease, indicating that the feline enzyme is more specific, since it requires more of its subsites to be occupied (*45*).

We investigate the origin of specificity of the HIV and FIV proteases by examining a model system involving a common inhibitor, LP149 bound to the two enzymes. This inhibitor exhibits specific binding thermodynamics with respect to these two enzymes and there is a crystal structure of LP149 bound to FIV protease (*46*). A major challenge in our investigation is that no crystal structure exists for LP149 bound to HIV protease and we must generate a reasonable model for this complex. We then use an empirical binding free energy partition to determine the energetic basis for binding specificity between HIV and FIV protease in complex with a common inhibitor: LP149.

4.2 Modeling HIV-LP149 complexes. The first step was to generate reasonable ensembles of HIV protease complexed with LP149. HIV protease adopts one of two conformations depending on whether or not a ligand is bound. In the unbound state the two "flap" regions, each consisting of two β-strands connected by a hairpin loop, are open, making the active site more accessible to the environment (*51*). In the bound state these flaps are closed and, in most cases, connected to the ligand through a tetra-coordinated water molecule. Since the backbone conformations of HIV protease molecules in complex with ligands are highly similar, we build a model of HIV protease complexed to a novel inhibitor by starting with the canonical closed backbone conformation of HIV protease and modifying the positions of the side chains to accommodate the new inhibitor. Such restraints were applied in both the HIV and FIV structure refinements. The 24 HIV protease complexes listed in Table V were extracted from the PDB and their native ligands removed.

Table V. PDB accession codes of HIV protease complexes

1aaq, 1dif, 1hbv, 1hef, 1heg, 1hih, 1hiv, 1hos, 1hps, 1hpv, 1hpx, 1hte, 1htf, 1htg, 1hvi, 1hvj, 1hvk, 1hvl, 5hvp, 9hvp, 1hvr, 1hvs, 4phv, 1sbg

All complexes have approximately the same backbone conformation and different side chain conformations. These structures are used as starting positions in the search for a new conformation bound to LP149. The inhibitor is docked in the active site using a least squares fit of the FIV protease complex backbone with each HIV protease closed conformation. The starting conformation of the inhibitor is the same in the FIV crystal structure and HIV protease models. The HIV protease complex models and the FIV crystal structure are minimized under successively reduced harmonic restraints to remove aberrant vdW clashes, while perturbing the rest of the structure as little as possible.

In order to equilibrate the side chains in the presence of the new inhibitor, a simulated annealing approach (*52*) utilizing the CHARMM package (*18*) is employed. The 24 HIV protease models are annealed at 1400K for 6 ps and cooled from 1400K to 0K over 84 ps to allow the side chains to adjust to LP149 within the active site. In order to generate a comparable ensemble of structures, the FIV protease crystal structure is subjected to the identical protocol using 24 distinct initial conditions. A weak harmonic restraint, inversely proportional to the B-factor, is placed on backbone atoms to permit significant movement at high temperatures while guiding the backbone to the initial canonical conformation at the end of the simulation. The resulting structures are minimized under the same conditions to bring each complex to the closest local minimum.

This protocol was validated using HIV protease complexed with indinavir (PDB accession code 1hsg). The structures resulting from HIV protease complex models and those resulting from the original crystal complex were highly similar both structurally and energetically (Table VI). Thus, our protocol, tested on the HIV/indinavir complex, is demonstrated to produce models of similar quality as the original, known, crystal structure when it is subjected to the same refinement procedure.

Table VI. Structural and Energetic Comparison of Models of the HIV Protease-Indinavir Complex Based on our Protocol and Derived from the Crystal Structure

Component	Model	Crystal Structure
Mean RMSD from Crystal	1.19 Å	1.08 Å
Mean Interaction Energy	-92 kcal	-96 kcal
Mean PB Energy	23 kcal	24 kcal
Mean Internal Energy	41 kcal	40 kcal
Mean Cavity Energy	1.5 kcal	1.6 kcal
Total Energy	-28 kcal	-29 kcal

4.3 Energy Function for Binding Energy Analysis. The ensemble of the complexes were analyzed using an empirical free energy function containing important energetic components of the binding free energy. The function includes the interaction energy between the ligand and the enzyme, the internal energy of the inhibitor, and the hydrophobic and electrostatic components of the solvation free energy. The interaction energy and internal energy are computed using the CHARMM force field (*18*). The hydrophobic solvation energy was computed as a linear function of the surface area using a coefficient fit to experimental solvent transfer results for hydrocarbons (*53*). The electrostatic component of the solvation energy was computed by solving the finite difference approximation of the linearized Poisson-Boltzmann equation (*37*) using DelPhi. This function permits a detailed investigation of the energetic basis for binding specificity. Two of the terms, the interaction energy and the electrostatic solvation free energy, are linear terms that can be separated into atomic, and thus active site residue contributions. This enhances our ability to investigate the molecular as well as the energetic basis of binding specificity.

Some assumptions were employed in the energetic analysis in order to make the necessary calculations tractable. Since no crystal structure is available for the open state of FIV protease, an alternative reference state was chosen for the unbound conformation. We use the closed conformations of both HIV and FIV proteases in the absence of ligand to represent the unbound state. By using this reference state, the change in the internal energy of the protein, which would be subject to intolerable errors using a modeled open conformation, is eliminated from the calculation. This choice of reference state also affects the solvation components of the energy function. The hydrophobic solvation energy term is small and very similar in both HIV and FIV protease and may be ignored. The PB solution term, accounting for the electrostatic component of the solvation free energy, is significant and may be under-estimated using this reference state. However, the same approximation is applied consistently to HIV and FIV protease and the errors arising from this approximation are assumed to cancel. Furthermore, we note that the use of the closed conformations for reference states is analogous to assuming that the *change* in each of the components of the binding free energy noted above is the same for HIV and FIV.

4.4 Energetic Basis of Binding Specificity. Analysis of the distributions of the energy components for the HIV and FIV protease complex ensembles yielded an interesting and surprising result. The internal energy of the inhibitor seems to be the primary component responsible for the difference in binding specificity of LP149 for HIV and FIV proteases. The other energetic components were similar for the HIV and FIV protease ensembles, whereas the internal energy showed a significant (14 kcal/mol) preference for the HIV complex. Table VII shows the mean value of each energy component of the HIV and FIV protease ensembles.

Table VII. Energetic comparison of HIV and FIV protease-LP149 complexes

Component	HIV (kcal/mol)	FIV (kcal/mol)
Mean Interaction Energy	-127	-121
Mean PB Energy	25	24
Mean Internal Energy	-33	-19
Mean Cavity Energy	-1.6	-1.8
Total Energy	-134	-115

The basis for the large difference in internal energy appears to be the ability of LP149 to find a low energy conformation within the HIV protease active site starting from the higher energy conformer present in the FIV protease crystal structure. Crowding within a hydrogen-bonded ring structure (also present in the original crystal structure) seems to be a major component of the internal energy difference. The average per-atom contributions to the vdW component of the intra-ligand energy are shown in Figure 2 along with a diagram of LP149, illustrating the regions of high energy vdW interactions. Although most of the structural diversity appears to be within the terminal naphthalene rings, the only significant difference in the vdW component of internal energy results from the hydrogen-bonded ring structure in the P3 position. Although no obvious change in conformation is apparent, even in the HIV protease conformation the energy is neutral or slightly unfavorable. This suggests that this region of LP149 is against a repulsive vdW wall, where the energy is proportional to r^{-12}, and a small change in the structure can result in a large change in energy.

This result is consistent with kinetic studies of FIV protease using HIV protease substrates. A series of HIV protease substrates, based on the same natural substrate as LP149, were investigated for kinetic properties within FIV protease (*46*). One substrate was distinguished by its high k_{cat}/K_m as well as its significantly faster cleavage relative to HIV protease. This substrate contained a threonine substituted for the glutamic acid side chain analogous to the one in LP149 critical for the strained hydrogen-bonded ring.

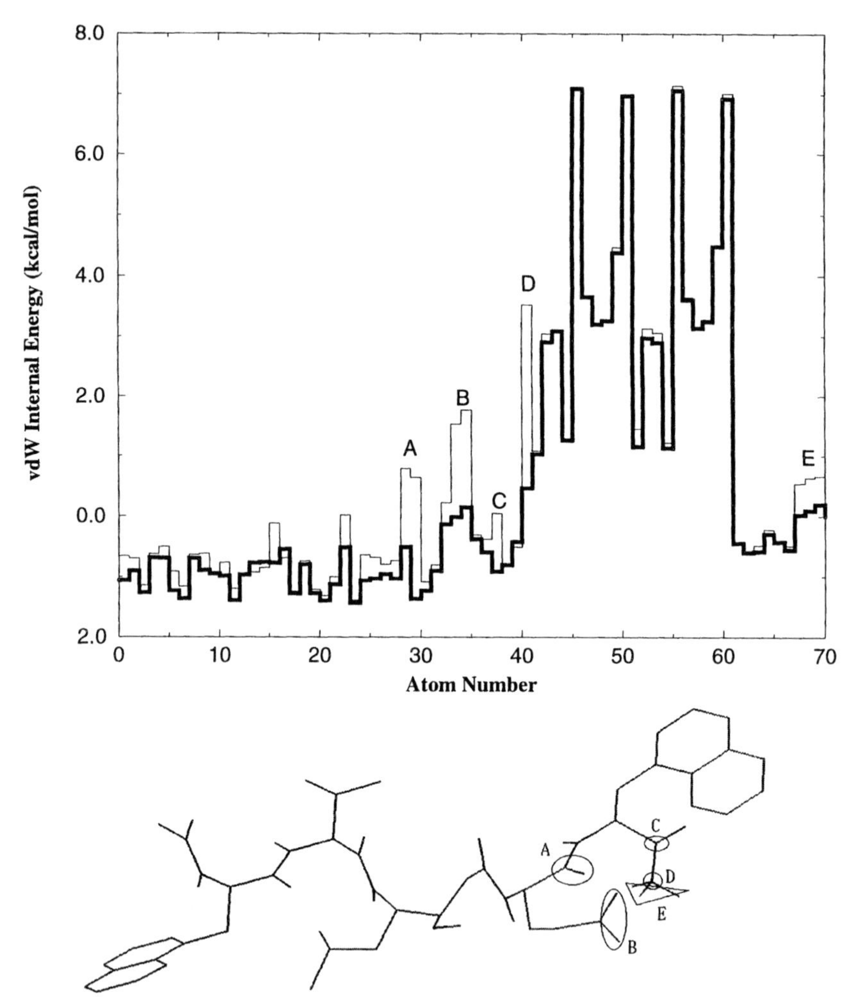

Figure 2. Per atom vdW internal energy. Heavy line is HIV; thin line is FIV. Regions of significant difference are labeled A through E and are correspondingly indicated on the structure of LP149.

This result is also consistent with work published (*54*) while this manuscript was under review. This paper probed the S3 and S3' subsite specificities in FIV and HIV proteases using a variety of peptide-based competitive inhibitors. The result was that small residues in the P3 and P3' positions were significantly favored in FIV protease. This experimental study is complementary to our work where we have examined the energetics of HIV and FIV protease binding a single inhibitor. The conclusions reached by both methods are compatible: small sidechains would be favored in the P3 and P3' positions due to overcrowding in the S3 and S3' subsites of FIV protease.

Although the total interaction and PB energy terms did not show a strong preference for the HIV or FIV protease structures, this does not preclude important differences at the level of individual residues. Significant differences in the average energy at the same relative backbone position in HIV and FIV protease could indicate sites important in conferring binding specificity. The active site residue interaction energy profiles for the HIV and FIV complex ensembles indicate that Asp 30 in HIV protease interacts very differently than the corresponding Ile 35 in FIV protease. As expected, this interaction is primarily electrostatic in the HIV complex, while a less favorable, primarily vdW interaction is observed in FIV protease. There is a significant difference between the mean interaction energy for this position in the two ensembles as well as a distinct separation between the distributions themselves. Based on this, we hypothesize that this position is linked to the specificity difference between HIV and FIV protease.

The same position also demonstrates a strong separation between HIV and FIV protease ensembles in the context of the electrostatic solvation free energy. As expected, the desolvation penalty is much more severe for the aspartate in the HIV protease than the corresponding isoleucine in FIV protease. Again, the distinct energy distributions and significant difference in the mean values between the two ensembles suggests that this effect is not an artifact, but a real difference in the binding energetics of these two enzymes. Although the solvation penalty associated with the Asp30/Ile35 opposes the favorable interaction energy (Table VIII), it is not enough to cancel the affect and this position is expected to favor the HIV protease / LP149 complex.

Table VIII. Interaction and electrostatic desolvation energy for the Asp30/Ile35 backbone position of HIV and FIV proteases

Protein	Interaction Energy (kcal/mol)	PB Energy (kcal/mol)
HIV Protease (Asp 30)	-9.1	2.8
FIV Protease (Ile 35)	-4.3	1.0

An understanding of binding specificity in closely related protein systems is an important goal for developing general principles of rational drug design and for understanding mechanisms of drug resistance. The HIV and FIV proteases embody the properties of close homology and distinct specificity, while also presenting the opportunity for understanding an important therapeutic target. The model developed here for understanding specificity within this system gives a surprising result, which is nevertheless consistent with experimental data. The further development of understanding how small changes in sequence can lead to dramatic differences in specificity may be crucial for neutralizing the threat of drug resistance.

5. λ-Dynamics: A Novel Approach for Evaluating Ligand Binding Affinity

An integral part in the design of new therapeutics is the search for lead compounds and their subsequent refinement. Computational approaches that rapidly evaluate the relative binding free energy of ligands to a common protein receptor play an important role in this process. Conventional free energy calculation methods, such as free energy perturbation (FEP) and thermodynamic integration (*55, 56*), evaluate the relative binding free energy of two ligands according to the following thermodynamic cycle:

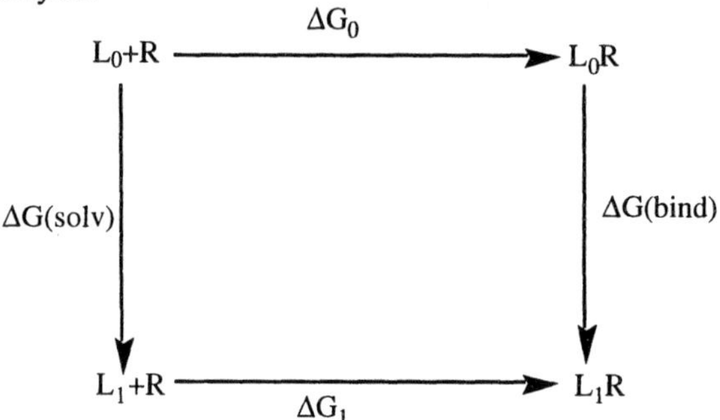

where L_0 and L_1 are the free ligands in aqueous solution. L_0R and L_1R are the corresponding ligands complexed with protein receptor R. The relative binding free energy of the two ligands, $\Delta\Delta G(\text{bind}) = \Delta G(\text{bind}) - \Delta G(\text{solv})$, is the difference between the relative free energy of the ligands in the complexed state and that of the free ligands. The free energy difference on each half of the thermodynamic cycle is calculated from the Hamiltonian, $H(\lambda,x) = \lambda H_1(x) + (1-\lambda)H_0(x)$. Simulations at a set of λ values ($0 < \lambda < 1$) are performed to transform the initial state, "0", slowly to the final state, "1". A free energy map along the coordinate λ is thus constructed and the free energy difference between the end states is obtained. These methods have been successfully applied to assess the relative binding free energy of protein-ligand systems (*57, 58*), but they are computationally expensive. A typical calculation could take days to weeks to complete, which is too long to be useful in the drug design process.

Alternative methods based on favorable interaction are commonly used in drug design to approximate the free energy. Although such approaches are relatively rapid in evaluating compounds, they can be inaccurate since the entropy contribution to the free energy is ignored. Because of its importance in chemical and biological applications, the development of new methodology for free energy calculations is an area of active research. Recently a semiempirical method that uses the differences in the average ligand-environment interaction energies between the bound and free states and a linear scaling procedure has been developed and was applied to the binding of sulfonamide inhibitors to human thrombin with promising results (*59*). In the previous section, we described a continuum solvent based model to estimate relative binding free energies for a ligand binding to HIV and FIV proteases. While such empirical approaches are desirable, and can be useful, further development of methods based on more rigorous free energy methods is also of import.

5.1 The λ-dynamics method. We have developed, from first principles, a free energy based approach to evaluate relative binding affinity. For a set of ligands and a common protein receptor, we construct the potential function of the system as the following (*60*):

$$V(\{\lambda\},(x)) = \sum_{i=1}^{L} \lambda_i^2 (V_i(x) - F_i) \qquad \left(\sum_{i=1}^{L} \lambda_i^2 = 1 \right) \tag{7}$$

Here $V_i(x)$ is the interaction energy involving protein and ligand i, L is the total number of ligands, λ_i is the coupling parameter, F_i is the reference free energy or biasing potential.

The dynamics of the system is governed by the extended Hamiltonian (*60*)

$$H(\{\lambda\},x) = T + T_{\{\lambda\}} + \sum_{i=1}^{L} \lambda_i^2 (V_i(x) - F_i) \tag{8}$$

Here the λs are treated as fictitious particles with masses m_λ and the method is thus called λ-dynamics. The free energy difference between molecules i and j, with reference free energy F_i and F_j respectively, is given by:

$$\begin{aligned} \Delta\Delta A_{ij} &= \Delta A_i - \Delta A_j \\ &= -\frac{1}{\beta}\left[\ln \int dx \exp\left(-\beta V_i(x) - \beta F_i\right)\right] + \frac{1}{\beta}\left[\ln \int dx \exp\left(-\beta V_j(x) - \beta F_j\right)\right] \\ &= -\frac{1}{\beta}\ln\frac{P\left(\lambda_i = 1, \{\lambda_{m\neq i}\} = 0\right)}{P\left(\lambda_j = 1, \{\lambda_{l\neq j}\} = 0\right)} \end{aligned} \tag{9}$$

where $P(\lambda_i = 1, \lambda_{m\neq i} = 0)$ is the probability that the hybrid system is in a state dominated by molecule I, A is the Helmholtz free energy and β is $1/k_BT$, where k_B is the Boltzmann constant and T is temperature. Therefore the free energy difference between two molecules can be obtained from the ratio of the probabilities of the ligands in the λ=1 state.

The λ-dynamics method is able to evaluate the relative binding free energy of multiple ligands simultaneously: short simulations enable one to obtain qualitative ordering of the binding affinities, while longer simulations provide quantitative results. When designing pharmaceutical agents, one is interested in identifying, from a pool of slightly different compounds, those with favorable binding affinity. Generally, the detailed value of the free energy change is not of interest. When comparison with experiment is required, then more quantitative results are required. The λ-dynamics method can perform both tasks.

5.2 Model system and simulation protocol. The system under study is a set of benzamidine inhibitors bound to trypsin. The inhibitors used are benzamidine, *p*-amino-benzamidine, *p*-methyl-benzamidine, and *p*-chloro-benzamidine. CHARMM version 22 parameter and topology files(*18*) were used except for the charges of the inhibitors, which were derived from the Quanta charge template method and modified slightly to confer identical charges on invariant atoms. The hybrid trypsin-inhibitor system was capped with a 24 Å sphere of TIP3P water (*61*) centered at the active site. The system was partitioned into a reaction region and a buffer region with a deformable boundary (*62*). The stochastic boundary MD method (*63*) was used throughout the simulations. A 30 ps simulation was performed before each calculation to allow the system to equilibrate.

The applications outlined in sections 5.3 and 5.4 demonstrate the use of λ-dynamics method for free energy calculations (*64, 65*).

5.3 Screening Calculations. In the screening calculations, where one is only interested in the ranking of the ligands, the reference free energy F_i in Eq. (7) corresponds to the free energy of the unbound ligands in solution. Its value is precalculated. This may be easily done by methods based on continuum solvation models such as the Poisson-Boltzmann (*66*) and generalized Born methods (*67, 68*). Since the reference free energy F_i is incorporated into the Hamiltonian of Eq. (7), the resulting free energy change from the simulations according to Eq. (8) gives the relative free energy change (ΔΔG), as was demonstrated in Eq. (9). The binding affinity is related to the probability of each compound to be in the λ = 1 state. Higher population indicates a more favorable binding free energy. Figure 3 shows the running average of each λ.

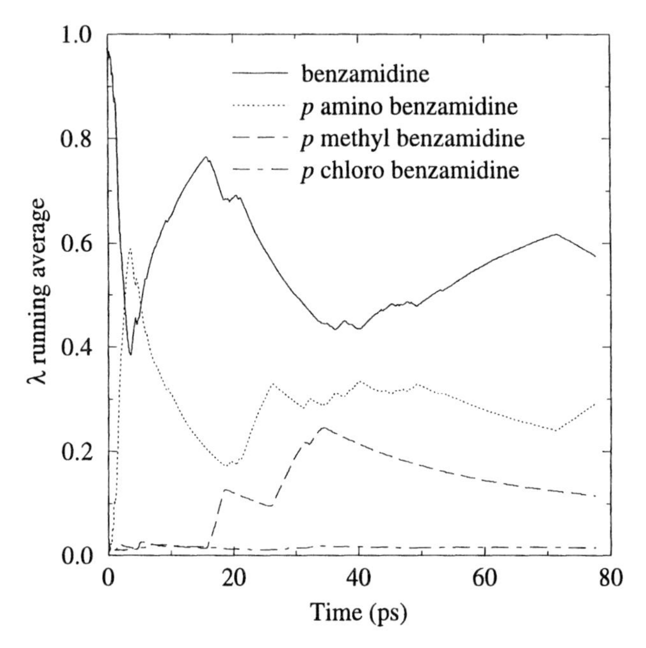

Figure 3. The cumulative λ running average of benzamidine derivatives complexed with trypsin as a function of simulation time.

Because the ligands all compete for the λ = 1 state and do not remain at an intermediate value of λ, the average value of λ reflects the probability of being in the λ = 1 state. The ordering of the ligands based on binding affinity is: benzamidine > *p*-amino-benzamidine > *p*-methyl-benzamidine > *p*-chloro-benzamidine. This ranking is clearly apparent after 50 ps of simulation time.

A validation calculation using the standard FEP method gives the binding free energies relative to benzamidine, 0.4 kcal/mol for *p*-amino-benzamidine, 2.3 kcal/mol for *p*-methyl-benzamidine, and 2.2 kcal/mol for *p*-chloro-benzamidine. Although the binding affinity between benzamidine and *p*-amino-benzamidine differs by only 0.4 kcal/mol, the λ-dynamics method clearly distinguishes the two ligands and provides the correct ranking. While this ranking differs from the experimental findings (*69*), the model calculations are converged using the FEP method and thus represent the correct answer for this model. Consequently, the screening calculations correctly reflect the relative binding affinities for our model. Additional simulations with different initial λ coordinates and velocities were also examined and resulted in the same ranking, indicating that the ligands were not trapped in local minima.

5.4 Precise Free Energy Changes. We also applied the λ-dynamics method to obtain the specific change in free energies for this system. In this situation, F_i is taken as a biasing potential. By properly choosing $\{F\}$, the barrier between different states along the reaction coordinates $\{\lambda\}$ is reduced and therefore one can completely sample the $\{\lambda\}$ space within a single simulation. An iterative procedure has been developed to improve sampling of the phase space and therefore make free energy calculations converge more rapidly (*65*). This procedure employed feedback from previous calculations to improve the bias of the current

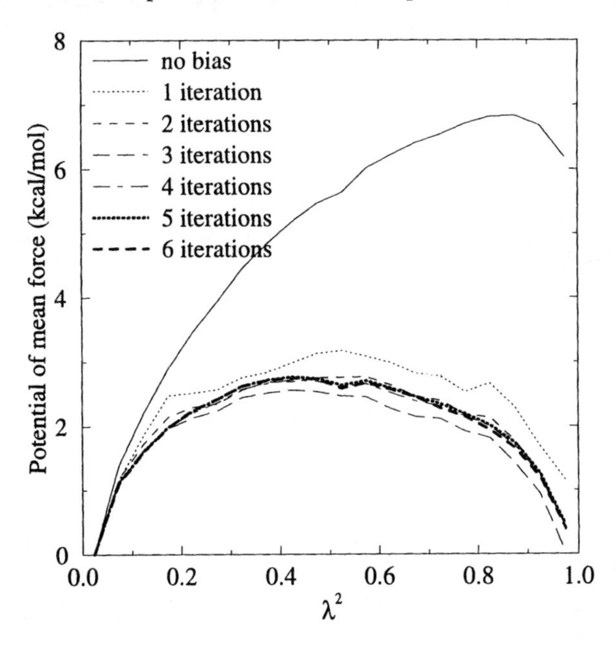

Figure 4. Potential of mean force along coordinate λ between *p*-amino-benzamidine (λ = 0) and *p*-methyl-benzamidine (λ = 1) in the unbound state. The barrier is reduced from 7 kcal/mol to 3 kcal/mol when a biasing potential is applied.

simulation. The optimal estimate of these biasing potentials is achieved by the multiple reaction coordinate WHAM technique (*70*). In these calculations, biasing potentials are derived from constant values that correspond to the estimated free energy of each species. Figure 4 illustrates the free energy surface between *p*-amino-benzamidine and *p*-methyl-benzamidine in the unbound state. The barrier between the two states decreases significantly when the biasing potential is applied. The calculations also converge rapidly. After the first iteration the free energy curve is already converged within statistical uncertainties. The simulation time is 150 ps per iteration using the λ-dynamics method. While for FEP calculations, a total of 240 ps simulation time (3 windows) was required to achieve similar precision.

5.5 A Novel Approach for Evaluating Ligand Binding Affinity. In summary, the λ-dynamics method may be used either to identify rapidly ligands with favorable binding free energy or to obtain specific changes in free energy. Since it screens the binding free energy of the ligands instead of the interaction energy, it provides an accurate assessment of relative binding affinity. Species whose binding free energy differ by more than a few kcal/mol from the most favorable binder can be screened out within a few tens of picoseconds of simulation, because they will not compete. In other words, they never reach the $\lambda = 1$ state. The total computation time is not expected to increase with the total number of ligands because only the few favorable binders are able to compete for the $\lambda = 1$ state. In contrast, conventional calculations of the relative binding free energy of two ligands would typically take hundreds of picoseconds of simulation time and increases with the number of ligands. Longer simulation, in combination with the iterative procedure, provides quantitative free energy changes. Although one can not generalize based on a single application, our results suggest that for detailed calculations where the specific change in free energy is required, the simulation time using λ-dynamics is about half of that using FEP. The efficiency could be further increased with improved biasing potentials.

Acknowledgments

Helpful discussions with Jeffrey Skolnick, Andrzej Kolinski, Felix Sheinerman and Angel Ortiz are acknowledged. This work was supported by NIH GM37554 and RR06009.

Literature Cited

(*1*) Ajay; Murcko, M. A. *J. Med. Chem.* **1995,** *38*, 4953.
(*2*) Jones, G.; Willet, P.; Glen, R. C. *J. Comput.-Aided Mol. Des.* **1995,** *9*, 532.
(*3*) Jones, G.; Willet, P. *Curr. Opin. Biotech.* **1995,** *6*, 652.
(*4*) Judson, R. S.; Jaeger, E. P.; Treasurywala, A. M. *J. Mol. Struct.* **1994,** *308*, 191.
(*5*) Judson, R. S.; Tan, Y. T.; Mori, E.; Melius, C.; Jeager, E. P.; Treasurywala, A. M.; Mathiowetz, A. *J. Comp. Chem.* **1995,** *16*, 1405.
(*6*) Di Nola, N.; Roccatano, D.; Berendsen, H. J. C. *Proteins: Struct. Func. Genetics* **1994,** *19*, 174.
(*7*) Clark, K. P.; Ajay *J. Comp. Chem.* **1995,** *16*, 1210.
(*8*) Caflisch, A.; Fischer, S.; Karplus, M. *J. Comp. Chem.* **1997,** *18*, 723.
(*9*) Goodsell, D. S.; Olson, A. J. *Proteins: Struct. Func. Genetics* **1990,** *8*, 195.
(*10*) Kuntz, I. D.; Blaney, J. M.; Oatley, S. J.; Langridge, R.; Ferrin, T. E. *J. Mol. Biol.* **1982,** *161*, 269.
(*11*) Fisher, E.; Thierfelder, H. *Berl. Dtsh, Chem. Ges.* **1894,** *27*, 2985.
(*12*) Koshland, D. E. *Proc. Natl. Acad. Sci. U. S. A.* **1958,** *44*, 98.
(*13*) Marquart, M.; Walter, J.; Deisenhofer, J.; Bode, W.; Huber, R. *Acta Crystallogr.* **1983,** *39*, 480.
(*14*) Padlan, E. A.; Cohen, G., H.; Davies, D. R. *Ann. Immunol. (Paris)* **1985,** *136*, 271.
(*15*) Weis, W. I.; Brünger, A. T.; Skehel, J. J.; Wiley, D. C. *J. Mol. Biol.* **1990,** *212*, 737.

(*16*) Noble, M. E. M.; Wierenga, R. K.; Lambeir, A. M.; Opperdoes, F. R.; Thunnissen, M. W. H.; Kalk, K. H.; Groendijk, H.; Hol, W. G. J. *Proteins: Struct. Func. Genetics* **1991,** *10*, 50.
(*17*) Weber, P. C.; Ohlendorf, D. H.; Wendoloski, J. J.; Salemme, F. R. *Science* **1989,** *243*, 85.
(*18*) Brooks, B. R.; Bruccoleri, R. E.; Olafson, B. D.; States, D. J.; Swaminathan, S.; Karplus, M. *J. Comp. Chem.* **1983,** *4*, 187.
(*19*) Bernstein, F. C.; Koetzle, T. F.; Williams, G. J. B.; Meer, E. F.; Brice, M. D.; Rodgers, J. R.; Kennard, O.; Shimanouchi, T.; Tasumi, M. *J. Mol. Biol.* **1977,** *112*, 535.
(*20*) Mirny, L. A.; Shakhnovich, E. I. *J. Mol. Biol.* **1996,** *264*, 1164.
(*21*) Gutin, A. M.; Abkevich, V. I.; Shakhnovich, E. I. *Proc. Natl. Acad. Sci. U. S. A.* **1995,** *92*, 1282.
(*22*) Godzik, A. *Protein Engng.* **1995,** *8*, 409.
(*23*) Godzik, A.; Kolinski, A.; Skolnick, J. *J.Comput.-Aided Mol. Des.* **1993,** *7*, 397.
(*24*) Jones, D.; Thornton, J. *J. Comput.-Aided Mol. Des.* **1993,** *7*, 439.
(*25*) Luthy, R.; Bowie, J. U.; Eisenberg, D. *Nature* **1992,** *356*, 83.
(*26*) Bowie, J. U.; Luthy, R.; Eisenberg, D. *Science* **1991,** *253*, 164.
(*27*) Miranker, A.; Karplus, M. *Proteins Struct. Func. Genetics* **1991,** *11*, 29.
(*28*) Clark, D. E.; Westhead, D. R.; Sykes, R. A.; Murray, C. W. *J. Comput.-Aided Mol. Des.* **1996,** *19*, 397.
(*29*) Wang, S.; Zaharewitz, D. W.; Sharma, R.; Marquez, V. E.; Lewin, N. E.; Du, L.; Blumberg, P. M.; Milne, G. W. *J. Med. Chem.* **1994,** *37*, 4479.
(*30*) Kiyama, R.; Honma, T.; Hayashi, K.; Ogawa, M.; Hara, M.; Fujimoto, M.; Fujishita, T. J. *J. Med. Chem.* **1995,** *38*, 2728.
(*31*) Hodgkin, E. E.; Miller, A.; Whittaker, M. *J. Comput.-Aided Mol. Des.* **1993,** *7*, 515.
(*32*) Dammkoehler, R. A.; Karasek, S. F.; Shands, E. F.; Marshall, G. R. *J. Comput.-Aided. Mol. Des.* **1989,** *3*, 3.
(*33*) Dammkoehler, R. A.; Karasek, S. F.; Shands, E. F.; Marshall, G. R. *J. Comput.-Aided Mol. Des.* **1995,** *9*, 491.
(*34*) Bohm, H.-J.; Klebe, G. *Angew. Chem. Int. Ed. Engl.* **1996,** *35*, 2588.
(*35*) Nicklaus, M. C.; Shaomeng, W.; Driscoll, J. S.; Milne, W. A. *Bioorg. & Med. Chem.* **1995,** *3*, 411.
(*36*) Jain, A. K.; Dubes, R. C. *Algorithms for Clustering Data*; Prentice Hall: Englewood Cliffs, New Jersey, 1988.
(*37*) Warwicker, J.; Watson, H. C. *J. Mol. Biol.* **1982,** *157*, 671.
(*38*) Edinger, S. R.; Cortis, C.; Shenkin, P. S.; Friesner, R. A. *J. Phys. Chem. B* **1997,** *101*, 1190.
(*39*) Cohen, M. L. *Science* **1992,** *257*, 1050.
(*40*) Ridkey, T.; Leis, J. *J. Biol. Chem.* **1995,** *270*, 29621.
(*41*) Sardana, V. V.; Schlabach, A. J.; Graham, P.; Bush, B. L.; Condra, J. H.; Culberson, J. C.; Gotlib, L.; Graham, D. J.; Kohl, N. E.; LaFemina, R. L.; Schneider, C. L.; Wolanski, B. S.; Wolfgang, J. A.; Emini, E. A. *Biochemistry* **1994,** *33*, 2004.
(*42*) Maschera, B.; Darby, G.; Palú, G.; Wright, L. L.; Tisdale, M.; Myers, R.; Blair, E. D.; Furfine, E. S. *J. Biol. Chem.* **1996,** *271*, 33231.
(*43*) Spratt, B. G. *Science* **1994,** *264*, 388.
(*44*) Perach, M.; Rubinek, T.; Hughes, S. T.; Hizi, A. *J. Mol. Biol.* **1997,** *268*, 648.
(*45*) Schnölzer, M.; Rackwitz, H.-R.; Gustchina, A.; Laco, G. S.; Wlodawer, A.; Elder, J. H.; Kent, S. B. H. *Virology* **1996,** *224*, 268.

(*46*) Wlodawer, A.; Gustchina, A.; Reshetnikova, L.; Lubkowski, J.; Zdanov, A.; Hui, K. Y.; Angleton, E. L.; Farmerie, W. G.; Goodenow, M. M.; Bhatt, D.; Zhang, L.; Dunn, B. M. *Nat. Struct. Biol.* **1995,** *2*, 480.
(*47*) Debouck, C.; Gorniak, J. G.; Strickler, J. E.; Meek, T. D.; Metcalf, B. W.; Rosenberg, M. *Proc. Natl. Acad. Sci. U. S. A.* **1987,** *84*, 8903.
(*48*) Kohl, N. E.; Emini, E. A.; Schleif, W. A.; Davis, L. J.; Heimbach, J. C.; Dixon, R. A.; Scolnick, E. M.; Sigal, I. S. *Proc. Natl. Acad. Sci. U. S. A.* **1988,** *85*, 4686.
(*49*) Miller, M.; Schneider, J.; Sathynarayana, B. K.; Toth, M. V.; Marshall, G. R.; Clawson, L.; Selk, L.; Kent, S., B.H.; Wlodawer, A. *Science* **1989,** *246*, 1149.
(*50*) Silva, A. M.; Cachau, R. E.; Sham, H. L.; Erickson, J. W. *J. Mol. Biol.* **1996,** *255*, 321.
(*51*) Spinelli, S.; Liu, Q. Z.; Alzari, P. M.; Hirel, P. H.; Poljak, R. J. *Biochimie* **1991,** *73*, 1391.
(*52*) Kirkpatrick, S.; Gelatt, J., C.D.; Vecchi, M. P. *Science* **1983,** *220*, 671.
(*53*) Sitkoff, D.; Sharp, K. A.; Honig, B. *J. Phys. Chem* **1994,** *98*, 1978.
(*54*) Lee, T.; Laco, G. S.; Torbett, B. E.; Fox, H. S.; Lerner, D. L.; Elder, J. H.; Wong, C. H. *Proc. Natl. Acad. Sci. U. S. A.* **1998,** *95*, 939.
(*55*) Zwanzig, R. W. *J. Chem. Phys.* **1954,** *22*, 1420.
(*56*) Kollman, P. *Chem. Rev.* **1993,** *93*, 2395.
(*57*) Brooks III, C. L.; Fleischman, S. H. *J. Amer. Chem. Soc.* **1990,** *112*, 3307.
(*58*) Bash, P.; Chandra, S.; Brown, F. K.; Langridge, R.; Kollman, P. A. *Science* **1987,** *235*, 574.
(*59*) Jones-Hertzog, D. K.; Jorgensen, W. L. *J. Med. Chem.* **1997,** *40*, 1539.
(*60*) Kong, X.; Brooks III, C. L. *J. Chem. Phys.* **1996,** *105*, 2414.
(*61*) Jorgensen, W. L.; Chandrasekhar, J.; Madura, J. D.; Impey, R. W.; Klein, M. L. *J. Chem. Phys.* **1983,** *79*, 926.
(*62*) Brooks III, C. L.; Karplus, M. *J. Chem. Phys.* **1983,** *79*, 6312.
(*63*) Brooks III, C. L.; Brunger, A.; Karplus, M. *Biopolymers* **1985,** *24*, 843.
(*64*) Guo, Z.; Brooks III, C. L. *J. Am. Chem. Soc.* **1998,** *120*, 1920.
(*65*) Guo, Z.; Brooks III, C. L.; Kong, X. *J. Phys. Chem. B* **1998,** *102*, 2032.
(*66*) Klapper, I.; Hagstrom, R.; Fine, R.; Sharp, K.; Honig, B. *Proteins: Struct. Funct. Genetics* **1986,** *1*, 47.
(*67*) Still, W. C.; Tempczyk, A.; Hawley, R. C.; Hendrickson, T. *J. Am. Chem. Soc.* **1990,** *112*, 6127.
(*68*) Hawkins, G. D.; Cramer, C. J.; Truhlar, D. G. *Chem. Phys. Lett.* **1995,** *246*, 122.
(*69*) Mares-Guia, M.; Nelson, D. L.; Rogna, E. *J. Am. Chem. Soc.* **1977,** *99*, 2231.
(*70*) Boczko, E. M.; Brooks, C. L. I. *J. Phys. Chem.* **1993,** *97*, 4509.

Chapter 3

New Free Energy Calculation Methods for Structure-Based Drug Design and Prediction of Protein Stability

Lu Wang, Mats A. L. Eriksson, Jed Pitera, and Peter A. Kollman

Department of Pharmaceutical Chemistry, University of California, San Francisco, CA 94143

We summarize some aspects of our recent studies of protein stability and protein-ligand interactions by a combination of several newly developed approximate free energy calculation methods and rigorous free energy calculation methods. The approximate free energy simulation methods employed in our studies include free energy derivatives (FED), pictorial representation of free energy changes (PROFEC), chemical Monte Carlo/molecular dynamics simulation (CMC/MD), Poisson-Boltzmann continuum electrostatics/solvent accessible area (PB/SA) and generalized Born approximation/solvent accessible area (GB/SA) methods. The thermostability of T4 lysozyme and the binding free energies of HIV-1 reverse transcriptase inhibitors were analyzed as test cases. It is shown that these different approaches are complementary to each other and when combined, can make predictions efficient, comprehensive and insightful. Potential applications of this strategy in structure-based drug design and protein engineering are discussed.

Calculations of free energies of ligand binding or protein stability can be very useful in drug design and protein engineering (1). The most rigorous approaches for calculating free energy changes, i.e., the free energy perturbation (FEP) or thermodynamic integration (TI) methods, have been limited in practical applications, due to their computationally intensive nature. Recently, there has been considerable interest in developing approximate, yet efficient free energy simulation methods. Notably, Åqvist *et al* (2) proposed the linear interaction energy approximation that correlates the binding free energies of ligands with their average interaction energies in protein and solvent. Radmer & Kollman (3) developed the pictorial representation of free energy changes (PROFEC) which can suggest modifications of a ligand to increase its binding affinity. To quickly rank the binding affinities of a series of ligands, Pitera & Kollman (4) introduced the chemical Monte Carlo/molecular dynamics (CMC/MD) technique and Kong & Brooks (5) have developed the λ dynamics simulation method. The potential of using free energy derivatives (FED) in estimation of free energy changes (6,7) and analog design (8) has been explored. In addition, Shen *et al* (9,10) advocated the Poisson-Boltzmann electrostatics/solvent accessible area (PB/SA) scheme for estimation of binding free energies of ligands. In general, these methods are much faster than FEP or TI and therefore more suited to practical applications. Due to their approximate nature, they may be less accurate. In this work, we tested some of these approximate methods in two cases. One is the thermostability problem of T4 lysozyme (11). Another is the binding free energies of HIV-1 reverse transcriptase (RT) TIBO inhibitors (12). For T4 lysozyme, we first used FED and PROFEC to suggest candidate modifications that may improve the stability of the protein and then analyzed an interesting qualitative prediction with a TI calculation. For the HIV-1 RT inhibitors, we first analyzed their binding affinities with the CMC/MD and PB/SA methods and later supported one of the results by TI calculations. As a comparison, the solvation free energies of the ligands were calculated with both PB/SA method and the generalized Born approximation/solvent accessible area (GB/SA) method (13). For both cases, we found that the approximate methods gave quite reasonable results and therefore may be useful in practical structure-based drug design and protein engineering applications.

Theory

Thermodynamic cycles. The following thermodynamic cycle was used to assess the effect of a modification on the stability of T4 lysozyme:

$$\begin{array}{ccc} & \Delta G_{u\rightarrow f} & \\ \text{T4}\,(f) & \Leftarrow & \text{T4}\,(u) \\ \Delta G_{f\rightarrow f*} \Downarrow & & \Downarrow \Delta G_{u\rightarrow u*} \\ \text{T4*}(f) & \Leftarrow & \text{T4*}(u) \\ & \Delta G^*_{u\rightarrow f} & \end{array} \quad (1)$$

where T4 and T4* stand for T4 lysozyme and a further modified enzyme. "f" and "u" stand for the folded and unfolded states of the enzyme. Because free energy is a state function, we have

$$\Delta\Delta G_{fold} = \Delta G^*_{u\rightarrow f} - \Delta G_{u\rightarrow f} = \Delta G_{f\rightarrow f*} - \Delta G_{u\rightarrow u*} \quad (2)$$

where $\Delta\Delta G_{fold}$ is the folding free energy difference or stability difference between T4* and T4. Direct calculations of $\Delta G^*_{u\rightarrow f}$ and $\Delta G_{u\rightarrow f}$ are difficult due to the large conformational changes involved. Fortunately, $\Delta G_{f\rightarrow f*}$ and $\Delta G_{u\rightarrow u*}$ can be readily determined by free energy calculation methods.

The following thermodynamic cycle was used to calculate the relative binding affinity of two ligands, L_1 and L_2, to the same protein, P:

$$\begin{array}{ccc} & \Delta G_{L_1,b} & \\ L_1 + P & \Rightarrow & L_1\bullet P \\ \Delta G_{L_1\rightarrow L_2,s} \Downarrow & & \Downarrow \Delta G_{L_1\rightarrow L_2,p} \\ L_2 + P & \Rightarrow & L_2\bullet P \\ & \Delta G_{L_2,b} & \end{array} \quad (3)$$

where $\Delta G_{L_1,b}$ and $\Delta G_{L_2,b}$ are the binding free energies of L_1 and L_2. $\Delta G_{L_1\rightarrow L_2,s}$ and $\Delta G_{L_1\rightarrow L_2,p}$ are the mutational free energies of $L_1\rightarrow L_2$ in the solvent and in the protein, respectively. The difference in binding free energy between L_1 and L_2,

$\Delta\Delta G_b$, can be expressed as

$$\Delta\Delta G_b = \Delta G_{L2,b} - \Delta G_{L1,b} = \Delta G_{L1\rightarrow L2,p} - \Delta G_{L1\rightarrow L2,s} \tag{4}$$

Similarly, direct calculations of $\Delta G_{L1,b}$ and $\Delta G_{L2,b}$ are difficult, but $\Delta G_{L1\rightarrow L2,s}$ and $\Delta G_{L1\rightarrow L2,p}$ can be calculated from simulations.

Free energy calculation with thermodynamic integration (TI) method. In this study, TI (1) was used to calculate the free energy changes. For the transformation of one state into another, a coupling parameter λ is introduced and the Hamiltonians of the two states are defined as H_0 (λ=0) and H_1 (λ=1). The free energy change of the transformation is expressed as the following integral (1)

$$\Delta G = \int_0^1 \langle \partial H(\lambda) / \partial\lambda \rangle_\lambda \, d\lambda \tag{5}$$

where $\langle \partial H(\lambda) / \partial\lambda \rangle_\lambda$ is an ensemble average at λ. In practice, a number of evenly-spaced windows with different λ values ranging from 0 to 1 are chosen and at each window, $\langle \partial H(\lambda) / \partial\lambda \rangle_\lambda$ is calculated by averaging over molecular dynamics trajectories. The kinetic energy contribution can be neglected and ΔG is estimated by

$$\Delta G \approx \sum_i \langle \partial V(\lambda) / \partial\lambda \rangle_{\lambda_i} \Delta\lambda \tag{6}$$

where λ_i is the λ value of the *i*th window and $\Delta\lambda$ is the interval between successive windows. According to the trapezoidal approximation, corrections are made to the contributions of the first and last windows (multiplying by 0.5). V is the potential function that describes the atomic interactions in the system.

Free energy derivatives (FED). The partial derivatives of free energy with respect to the nonbonded interaction parameters, q_i, ε_i and R_i^*, were calculated by the following equations (6,7):

$$\partial G / \partial q_i = \left\langle \sum_j q_j / (\varepsilon R_{ij}) \right\rangle = 1/q_i \langle V_{coul}(i) \rangle \tag{7}$$

$$\partial G / \partial \varepsilon_i = \left\langle \sum_j [1/(2\varepsilon_i)] \varepsilon_{ij} \left[\left(R_{ij}^* / R_{ij} \right)^{12} - 2 \left(R_{ij}^* / R_{ij} \right)^6 \right] \right\rangle$$
$$= [1/(2\varepsilon_i)] \langle V_{L-J}(i) \rangle \tag{8}$$

$$\partial G / \partial R_i^* = \left\langle \sum_j \varepsilon_{ij} \left(12 / R_{ij}^* \right) \left[\left(R_{ij}^* / R_{ij} \right)^{12} - \left(R_{ij}^* / R_{ij} \right)^6 \right] \right\rangle \tag{9}$$

where $\langle V_{coul}(i) \rangle$ and $\langle V_{L-J}(i) \rangle$ are the mean Coulombic and Lennard-Jones interaction energies of the *i*th atom with the rest of the system. To analyze the change of a protein's stability due to modifications of the properties of its *i*th atom, according to equation 2, the free energy derivatives need to be calculated for both the folded and the unfolded states of the protein and their difference indicates the stability change.

Pictorial representation of free energy changes (PROFEC). The PROFEC contour maps can be used to visualize how a protein's stability or a ligand's binding affinity changes when additional particles are added to a residue of the protein or the ligand (3). The contour map is generated by evaluating the insertion free energy of a test particle at various grid points near the residue of interest, using coordinates from molecular dynamics simulations (3). To analyze a protein's stability, according to equation 2, two contour maps for the folded and the unfolded states of the protein have to be generated and their difference map is used to indicate the stability change of the protein upon modification of the residue of interest. The free energy cost of adding a test particle at a grid point is calculated by (3)

$$\Delta G\,(i, j, k) = -RT \ln < \exp\,(-\Delta V(i, j, k)\,/\,RT >_0 \qquad (10)$$

where i, j and k are the coordinates of a grid point, ΔG (i, j, k) is the insertion free energy and ΔV(i, j, k) is the interaction energy between the test particle and the surrounding atoms.

Chemical Monte Carlo/molecular dynamics (CMC/MD). The CMC/MD method (4) has recently been developed for determination of relative binding free energies of a series of ligands to a common receptor. The method (described in detail in ref. 4) employs the MD method for generating a set of coordinates for one distinct chemical system and the MC method to sample the *chemical* space of the system. Applied to a protein-inhibitor system, the chemical space can be 5-10 different derivatives of the inhibitor. Each derivative is included in the simulated system but the potential function is "masked" so that only one ligand interacts with the protein and solvent at a time. The Monte Carlo steps consist of changes to the "masking" function, effectively changing the ligand being simulated. MD is used to propagate the coordinates of all the ligands. During the course of the MC/MD run, the probability (P_i) of each inhibitor 'i' is accumulated according to the Metropolis (14) criteria for accepting an inhibitor in an MC step:

$$\text{if } \Delta E_i \leq 0 \Rightarrow P_{acc} = 1, \qquad \text{if } \Delta E_i > 0 \Rightarrow P_{acc} = \exp(-\Delta E_i/RT) \qquad (11)$$

where ΔE_i is the difference in protein-inhibitor interaction energy between a randomly chosen ligand and the old ligand and P_{acc} is the acceptance probability. Prior to each MC step, the "Boltzmann" probabilities of each ligand 'i' is calculated by:

$$P_i = \exp(-\Delta E_i/RT)/\Sigma\exp(-\Delta E_i/RT) \qquad (12)$$

It can be shown that if an infinite number of MC steps were performed on a given Cartesian conformation, the resulting probability distribution would coincide with that calculated from equation 12. In the TIBO-HIV-1 RT systems studied here, we used the averaged P_i's from equation 12, since they also allow for estimations of the relative free energies of poorly sampled inhibitors. The relative free energy of the bound state for inhibitors 'j' and 'i' is then related to their ratio of probabilities according to:

$$\Delta G_j - \Delta G_i = -RT\ln P_j/P_i \qquad (13)$$

Solvent effects, i.e. differences in free energy of binding due to different free energies of solvation (ΔG_{solv}), can be taken into account by testing the acceptance against ΔE_i-$\Delta G_{solv,i}$, rather than ΔE_i in equation 11. Using ΔG_{solv} as a biasing potential in the MC step, the sampling mirrors the binding free energy - which is the relevant property when ranking inhibitors - rather than the free energy of the bound state.

The CMC/MD, as outlined above, was found to converge very slowly when applied to a series of HIV-1 RT TIBO inhibitors. In order to increase the convergence rate, a variant of this method was developed - herein called the "adaptive CMC/MD" method (J. Pitera, unpublished). Rather than sampling the chemical space according to the relative free energies of the inhibitors, the goal of the adaptive CMC/MD is instead to sample this space *evenly*. This can be achieved by introducing biasing offsets, $\Delta G_{offs,i}$, that for each ligand 'i' reflects its relative free energy in the bound state. An MC sampling by testing the acceptance against ΔE_i-$\Delta G_{offs,i}$, rather than ΔE_i in equation 11, would then result in an even sampling of all ligands, since all ΔE_i-$\Delta G_{offs,i}$ would be equal to zero. The offsets are solved iteratively. Starting with all $\Delta G_{offs,i}$=0, the probabilities of each ligand are calculated according to equation 12, averaged over a certain number of CMC/MD-cycles (a CMC/MD run). A first set of $\Delta G_{offs,i}$'s, relative to some arbitrarily chosen ligand, is estimated from equation 13, and is then used as biasing offsets in the next MC/MD run. The offsets are then adjusted after each CMC/MD run, by averaging the P_i's from equation 12, and add the adjusted offsets obtained from equation 13 to $\Delta G_{offs,i}$. When this procedure has converged, all P_i's are equal and the relative free energies of the bound state ($\Delta G_{bound,i}$) equals to -$\Delta G_{offs,i}$. $\Delta\Delta G_b$ can then be calculated by subtracting ΔG_{solv} from ΔG_{bound}. We estimated ΔG_{solv} by the GB/SA method discussed below.

Poisson-Boltzmann continuum electrostatics/solvent accessible area (PB/SA) and Generalized Born approximation/solvent accessible area (GB/SA) methods. In both the PB/SA and the GB/SA methods, a solvated protein or a small solute molecule is represented as a low dielectric cavity containing fixed charges and dipoles. The solvent water is represented as a medium of dielectric constant 80 which may contain ions. For the PB/SA method, the electrostatic field around a protein or a small solute molecule in the presence of salt is estimated by the solution to the linearized Poisson-Boltzmann equation (15)

$$\vec{\nabla}(\varepsilon(\mathbf{x})\vec{\nabla}\phi(\mathbf{x})) - \kappa^2\phi(\mathbf{x}) + 4\pi\rho(\mathbf{x}) = 0 \qquad (14)$$

where ϕ is the electric potential, ε is the dielectric constant, ρ is the fixed charge density and κ is the modified Debye-Huckel parameter which depends on the ionic strength and temperature of the solution. The Poisson-Boltzmann equation can be solved numerically by the finite difference method, in which the continuous functions are approximated by distinct values at points on a cubic grid (15). With the electrostatic potential obtained from solving the PB equation, the electrostatic interaction between a protein or a small solute with the solvent is expressed as (15)

$$\Delta G_{pol} = 1/2 \sum \phi_i q_i \qquad (15)$$

where ϕ_i is the potential on charge q_i and the sum is over the fixed charges. To estimate the electrostatic contribution to the hydration free energy of a molecule, two calculations, one for the molecule *in vacuo* and the other for the molecule in aqueous solution, should be performed and their difference in ΔG_{pol} gives the electrostatic contribution to the solvation free energy. $\Delta G_{L_1 \rightarrow L_2,s}$ and $\Delta G_{L_1 \rightarrow L_2,p}$ in equation 4 may be approximately estimated by the PB/SA or the GB/SA methods. For $\Delta G_{L_1 \rightarrow L_2,s}$, this involves the calculation of the solvation free energy difference of L_1 or L_2. According to equation 15, the calculation of the electrostatic contribution to $\Delta G_{L_1 \rightarrow L_2,s}$ is straightforward. The nonpolar contribution to $\Delta G_{L_1 \rightarrow L_2,s}$ can be calculated according to the following empirical linear relation which correlates the solvation free energies of nonpolar solutes with their solvent accessible area (16,17)

$$\Delta G_{npol} = \Sigma\ \sigma\ A \qquad (16)$$

where A is the solvent accessible area (in $Å^2$). σ is the empirical solvation parameter. We used σ = 5 cal/mol^{-1}Å^{-2} in this work. The solvent accessible areas were calculated with Connolly's MS program (18). For $\Delta G_{L_1 \rightarrow L_2,p}$, this involves the calculation of the solvation free energy difference between L_1•P and L_2•P. Direct and accurate calculation of this difference is difficult because the solvation free energies of the two complexes are large numbers and the estimated difference may have a large error. As a first order approximation, we estimated $\Delta G_{L_1 \rightarrow L_2,p}$ as the difference in interaction energy between L_1 and L_2 with the protein and the solvent in energy minimized conformations of the complexes. The polar contribution to $\Delta G_{L_1 \rightarrow L_2,p}$ was estimated as the difference in electrostatic interaction energies between the two ligands with the protein and the solvent, which were calculated with equation 15 by summing over the atoms of L_1 or L_2. The nonpolar contribution to $\Delta G_{L_1 \rightarrow L_2,p}$ was estimated as the difference in van der Waals interaction energy between the two ligands with the protein and the solvent.

The GB/SA method uses a similar relation for calculation of the nonpolar contribution to the solvation free energy. For the electrostatic part, it uses the so-called generalized Born approximation which express the interactions between the fixed charges and their interactions with the solvent as a sum of pairwise interactions (13). The generalized Born approximation is only valid for the Poisson equation which correspond to κ=0 in equation 14. Since it has not been tested extensively on protein systems, we only used it to calculate the solvation free energies of the HIV-1 RT inhibitors.

Methods

1. T4 lysozyme

The models of the folded and unfolded states. The T4 lysozyme we analyzed is a mutant that has an unnatural amino acid at position 133, S-2-amino-3-cyclopentylpropanoic acid (Cpe) (19), which differs from an alanine residue in that it has a cyclopentyl group attached at C_β on the side chain. Our previous simulation study indicated that replacement of the original Leu at 133 with Cpe will better stabilize the enzyme than with 19 other natural amino acids, a prediction which was confirmed experimentally (19).

The structure of the mutant T4 lysozyme with a Cpe was obtained by model building using standard geometries based on the structure of the wild type T4 lysozyme (11) from the Brookhaven Protein Data Bank and was subjected to energy minimizations before the molecular dynamics simulations. An 18 Å of cap TIP3P water (20) molecules centered around the $C_{\varepsilon 2}$ atom of Cpe133 was used in the simulations on the enzyme. 17 counterions (Na^+ or Cl^-) were added to keep the whole system neutral. Only residues within the sphere and the cap water molecules, which consists of about 1600 protein and counterion atoms and 260 water molecules, were allowed to move in the molecular dynamics simulations. The cap water molecules were kept from escaping by a weak repulsive potential (1.5 kcal/mol) at the surface of the sphere.

The unfolded state of the enzyme was represented by a terminally blocked solvated dipeptide, Ace-X-NMe, in which Ace and NMe are the acetyl and N-methylamide groups respectively and X is the unnatural amino acid residue. The backbone of the dipeptide was chosen to be in the extended state (-180°<ϕ<0°,

0°<ψ<180°). The side chain dihedral χ (N-C_α-C_β-C_γ) was chosen to be around 180°. The dipeptide was placed at the center of a box of 30 × 30 × 30 $Å^3$ filled with TIP3P water molecules under standard conditions. The number of water molecules is about 760. It should be noted that although different boundary conditions were used for the folded and unfolded states, because only $\Delta\Delta G$ contribute to the stability difference of the protein (equation 2), the error due to different boundary conditions is most likely canceled. Especially, the mutations we studied are nonpolar, in which the major contributions are short-range interactions, and we used the same cutoff radius (see below) in the simulations of the folded and unfolded states.

Force field parameters. The all atom force field developed by Cornell *et al* (21) were used. The atomic charges of the unnatural amino acid were obtained by fitting the electrostatic potential around the dipeptide model using the RESP method (22). The electrostatic potential was obtained by a single point *ab initio* quantum mechanical calculation using Gaussian94 (23) with 6-31G* basis set on a geometry generated by energy minimization with the AM1 method.

MD simulations. The MD simulations of PROFEC and CMC/MD were performed with the SANDER module of the AMBER4.1 program (24). The MD simulations of FED and TI were performed with the GIBBS module. Each simulation was performed with 2 fs time step, 8 Å cutoff radius and restrained temperature around 300K (25). The bond lengths were constrained by the SHAKE algorithm (26). For the simulations in water, periodic boundary conditions were applied and the pressure was controlled at 1 atm (25). The SETTLE algorithm was used to speed up the calculations on water molecules (27).

FED and PROFEC calculations. The free energy derivatives (FED) and the PROFEC contours were calculated from 300 ps and 100 ps MD simulations in the enzyme and in water, respectively. The $C_{\varepsilon 2}$ of Cpe133 was used as the origin of the grid. The two hydrogens attached to $C_{\varepsilon 2}$ were used to define the x-axis and the xy-plane. The parameters of the Lennard-Jones test particle are R*=2.0 Å and ε=0.15 kcal/mol, which are close to the van der Waals parameters of a tetrahedral carbon atom. The PROFEC results were visualized with UCSF MidasPlus (28) through a special delegate program written by R. J. Radmer.

Free energy calculations (TI). Each calculation was performed from $\lambda=0\rightarrow1$ (the forward change) and $\lambda\rightarrow0$ (the backward change). The average of the two results and their absolute difference were taken as the estimated ΔG and the hysteresis. The simulation time for each change ranges from 164 ps to 504 ps and the number of windows are from 41 to 126 (see Table II). For each window, the first 2 ps simulation was used as equilibration and the following 2 ps was used as sampling.

2. HIV-1 RT and its TIBO inhibitors

Force field parameters for the TIBO derivatives. Van der Waals (VDW) parameters of the chlorine atoms were taken from parameters used for chloroform (29) and the parameters for the sulfur atom (VDW, bond, angles, dihedrals and improper dihedrals) were adopted from a parameterization of thiobiotin (30). We used both the conformation of 8Cl-TIBO (R86183, see Table III) in complex with HIV-1 RT (31) as well as the A-form of the crystal structure of 9Cl-TIBO (R82913) (32) to estimate the partial atomic charges of 8Cl-TIBO, 9Cl-TIBO and unchlorinated TIBO (R82150). The two respective conformers were geometrically optimized using Gaussian94 (23) at the STO-3G level, each followed by a calculation of the electrostatic potential with the 6-31G* basis set. Atomic partial charges of the TIBO derivatives were fitted to the electrostatic potentials around the two structures using the RESP method (22). Comparison of the partial charges evaluated from the two conformers individually showed a very small difference and we therefore evaluated the partial charges of the remaining TIBO derivatives (Table III) using only the TIBO conformer of 8Cl-TIBO in HIV-1 RT.

Setup and equilibration of HIV-1 RT in complex with 8Cl-TIBO. Unresolved residues (modeled as alanines) as well as hydrogens were added to the 3.0 Å resolution crystal structure of 8Cl-TIBO in HIV-1 RT (31). After a short minimization of the hydrogens *in vacuo*, the complex was hydrated by immersing it in a 55 Å radius sphere of TIP3P-water (20). The solvent sphere and the protein-inhibitor complex were minimized to let the protein relax in an aqueous environment. All water molecules beyond the first hydration shell (i.e. at a distance > 3.5 Å from any protein atom) were then removed and counterions (11 Cl^-) were added to achieve electroneutrality. Protein residues with any atom closer than 12 Å from 8Cl-TIBO were chosen to be flexible in the simulations and all protein residues, water molecules and counterions further than 15 Å from any flexible residue were deleted. A 20 Å radius spherical cap of TIP3P-water, including the hydrating water molecules within the sphere from the previous step, was centered on TIBO and equilibrated for 50 ps at 300 K. The protein, 8Cl-TIBO and the hydrating water molecules outside the water cap were then kept rigid. Thereafter, the flexible residues (as defined above) and 8Cl-TIBO together with the cap of water molecules were then heated (50 ps) and equilibrated for 300 ps at 300 K. The simulations were carried out with the SANDER module of AMBER 4.1 (24) using the Cornell *et al.* force field (21). We applied a dual cutoff of 9 and 13 Å, respectively, where energies and forces due to interactions between 9 and 13 Å were updated with the same frequency as the non-bonded list, i.e., every 20 time steps. A time step of 2 fs was used and all bonds were constrained with the SHAKE algorithm (26). The temperature was maintained using the Berendsen method (25), with separate couplings of the solute and solvent to the heat bath because the relaxation times of the solute and solvent may be different (33).

Setup and equilibration of 8Cl-TIBO in solution. As starting conformer we chose the A-form from the crystal structure of 9Cl-TIBO (32), with a substitution of the atoms at positions 8 and 9. 8Cl-TIBO was then immersed in a box of TIP3P water with dimensions 34 × 33 × 29 $Å^3$. Keeping the inhibitor rigid, the water molecules were equilibrated at constant pressure for 100 ps. The TIBO atoms were then released and the system was equilibrated for 200 ps, using the same dual cutoff and time step as for 8Cl-TIBO in HIV-1 RT.

Adaptive CMC/MD. The method was applied to 8 different TIBO derivatives, shown in Table III. Each inhibitor was positioned in the equilibrated HIV-1 RT - 8Cl-TIBO complex (see above), by substituting and/or deleting atoms in 8Cl-TIBO. The inhibitors were then allowed to relax in the binding pocket by individually minimizing them, keeping everything but the inhibitor rigid. Due to problems with the SHAKE algorithm during the MD steps, the time step was reduced to 1.5 fs and one MC step was performed every 20 MD time steps. We applied the adaptive CMC/MD method for two sets of inhibitors, the $\Delta G_{offs,i}$'s (see the "Theory" section) were iteratively adjusted every 500 MC steps for set 1 and we shortened that interval to every 125 MC steps for set 2. The free energies of solvation for the TIBO derivatives were estimated from GB/SA (13) calculations, using the program MacroModel/BatchMin, version 4.5 (34). For these calculations, we used our RESP derived charges on the derivatives, which were minimized *in vacuo* prior to the calculations.

PB calculations. The 8 different TIBO - HIV-1 RT systems were further minimized, now with flexible residues, water molecules and counterions as in the MD simulations (see above). All water molecules and counterions were then removed and the PB calculations were carried out with the latest Delphi package (35,36), using a dielectric constant ε=2 for both the protein and the inhibitors and with a ionic strength of 0.13 M. For estimations of $\Delta G_{L1\rightarrow L2,s}$ (equation 3), the same structures as for the GB/SA calculations (above) was used, with ε=2 and an ionic strength of 0.13 M.

Free energy calculations (TI). For these calculations, which were carried out with the GIBBS module of AMBER4.1, we applied the same parameters and protocol as for the MD simulations (above). Starting with the equilibrated systems of 8Cl-TIBO in HIV-1 RT and in solution, respectively, the 8-chloro atom was perturbed into a hydrogen (R82150), using a window size ($\Delta\lambda$) of 0.02 (i.e. 51 windows in the λ-interval [0,1]). For TIBO in solution each window was equilibrated for 2 ps prior to a data collection time of 5 ps per window. The equilibration/data collection times for TIBO in HIV-1 RT were 3 ps and 8 ps, respectively.

Results

1. The stability of T4 lysozyme

Free energy derivatives and PROFEC. The free energy derivatives with respect to VDW radius (R*) of eight hydrogen atoms on the cyclopentyl ring of Cpe were calculated (Table I). The configurations of the hydrogens are defined as either pro-α or pro-β. The pro-α hydrogen is on the opposite face of the cyclopentyl ring from the C_β atom; the pro-β hydrogen is on the same face as the C_β atom. This definition of configurations is similar to that of the anomers of sugars. From Table 1, one sees that the free energy derivatives of HD12 (pro-α), HE12 (pro-β), HE21 (pro-α) and HD22 (pro-β) are negative, indicating that introducing some VDW group larger than hydrogen on either one of these sites may stabilize the protein. Because the free energy derivative of HE21 is the lowest, we focused our analyses around CE2 where HE21 is attached.

Table I. The free energy derivatives of Cpe (kcal/mol)[a]

Atom	dG/dR*,prot	dG/dR*,soln	Δprot-soln	configuration
HD11	7.7	6.1	1.6	pro-β
HD12	3.9	6.4	-2.5	pro-α
HE11	6.7	2.5	4.2	pro-α
HE12	4.5	6.3	-1.8	pro-β
HE21	2.4	6.3	-3.9	pro-α
HE22	7.1	2.5	4.6	pro-β
HD21	6.8	3.7	3.1	pro-α
HD22	6.4	8.9	-2.5	pro-β

[a] The free energy derivatives were obtained by 300 ps MD simulations in the enzyme and in water.

The PROFEC contour of zero VDW potential with $C_{\varepsilon 2}$ of Cpe as the origin is shown in Figure 1. Interestingly, the contour has the shape of a vase; its mouth faces the cavity and its neck embraces $C_{\varepsilon 2}$. The contour agrees with the free energy derivatives in that there is much more space for introducing a group at HE21 (negative derivatives) than at HE22 (positive derivatives). A natural proposal is the introduction of a methyl group at HE21 in the α configuration. In the following, we refer to this modification as Cpe→α–Mcpe and refer to the introduction of the methyl group at HE22 in the β configuration as Cpe→β-Mcpe. Figure 2 shows the superimposition of a methyl group at HE21 and at HE22. Obviously, the methyl

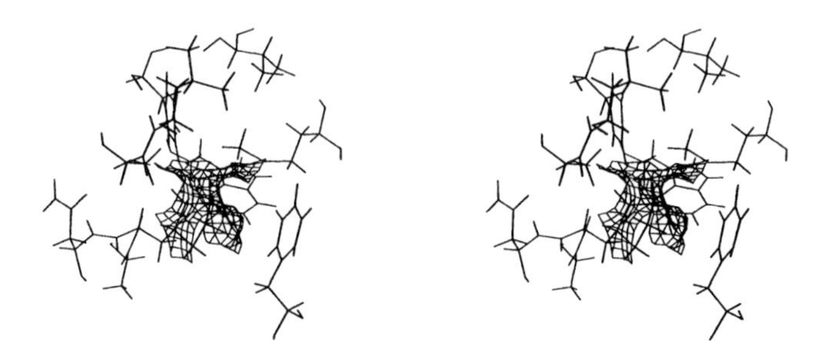

Figure 1. The PROFEC contour centered at $C_{\varepsilon 2}$ of Cpe133.

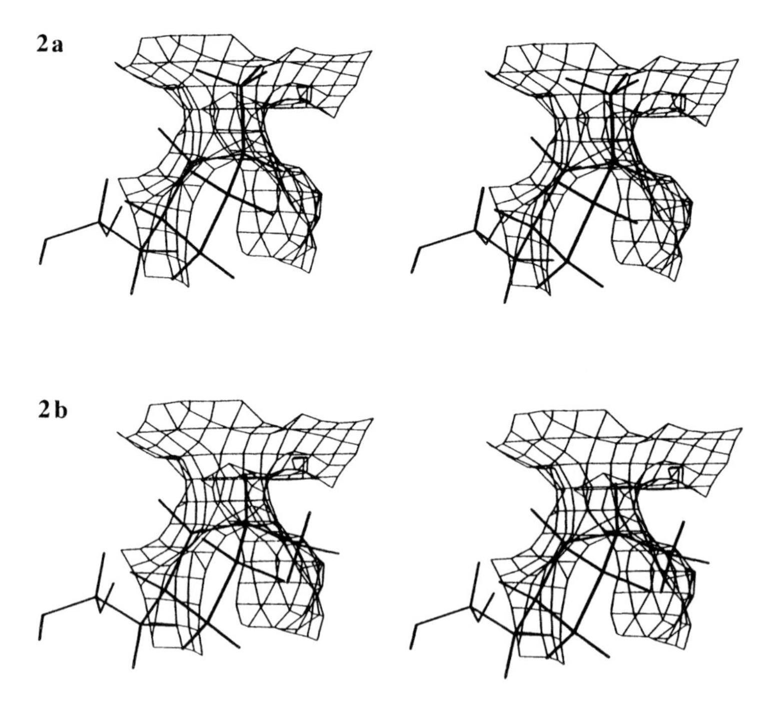

Figure 2. The superimposed structures of a methyl group at HE21 in the α configuration (a), and at HE22 in the β configuration (b).

group introduced at HE21 can fit very well in the cavity while the methyl group introduced at HE22 will collide with the wall of the vase belly. This suggests that adding a methyl group at HE21 in α configuration will improve the stability of the enzyme with a Cpe at position 133.

Table II. The ΔG of Cpe→Mcpe in water and in the enzyme (kcal/mol)[a]

medium	config.	time	forward ΔG	backward ΔG	average ΔG
protein	α	164 ps	-1.92	-2.20	-2.06
protein	α	324 ps	-2.70	-2.59	-2.65
protein	α	488 ps	-2.64	-2.56	-2.60
protein	β	164 ps	-0.24	0.30	0.03
protein	β	324 ps	-0.66	-0.22	-0.44
water	α	164 ps	-1.18	-0.94	-1.06
water	α	324 ps	-0.51	-0.36	-0.44
water	α	504 ps	-0.86	-0.85	-0.86
water	β	324 ps	-0.66	-0.22	-0.44

[a] Each window has 2 ps / 2 ps for equilibration and sampling. The 164, 324 and 504 ps simulations have 41, 81 and 126 windows respectively. The 488 ps simulation: 41 windows for the electrostatic contribution and 81 windows for the VDW contribution.

Free energy calculations (TI). We have calculated the free energy changes for introducing a methyl group in the α (at HE21) and β (at HE22) configurations in water and in the T4 lysozyme (Table 2). One sees that the calculated average ΔGs are not sensitive to the length of simulation time and the hystereses are ≤ 0.5 kcal/mol. For the simulations in the enzyme, we used the results of 324 ps for forward or backward change as the estimated free energy changes: ΔG (Cpe→α-Mcpe) = -2.65 kcal/mol and (Cpe→β-Mcpe)= -0.44 kcal/mol. For the simulations in water, Cpe→α-Mcpe and Cpe→β-Mcpe should give similar results as is seen from the results of 324 ps for both configurations. Since the average ΔG of Cpe→α-Mcpe fluctuates around -0.8 kcal/mol for different lengths of simulation time, -0.8 kcal/mol is taken as the estimated average ΔG in water. Based on these results, Cpe→α-Mcpe will stabilize the enzyme by 1.8 kcal/mol while Cpe→β-Mcpe will destabilize the enzyme by 0.4 kcal/mol. Therefore, the free energy calculations support the predictions of PROFEC.

2. Relative binding affinity to HIV-1 RT for a series of TIBO inhibitors

Adaptive CMC/MD and PB calculations. In Table IV, we present the relative ranking of the TIBO derivatives found in the CMC/MD and PB calculations. The numerical values of the relative free energies and full details will be presented elsewhere. The three best binding TIBO derivatives were also ranked as the best with the adaptive CMC/MD method. In this context, it should be noted that deviations in the rank order from the experimental results also might be due to an imperfect agreement between HIV-1 RT activity and binding affinity, caused by differences in cell penetration ability and metabolic stability between the TIBO derivatives. The experimental EC_{50} values (Table III) of the three next inhibitors, ranked 4 to 6, are very close to each other and they are also ranked 4 to 6 with the CMC/MD method. In the PB calculations, $\Delta\Delta G_b$ for R84914 has significantly been overestimated and its rank order is thus too favorable, which shifts the rank order of these three derivatives downwards. Omitting R84914 in the ranking according to PB calculations, the agreement with experiment is very good and the order is reversed for only two inhibitors (R87027 and R84674). Finally, the two inhibitors with the highest $\Delta\Delta G_b$'s

(R84914 and R80902) are also correctly ranked with both methods. The poor binding of R80902, which has an oxygen instead of a sulfur atom at position 2 (Table III), is mainly due to its favorable free energy of solvation (ΔG_{solv}). According to the GB/SA method, we estimated ΔG_{solv} for R80902 around 2 kcal/mole lower than for the other TIBO derivatives. Both the adaptive CMC/MD and the PB calculations have very reasonable average errors in the $\Delta\Delta G_b$ compared to experiment (0.71 kcal/mole and 1.12 kcal/mole, respectively).

Table III. The selected set of TIBO derivatives (see also Figure 3).

Compound	R_1	R_2	R_3	R_4	EC_{50} [a]
R86183	8-Cl	S	H	$-CH_2-CH=C(CH_3)_2$	4.6
R82913	9-Cl	S	H	$-CH_2-CH=C(CH_3)_2$	33
R82150	H	S	H	$-CH_2-CH=C(CH_3)_2$	44
R80902	H	O	H	$-CH_2-CH=C(CH_3)_2$	4200
R84674	$8-CH_3$	S	H	$-CH_2-CH=C(CH_3)_2$	14
R84963	H	S	$-CH_3$ (*trans*)[b]	$-CH_2-CH=C(CH_3)_2$	39
R84914	H	S	$-CH_3$ (*cis*)[b]	$-CH_2-CH=C(CH_3)_2$	790
R87027	8-Cl	S	H	$-CH_2-CH=C(CH_2CH_3)_2$	5.1

[a] ref. 12.
[b] relative stereochemistry of the methyl groups at positions 5 and 7.

Table IV. Rank order of $\Delta\Delta G_{b,i}$ [$=\Delta G_{b,i}-\Delta G_{b,R86183} - (\Delta G_{solv,i}-\Delta G_{solv,R86183})$] for the TIBO derivatives from adaptive CMC/MD, PB calculations and experiments (12).

Compound	adaptive CMC/MD rank	PB-calculations rank	experimental $\Delta\Delta G_{b,i}$ [a]	experimental rank
R86183	**3**	**1** (1)[b]	0	**1** (1)
R87027	**1**	**4** (3)	0.06	**2** (2)
R84674	**2**	**2** (2)	0.66	**3** (3)
R82913	**6**[c]	**5** (4)	1.17	**4** (4)
R84963	**4**[c]	**6** (5)	1.27	**5** (5)
R82150	**5**[c]	**7** (6)	1.34	**6** (6)
R84914	**7**[c]	**3**	3.05	**7**
R80902	**8**	**8** (7)	4.04	**8** (7)

[a] calculated from $\Delta\Delta G_{b,i}=-RT\ln(EC_{50,i}/EC_{50,R86183})$, T=298 K, R=1.986 cal/K/mole.
[b] The ranking in parenthesis is when excluding R84914.
[c] Average rank order from two sets of MC/MD-runs. Set 1 consists of the following derivatives with known experimental binding affinities: R86183, R81913, R84963, R82150, R84914 and R80902 and was run for 450 ps. Set 2 consists of R86183, R87027, R84674, R81913, R84963, R82150 and R84914 and was run for 560 ps.

Free energy calculations (TI). Perturbing 8Cl-TIBO (R86183) into 8-H TIBO (R82150), yields a relative binding free energy (-1.9±0.5) in close agreement with experiment (12) (-1.34). The perturbation in water is well converged as seen from the close values from the forward and reverse runs (Table V). However, in spite of the extended equilibration/data collection times (see "Methods") for the inhibitor in HIV-1 RT, these perturbations show a considerable hysteresis. Decomposition of the free energies shows that the van der Waals contribution is responsible for the slow convergence, differing by 1.4 kcal/mole between the forward and reverse runs. The same decomposition also reveals that the better binding of R86183 compared with R82150 is almost entirely due to a more favorable van der Waals contribution to the

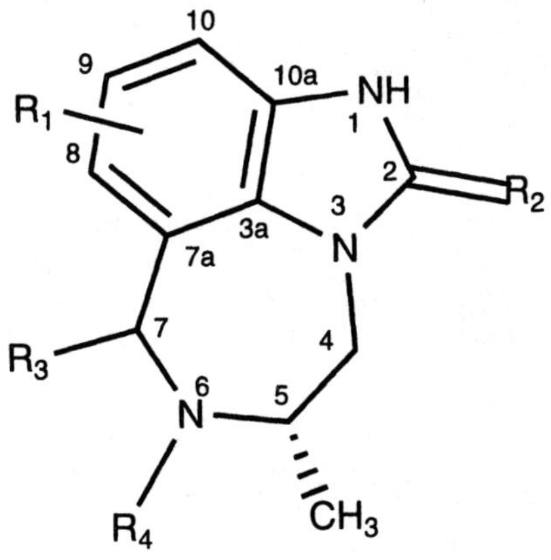

Figure 3. TIBO derivatives. The substitutions are explained in Table III.

free energy for the former inhibitor. Considering that the relatively non-polar TIBO is positioned in a hydrophobic and aromatic binding pocket, it is not surprising that van der Waals forces dominate the binding interaction.

Table V. Free energy calculations, R82150→R86183. The numbers are in kcal/mole and error estimates are within parenthesis.

	$\Delta G_{L1\to L2}$[a]			$\Delta G_{L1\to L2,P}$[a]			$\Delta\Delta G_b$[a]
	forward	reverse	average	forward	reverse	average	
ΔG_{tot}[c]	1.53	1.69	1.61	-0.78	0.13	-0.32	-1.9 (0.5)[b]
ΔG_{elec}	2.52	2.70	2.61	2.71	2.20	2.45	-0.16 (0.26)
ΔG_{vdw}	0.70	0.70	0.70	-1.65	-0.27	-0.96	-1.7 (0.7)
ΔG_{pmf}	-1.69	-1.71	-1.70	-1.83	-1.73	-1.73	-0.03 (0.11)

[a] see equations 3-4.
[b] experimental value: -1.34 kcal/mole (12).
[c] ΔG_{tot} is the total free energy, ΔG_{elec} and ΔG_{vdw} are the electrostatic and van der Waals contributions, respectively, and ΔG_{pmf} is the bond potential of mean force contribution.

Conclusions

We have studied the stability of T4 lysozyme and the binding free affinities of the RT TIBO inhibitors by several approximate and efficient free energy calculation methods and the rigorous TI method. For T4 lysozyme, the FED and PROFEC are useful for suggesting promising sites and candidate modifications to improve the stability of the protein. The results were supported by the TI calculations. The combination of FED and PROFEC appears to be very efficient and effective in making predictions. By using FED, one can quickly find the promising sites and then PROFEC can be used on these sites to suggest candidate modifications. Compared with TI, FED and PROFEC calculations are much faster: only a few hundred ps of MD simulation is required to obtain reasonable results. The information generated by FED and PROFEC can be quite comprehensive because in many cases it is appropriate to calculate the free energy derivatives for many atoms in a single MD simulation and use the same trajectories to construct the PROFEC contours. For the RT inhibitors, both the chemical MC/MD method and the PB calculations are able to rank TIBO derivatives in good agreement with experiment. Since these methods are quite different in nature, each with their own set of approximations, they serve as a good complement to each other. That is, if both methods predict the same rank order, the reliability of this prediction will significantly increase. Considerable simulation times (1.1 ns) were required for a reasonable estimate of the relative free energies of just *two* TIBO derivatives bound to RT in our TI calculations. In that perspective, the two more approximate methods worked surprisingly well and produced valuable information with substantially less effort and time. However, one sees that by decomposing of free energy differences into components, the TI calculations can provide important insights into the nature of TIBO-RT interactions.

In summary, the approximate methods used here are able to make quite reasonable predictions with much less computational cost than the rigorous TI calculations. Therefore, they are valuable at least as fast screening tools in the last stages of structure-based drug refinement and protein engineering. The rigorous free energy calculations (TI, or FEP) are computationally expensive, but can be used to support the predictions of the approximate methods and help to gain insights into the nature of molecular interactions. We envision a hierarchy of computational methods that can be applied to these sorts of problems. First, FED and PROFEC can be used to suggest lead plausible modifications for lead optimizations. Second, CMC/MD or PB calculations can be used to rank the binding affinities or stabilities of many ligands or mutants in a short time. Finally, traditional free energy methods (TI, or FEP) can be used to analyze a few particularly interesting cases.

Acknowledgment

L. Wang and M. A. L. Eriksson contributed equally to this work. This work was supported by the NIH (GM-29072, P. A. Kollman; GM-56531, P. Ortize de Montellano; GM-56609, E. Arnold). We acknowledge the use of the facilities of UCSF computer graphics laboratory, supported by NIH P41-RR01081 (T. Ferrin, P.I.). Mats Eriksson acknowledges a postdoctoral grant from the Swedish Natural Research Council (NFR).

References

(1) Kollman, P. A. *Chem. Rev.* **1993**, *93*, 2395-2417.
(2) Åqvist, J.; Medina, C.; Samuelsson, J.-E. *Protein Eng.* **1994**, *7*, 385-391.
(3) Radmer, R. J.; Kollman, P. A. *J. Comput.-Aided Mol. Design.* **1997**, accepted for publication.
(4) Pitera, J.; Kollman, P. A. *J. Am. Chem. Soc.* **1997**, submitted for publication.
(5) Kong, X.; Brooks III, C. L. *J. Chem. Phys.* **1996**, *105*, 2414-2423.
(6) Gerber, P. R.; Mark, A. E.; van Gunsteren, W. F. *J. Comput.-Aided Mol. Design* **1993**, *7*, 305-323.
(7) Cieplak, P.; Pearlman, D. A.; Kollman, P. A. *J. Chem. Phys.* **1994**, *101*, 627-633.
(8) Pang, Y. P.; Kollman, P. A. *Perspect. Drug Discov. Design* **1995**, *3*, 106-122.
(9) Shen, J.; Quiocho, F. A. *J. Comput. Chem.* **1995**, *16*, 445-448.
(10) Shen, J.; Wendoloski, J. *J. Comput. Chem.* **1996**, *17*, 350-357.
(11) Weaver, L. H.; Matthews, B. M. *J. Mol. Biol.* **1987**, *193*, 189-199.
(12) Pauwels, R.; Andries, K.; Debyser, Z.; Kukla, M. J.; Schols, D.; Breslin, H. J.; Woestenborghs, R.; Desmyter, J.; Janssen, M. A. C.; de Clercq, E.; Janssen, P. A. J. *Antimicrob. Agents. Chemother.* **1994**, *38*, 2863-2870.
(13) Still, C. W.; Tempczyk, A.; Hawley, R. C.; Hendrickson, T. *J. Am. Chem. Soc.* **1990**, *112*, 6127-6129.
(14) Metropolis, N. R.; Rosenbluth, M. N.; Teller, A. H.; Teller, E. *J. Chem. Phys.* **1953**, *21*, 1087-1092.
(15) Sharp, K.; Honig, B. *Ann. Rev. Biophys. Biophys. Chem.* **1990**, *19*, 301.
(16) Hermann, R. B. *J. Phys. Chem.* **1971**, *76*, 2754.
(17) Nozaki, Y.; Tanford, C. H. *J. Biol. Chem.* **1971**, *246*, 2211.
(18) Connolly, M. L. *Science* **1983**, *221*, 709-713.
(19) Mendel, D.; Ellman, J. A.; Chang, Z.; Veenstra, D. L.; Kollman, P. A.; Schultz, P. G. *Science* **1992**, *256*, 1798-1802.
(20) Jorgensen, W. L.; Chandrasekhar, J.; Madura, J. D.; Impey, R. W.; Klein, M. L. *J. Chem. Phys.* **1983**, *79*, 926-935.
(21) Cornell, W. D.; Cieplak, P.; Bayly, C. I.; Gould, I. R.; Mertz, K. M.; Ferguson, D. M.; Spellmeyer, D. C.; Fox, T.; Caldwell, J. W.; Kollman, P. A. *J. Am. Chem. Soc.* **1995**, *117*, 5179-1597.
(22) Bayly, C. I.; Cieplak, P.; Cornell, W. D.; Kollman, P. A. *J. Phys. Chem.* **1993**, *97*, 10269-10280.
(23) Frisch, M. J.; Trucks, G. W.; Schlegel, H. B.; Gill, P. M. W.; Johnson, B. G.; Robb, M. A.; Cheeseman, J. R.; Keith, T.; Petersson, G. A.; Montegomery, J. A.; Rahavachari, K.; Al-Laham, M. A.; Zakrzewski, V. G.; Ortiz, J. V.; Foresman, J. B.; Peng, C. Y.; Ayala, P. Y.; Chen, W.; Wong, M. W.; Andres, J. L.; Replogle, E. S.; Gomperts, R.; Martin, R. L.; Fox, D. J.; Binkley, J. S.; Defrees, D. J.; Baker, J.; Stewart, J. P.; Head-Gordon, M.; Gonzalez, C.; Pople, J. A. *Gaussian 94, Revision B.3* **1995**, Gaussian Inc., Pittsburg, PA.

(24) Pearlman, D. A.; Case, D. A.; Caldwell, J. W.; Ross, W. S.; III, T. E. C.; Ferguson, D. M.; Seibel, G. L.; Singh, U. C.; Weiner, P.; Kollman, P. A. *AMBER, version 4.1* **1995**, University of California at San Francisco.
(25) Berendsen, H. J. C.; Postma, J. P. M.; van Gunsteren, W. F.; DiNola, A.; Haak, J. R. *J. Chem. Phys.* **1984**, *81*, 3684-3690.
(26) Ryckaert, J. P.; Ciccotti, G.; Berendsen, H. J. C. *J. Comput. Phys.* **1977**, *23*, 327-341.
(27) Miyamoto, S.; Kollman, P. A. *J. Comput. Chem.* **1992**, *13*, 952-962.
(28) Ferrin, T. E.; Huang, C. C.; Jarvis, L. E.; Langridge, R. *J. Mol. Graph.* **1988**, *6*, 13-27.
(29) Fox, T.; Thomas, B. E.; McCarrick, M.; Kollman, P. A. *J. Phys. Chem.* **1996**, *100*, 10779-10783.
(30) Miyamoto, S.; Kollman, P. A. *Proteins* **1993**, *16*, 226-245.
(31) Ding, J.; Das, K.; Moereels, H.; Koymans, L.; Andries, K.; Janssen, P. A. J.; Hughes, S. H.; Arnold, E. *Struct. Biol.* **1995**, *2*, 407-415.
(32) Liaw, Y. C.; Gao, Y. G.; Robinson, H.; Wang, A. H. J. *J. Am. Chem. Soc.* **1991**, *113*, 1857-1859.
(33) Guenot, J.; Kollman, P. A. *Prot. Sci.* **1992**, *1*, 1185-1205.
(34) Mohamadi, F.; Richards, N. G. J.; Guida, W. C.; Liskamp, R.; Lipton, M.; Caufield, C.; Chang, G.; Hendrickson, T.; Still, W. C. *J. Am. Chem. Soc.* **1990**, *11*, 440-467.
(35) Gilson, M. K.; Sharp, K. A.; H., H. B. *J. Phys. Chem.* **1988**, *9*, 327-335.
(36) Honig, B.; Nicholls, A. *Science* **1995**, *268*, 1144-1149.

Chapter 4

Binding Evaluation Using the Finite Difference Solution to the Linearized Poisson–Boltzmann Equation and Solvation Entropy Correction

Jian Shen

Computational Chemistry, Hoechst Marion Roussel, Inc., Route 202-206, Bridgewater, NJ 08807

The finite difference solution to the linearized Poisson-Bolzmann equation is used to compute the electrostatic binding free energy of various receptor-ligand systems including L-arabinose-binding protein, thermolysin, thrombin, collagenase, etc. The non-electrostatic binding free energy is approximated using the empirical solvation entropy correction. Besides the balanced speed and accuracy, the method provides detailed mechanistic insights of the binding interactions and thus the rationales for both lead optimization and lead generation. Practical concerns and limitations of the method are also discussed.

The determination of receptor-ligand binding affinity (in terms of K_i, K_d, IC_{50} or binding free energy) has been an integral part of new drug discovery. Enhancing binding affinity is often a top priority in lead generation and lead optimization. Chemical modifications to improve other pharmacological properties also require maintaining a predefined binding affinity. Medicinal chemists often estimate binding affinities of prospective drug molecules based on statistical analysis or their own intuition for selective syntheses. If the structural information of the receptor is available, many suggestive ideas about binding improvements can be simulated by visualizing the 3D graphics. However, the outcome of a new compound usually remains chancy. On the other hand, the structural information provides essential input for free energy evaluation based on statistical mechanics (*1*,*2*) and promises an ultimate solution to predict the binding free energy and other thermodynamic properties (*3*).

Indeed, many developing methods (*4*) based on the structure of receptors have various success stories. To be able to impact on pharmaceutical research, a binding energy calculation method has to be reliable, fast and comprehensible. First, a qualitatively correct prediction for a variety of receptor-ligand binding systems is essential to real application. The numerical values are also required to be quantitatively close to experiments and reproducible. Second, the method needs to generate results quickly enough for making a synthetic decision, which usually means within a day or two. Third, the analysis of calculated results should offer an explanation of the ligand-receptor interaction, and thus suggests new directions to improve binding.

Among several promising computational technologies, the finite difference solution to the linearized Poisson-Bolzmann equation and solvation entropy correction (FDPB + SEC)(*5*) has unique features in terms of accuracy, speed and mechanistic insight for binding energy calculation. Because of no cut-off for electrostatic interaction and inclusion of solvent effect, FDPB is considered to be one of the most accurate methods to compute electrostatic energy(*6*). Studies (*7*) show that the numerical error in the calculation can be reduced to just a few tenths of kcal/mol when appropriate settings are used. Furthermore, the computing speed has been improved in parallel with the computer technology. A typical protein-ligand binding energy calculation (consisting of six FDPB calculations on a 100^3 grid) can be completed within one hour CPU on a R4000 processor (SiliconGraphics) while the cost is only half of that on a R10000. The computing speed is no longer a practical concern because the manual preparation of correct inputs usually takes more time. Finally, the method provides components of calculated binding energy difference such as solvation and hydrophobic binding. Analyzing these components can direct new molecular modification for improving binding and other pharmacological properties of ligands.

The principle of using FDPB for electrostatic binding energy calculation was described by Gilson and Honig (*8*). The early applications were focused on the effect of salt concentration on binding energy. The first published attempt by Karshikov et al.(*9*) to calculate hirudin-thrombin binding energy was not successful. One common opinion about the method was its inaccuracy due to the grid representation of molecules (both partial charges and dielectric boundaries). However, improved techniques in FDPB have essentially eliminated the problem. The method has been successfully applied to several binding complex systems including sulfate-binding protein (*10*), L-arabinose-binding protein (ABP), thermolysin (*11*), carbonic anhydrase (*12*) and isocitrate dehydrogenase (*13*).

In this paper, we first review the basics of the FDPB + SEC approach. Then, several applications to ligand-protein systems including ABP, thermolysin, thrombin and collagenase are presented. These systems are either therapeutic targets or drug binding models. ABP is one of periplasmic receptors, which function as uptake of a variety of nutrients. There is increasing interest in carbohydrate-related drug discovery. Thermolysin, a zinc endopeptidase, has served as a model for angiotensin-converting enzyme. Recently, it has been found to degrade amyloid, and thus is implicated as a potential therapeutic for Alzheimer's disease. Thrombin is a well-known target for cardiovascular disease. Collagenase is a member of matrix metalloproteases (MMPs). These enzymes play a cardinal role in the breakdown of extracellular matrix and are involved in a variety of biological and pathological processes. In addition to their therapeutic value, these receptor-ligand systems have multiple high-resolution 3D structures determined by x-ray crystallography. The specific methods used to treat problems associated with electrostatic, nonelectrostatic, H-bond, conformation, and the origin of electrostatic binding of proteins are described below. Finally, certain underlying assumptions and limitations of the method are discussed.

Theory and Method

For noncovalent binding of a ligand, L_0 (reference system), to a receptor, P

$$L_0 + P \rightarrow L_0P,$$

the associated binding free energy, ΔG_0, can be partitioned into electrostatic and nonelectrostatic components, ΔG_e and ΔG_n, respectively:

$$\Delta G_0 = \Delta G_e + \Delta G_n. \quad (1)$$

Similarly, we have ΔG_1 for a modified ligand, L_1. Thus, the binding energy difference, $\Delta\Delta G$, between the reference system and the modified system is:

$$\Delta\Delta G = \Delta G_1 - \Delta G_0 = \Delta\Delta G_e + \Delta\Delta G_n. \quad (2)$$

The reason to calculate $\Delta\Delta G$ is that the model to calculate $\Delta G_e + \Delta G_n$ may not include every significant interaction in a binding process. We assume that the neglected interactions are identical in both the reference and modified systems. Therefore, they will cancel in $\Delta\Delta G$. Knowledge of the relative binding energy is sufficient for most practical application.

Electrostatic Calculation. The electrostatic binding energy, ΔG_e, is calculated through the following thermodynamic cycle:

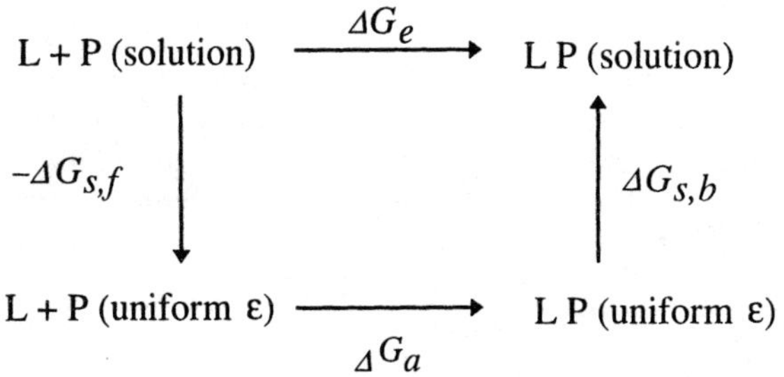

L + P (solution)	$\xrightarrow{\Delta G_e}$	L P (solution)
$-\Delta G_{s,f}$ ↓		↑ $\Delta G_{s,b}$
L + P (uniform ε)	$\xrightarrow[\Delta G_a]{}$	L P (uniform ε)

$$\Delta G_e = \Delta G_{s,b} - \Delta G_{s,f} + \Delta G_a \quad (3)$$

$$\Delta\Delta G_e = \Delta\Delta G_{s,b} - \Delta\Delta G_{s,f} + \Delta\Delta G_a \quad (4)$$

where ε is the dielectric constant of the system including the binding molecules and their environment. In this approach, two solvation energies ($\Delta G_{s,b}$ and $\Delta G_{s,f}$) are obtained from four electrostatic calculations using FDPB. The binding interaction in the uniform ε state (also called assembly energy or coulombic energy), ΔG_a, is analytically calculated using Coulomb's law. In principle, the ΔG_e can be obtained by computing electrostatic energies of L + P and LP without bypassing. However, such calculated values are difficult to interpret or analyze. The value of this approach will become apparent later in this article. The other advantage of this approach is the reduction of error even though this is not so critical.

One of the most controversial parameters used in continuum models of proteins is the dielectric constant, which ranges from 2 up to 20 in many applications. It is used to account for the polarizability of electrons, atomic dipoles, macrodipoles, and probably the uncertainties of coordinates of a protein. Because it cannot be measured experimentally nor calculated with ab initio, this empirical value is determined by comparing calculated and experimental electrostatic interaction energy. After testing several values (*7*), we found that a constant of 3 is good for the binding energy calculation with the force fields across several protein systems. Due to the solute dielectric (i.e., 3 in our work), a dielectric constant of 3 is used for the environment of the uniform ε state to simplify the calculation of ΔG_a.

The Finite Difference Solution to the Linearized Poisson-Boltzmann Equation (FDPB). The electrostatic free energy (G_e) of a molecular system is calculated through

$$G_e = 0.5 \Sigma q_i \phi \tag{5}$$

where q_i, a charge of atom i, and ϕ, the electrostatic potential, are related though the linearized Poisson-Boltzmann (LPB) equation:

$$-\nabla \bullet \varepsilon \nabla \phi + \varepsilon \kappa^2 \phi = \rho \tag{6}$$

where κ is the inverse Debye-Huckel screening length and ρ, the charge density, relates to q ($q = \int \rho \, dV$). This equation is solved numerically using the finite difference method. In this method, the atomic charges and dielectric of a molecular system are partitioned into 3D grids. The associated potential grid is obtained by iterative solution of the LPB equation. Available commercial software includes UHBD (*14*) and DelPhi (*15*).

Non-electrostatic energy calculation. It has been assumed that non-electrostatic binding energy difference is due to the solvation entropy difference. Indeed, the solvation entropy dominates the solvation free energy of nonpolar molecules such as alkanes. Although the free energy calculated by solving the PB equation has an entropy component (electrostatic) in terms of the temperature derivative of dielectric, it does not include solvent-solute correlation functions (*16*) that constitute the solvation entropy.

In an early binding energy application (*11*), we tried to estimate the nonelectrostatic energy by counting the number of waters displaced from a binding site of an enzyme assuming that the release of one water molecule equals 3/2 kT (translation degree of freedom). Although it can account for most hydrophobic modifications in a series of thermolysin inhibitors, the treatment lacks strong theoretical basis and is too arbitrary to handle a volume smaller than a water molecule.

The molecular surface area, or solvent accessible surface area (SASA), has long been recognized and used as a correlation to solvation free energies. Sitkoff et al (*17*) derived a single surface tension coefficient to account for nonpolar solvation energies in FDPB calculation. Because the coefficient is optimized along with the force field parameters, it can be complicated when applied to macromolecular systems.

Solvation entropy correction (SEC). In seeking a better way to calculate $\Delta\Delta G_n$, we(*5*) found that the average solvation entropic energy ($-T\Delta S$) is 40 ± 5 cal/mol Å^2 for a unit SASA of both polar and nonpolar molecules(*18*). This value is also close to 30 cal/molÅ^2 derived from a simulation (*19*). Thus, our $\Delta\Delta G_n$, or the solvation entropy correction (SEC) is derived as:

$$\Delta\Delta G_n = 0.04\ (\Delta A_b - \Delta A_f)\ \text{(kcal/mol)} \tag{7}$$

where ΔA_f and ΔA_b are water accessible surface area changes of the free state and the bound state in two binding complexes, respectively.

Calculation Setup. Protein polar hydrogens are added to the x-ray structures studied in this work and energetically minimized using the program X-PLOR (*20*) or CHARMm (*21*). The coordinates of the modified inhibitors are modeled based on their reference structures. All calculations are done using UHBD with a 100x100x100 grid. The solvent region in the FDPB calculation is determined using the accessible surface of a probe of 1.1Å radius and modeled with a dielectric of 78. An ionic strength of 0.01M and an ion excluding shell of 1.4Å represent the solution. A dielectric constant of 3.0 with a smooth dielectric boundary is used to define the interior of ligands and proteins. A single focusing step is applied to reduce the grid spacing from the initial 1Å to the final 0.25Å around the binding site. GROMOS (22) charges are used for proteins and some ligands. Other charges on ligands are either from published data or ab initio calculations (*10,11*).

Applications

Electrostatic binding. For a given receptor, two binding ligands may differ electrostatically (i.e., their partial charge distributions vary but there is no significant change in size). An example is the substitution of an H-bond donor or acceptor like a hydroxyl of D-galactose (Gal) as shown in Table I. Compared with the Gal, the

Table I. Sugar Binding Free Energy Differences and Their Components for ABP (kcal/mol)[a]

Sugar	$\Delta\Delta G_a$	$\Delta\Delta G_s{}^b$	$\Delta\Delta G_e$	$\Delta\Delta G_{exp}$
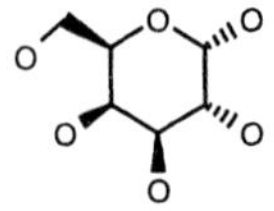 D-galactose	0.0	0.0	0.0	0.0
L-arabinose	4.50 4.69	-5.78 -5.91	-1.28 -1.22	-0.5
D-fucose	4.64 5.05	-3.10 -3.43	1.53 1.62	1.62
6-F-deoxyl-Gal	5.81 6.02	-2.46 -2.41	3.35 3.62	2.6
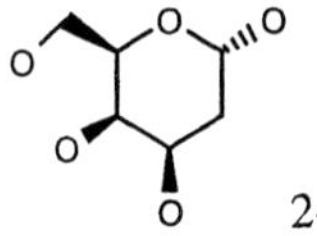 2-deoxyl-Gal	5.78 5.89	-3.20 -4.39	2.58 1.50	3.6
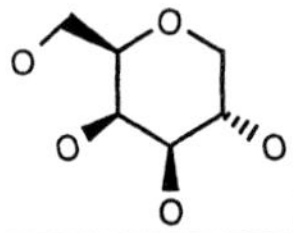 1-deoxyl-Gal	1.64 7.23	-0.32 -3.12	1.32 4.11	1.8

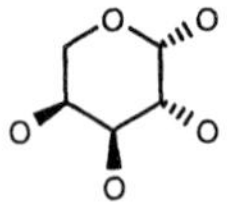

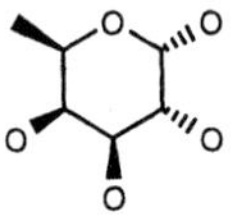

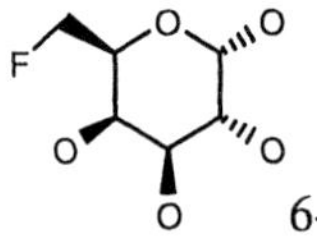

a. The sugar structures are shown as α form. The first calculated value in each column is for α anomer and the second one for β anomer, respectively.

b. $\Delta\Delta G_s = \Delta\Delta G_{s,b} - \Delta\Delta G_{s,f}$.

lack of one hydroxyl in other sugars results in various binding energy differences from -0.5 to 3.6 kcal/mol. Although a few complex structures of ABP-sugar have been solved by x-ray crystallography (*23*), it is still difficult to predict and quantify these energetic differences.

The calculation (*10*) using FDPB was quite satisfactory regarding qualitative trends and quantitative values. With reference to Gal binding, it correctly reproduced better binding for arabinose (Ara) and smaller affinities for the rest. Since two forms (depending on the configuration of 1-OH) of Gal were found in the x-ray structure, both were used as references to calculate the corresponding modified sugars. The populations of two anomers may not be equal in both free and bound states. In fact, only one anomer of bound arabinose can be found in a mutant ABP crystal structure (Quiocho, F. A. et al. Baylar College of Medicine, unpublished data). Nevertheless, the quantitative discrepancies between calculated sugar binding energy differences and experiments are generally less than 1 kcal/mol. Further improvements can be made by including the non-electrostatic contribution. As previously analyzed, this term generally favors a large group. Because Ara, Fucose (Fuc), 1-deoxy-Gal and 2-deoxy-Gal are smaller than Gal is, this contribution will make their binding energies more positive. 6-F-deoxyl-Gal is considered bigger than Gal is due to the greater van der Waals radius of fluorine and the longer C-F bond length. Thus, its binding energy should be adjusted negatively. This work will be carried out in the future in our lab and is expected to bring the results closer to the experiments.

The calculation also provides an important mechanistic explanation of why Ara is a better ligand and Fuc is a worse one than Gal is. Both Fuc and Ara have less binding contribution from the coulombic energy (positive $\Delta\Delta G_a$) than Gal does due to the lack of a hydroxyl (H-bonded to Asn 89 of ABP). However, Ara is much more solvated than Gal (negative $\Delta\Delta G_S$) in bound state due to the lack of 6-CH_2OH. This favorable solvation outweighs the loss of a hydrogen bond and makes Ara the tightest substrate for ABP. In contrast, the solvation contribution for Fuc is not big enough to compensate for the loss in coulombic energy, which dominates the overall change. The solvation difference from the calculation is consistent with the x-ray crystallographic observation (*24*) that a bound water, water 311, shifts toward the bound Ara more than toward the bound Fuc.

Hydrophobic interactions. Omission of non-electrostatic differences in sugar binding does not affect qualitative results because the size changes among these sugars are small, mostly one non-hydrogen atom. However, the non-electrostatic solvation cannot be neglected in other cases such as the calculation of a series of thermolysin inhibitors (*25*). These peptidyl analogues have different hydrogen bond

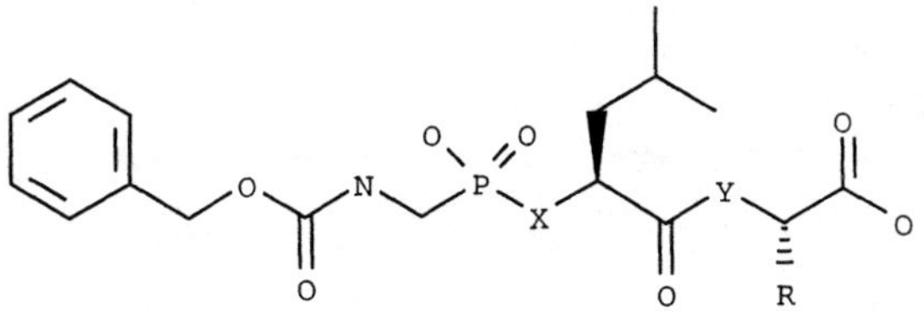

Phosphorus-Containing Inhibitors Cbz-Gly-ψ(PO_2)-X-Leu-Y-R (ZG^P(X)L(Y)R), where X = NH, O or CH_2; Y = NH or O; R = Leu, Ala, Gly, Phe, H or CH_3.

donor and acceptor on the backbone as well as varied P2' amino acid residues. The difference in molecular size between the reference ligand and a modified ligand ranges from two to eight non-hydrogen atoms, which challenges any binding energy calculation method. By combining FDPB with SEC, we (*5*) demonstrated that the calculated binding energy differences correlate very well to the corresponding experimental values as illustrated in Figure 1.

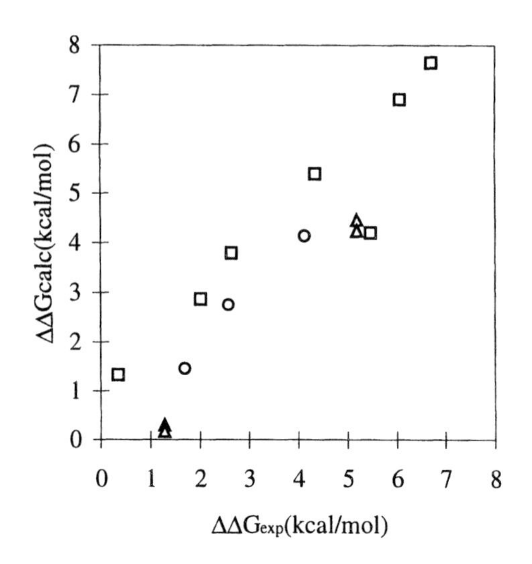

Figure 1. The calculated binding energy difference vs. experiments for 13 phosphorus-containing inhibitors, (ZG^P(X)L(Y)R). The symbol circle is for ZG^P(X)L(Y)L, square for ZG^P(X)L(Y)R (R ≠ Phe (F)) with solvation energy correction (SEC) and triangle for ZG^P(X)LF (two conformations) without SEC. The calculated binding energy differences are well correlated to the experiments (r = 0.90) at the same energy scale. (reproduced with permission from ref. 4. Copyright 1995 The Protein Society)

The need to include SEC can be seen more clearly in Figure 2, where another series of thermolysin inhibitors, ZR^P(O)LA, was calculated (*26*). The electrostatic binding, which prefers a small hydrophobic group, has an opposite trend compared with that of the experiment. Understandably, the electrostatic solvation of the complexes dominates $\Delta\Delta G_e$ because $\Delta\Delta G_a$ is zero and $\Delta\Delta G_{s,f}$ is small. The addition of $\Delta\Delta G_n$, which favors a large hydrophobic group, brings the calculated energy difference trend in line with the experiment.

H-bond strength. In many cases, the initial hits in an enzyme-targeted project are peptidyl inhibitors, which usually suffer from low oral bioavailability and quick degradation. While further screening may generate new hits, designing a new scaffold based on the peptide hits is an alternative approach in lead generation. In

MMP project, we reasoned that each amide or carbonyl of a bound peptide inhibitor does not contribute equally to the binding. The weakest H-donor or acceptor might be a potential moiety for modifications.

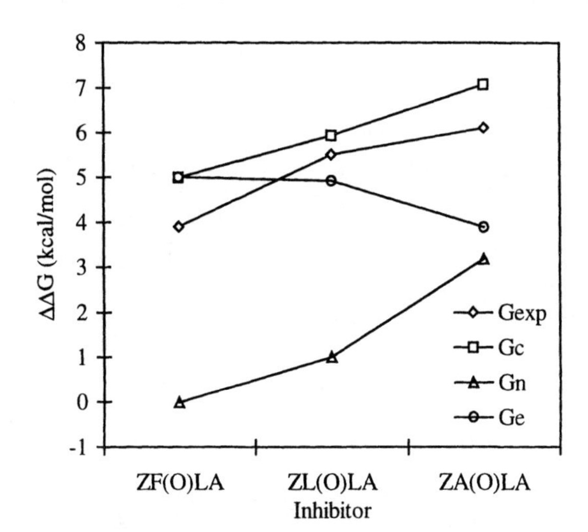

Figure 2. Calculated and experimental binding energy differences for ZR^{P}(O)LA.

The x-ray studies of collagenase(*27*) as well as stromelysin(*28*) show that all amides and carbonyls of a hydroxamate inhibitor are H-bonded to the protein active site. In order to predict each H-bond strength, we artificially turned off the partial charges on six NH and CO groups one at a time and calculated the binding energies using the FDPB method. Due to the uncertainties of the proton position between the active site Glu 219 and the hydroxamate, and of ionization of the hydroxamate, three binding mechanisms were assumed. Thus, differently protonated inhibitors and enzymes were subjected to this study, as shown in the Figure 3 legend.

The results support our hypothesis that not all H-bonds in the peptidyl inhibitor are equally important. The P1 CO and P1' NH are the strongest H-bonds and P1' CO and P2' NH the weakest. Understandably, the different proton assignments only affect the nearby P1 CO and P1' NH. The three binding mechanisms result in basically the same conclusion, which also validates the numerical calculations.

Although elimination of the H-bonds will reduce the electrostatic binding affinity from this study, the overall binding can be enhanced by hydrophobic modification based on both experiments and solvation entropy analysis. For example, adding a methyl to a ligand may increase its surface area by about 40 $Å^2$, which equates to -1.5 kcal/mol of solvation entropy energy. This amount of energy could offset the loss of weak H-bond if other interactions remain the same. Directed by this theoretical prediction, our medicinal chemists have generated new scaffolds by modifying P1'-P2' and P2'-P3' peptide bonds. Similar chemistry approaches have been disclosed by chemists in Dupont Merck (C. Decicco et al., poster in 214th ACS national meeting). These non-peptidyl inhibitors can achieve nM Ki binding and good oral bioavailability.

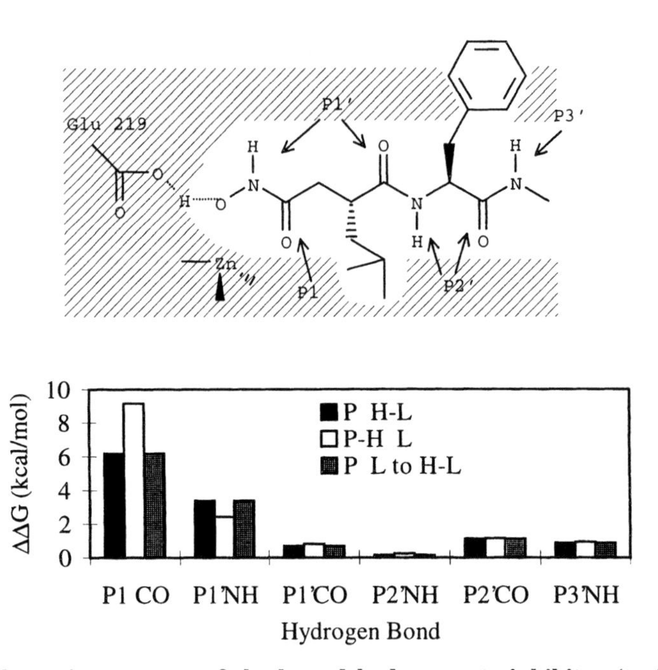

Figure 3. Schematic structure of the bound hydroxamate inhibitor (top) and relative H-bond strength of each NH and CO group (bottom). The ambiguous proton is indicated with two dash lines between Glu 219 and the inhibitor. The legend indicates whether a proton (H) is associated with Glu 219 of collagenase (P) or the ligand (L) in calculations.

Conformational Change. One of the most challenging tasks in molecular modeling is to determine the bioactive conformation of a ligand. If one ligand has been determined by biophysical methods, we usually assume that its analogs will bind to a receptor in an identical mode. This assumption is reasonable if the binding data and multiple x-ray structures can justify it as in the case of the ZG^{P}LR analogs. However, x-ray analysis (*29*) and binding kinetics (*30*) also prove that a modification can lead to different binding conformations, such as those of ZR^{P}LA (R $\neq$ G). In the best situation, computer-docking(*31*) may generate a correct bioactive conformation. The question is whether one can verify the binding mode and predict its binding energy.

The calculation of binding energy difference between conformationally different ligands is difficult for any detailed simulation method. To explore the limitation of the FDPB + SEC method in dealing with this problem, we (*26*) use three approaches to calculate the binding energy difference between ZF^{P}LA and ZG^{P}LL. Because both protein-inhibitor complexes (pdb4tmn and pdb5tmn) have been solved crystallographically, either one can be used as a reference protein structure. As shown in Figure 4, two direct calculations yield binding energy

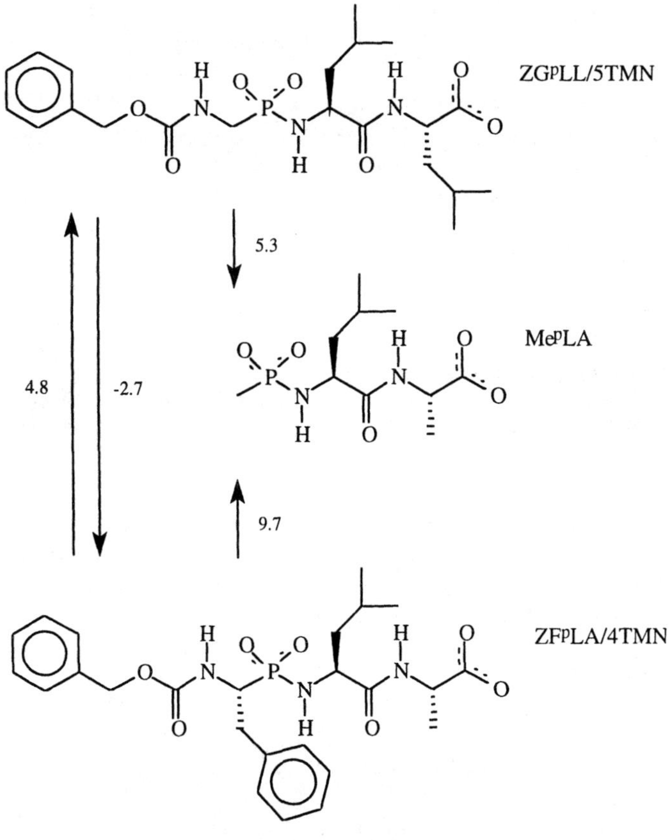

ZGPLL/5TMN

5.3

MePLA

4.8

-2.7

9.7

ZFPLA/4TMN

Figure 4. Structure of Me^{P}LA and schemes to compute energy difference between ZF^{P}LA and ZG^{P}LL. The arrows indicate calculations from a reference system to a modified system. The values (taken from Table 1) along the arrows are calculated binding energy differences (in kcal/mol) associated with each process.

difference of 4.79 and 2.74 kcal/mol, respectively. One of them closely matches the experimental binding energy difference of 2.94 kcal/mol. In these calculations, the modified systems are modeled by replacing the reference ligand with the one from the other x-ray structure while keeping the reference protein structures unchanged.

The reason for using only one protein structure is to reduce errors from the difference between the two protein structures. The third approach is composed of two calculations through a hypothetical inhibitor $Me^{P}LA$. Both pdb4tmn and pdb5tmn are used as the reference structures to calculate the energy differences from $ZF^{P}LA$ and $ZG^{P}LL$ to $Me^{P}LA$, respectively. This approach yields a binding energy difference of 4.4 kcal/mol between $ZF^{P}LA$ and $ZG^{P}LL$.

Although this approach produced a result no better than one of the direct calculations, it correctly modeled tighter binding of $ZF^{P}LA$ like the two other approaches. According to the analysis of decomposed energies, the enhancement of the $ZF^{P}LA$ binding is attributed largely to the solvation entropy or "hydrophobic force". The binding mode of the $ZG^{P}LR$ N-terminal moiety appears to be electrostatically unfavorable. The study suggests that increasing lipophilicity of that moiety may enhance $ZG^{P}LR$ binding.

Ionic Binding. When an ionic residue is identified at the active site of a receptor, it is natural to think of using a counter charge to enhance binding. Indeed, attaching a positively charged group, such as guanidinium or benzamidine(*32*), to an inhibitor seems necessary to achieve respectable binding to thrombin(*33*), which has a negatively charged Asp 189 at the active site. However, a similar rule does not work for collagenase, which has a positively charged arginine at the bottom of the S1'pocket. Unable to achieve significant binding using negatively charged P1' inhibitors, Singh et al.(*34*) attributed it to the desolvation penalty. If we look deeper, there is no desolvation problem for collagenase because experiments show that a hydrophobic residue can bind to the S1' site. The desolvation of the ligand should neither be an obstacle. In fact, an aspartate at P1 appears to be necessary to bind to interleukin 1-β converting enzyme (ICE)(*35*), which has two arginines at the active site. Is there any simple rule to allow us to predict ionic binding qualitatively? This question is related to the issue whether electrostatic interactions stabilize or destabilize proteins.

We carried out a series of FDPB calculations for thrombin and collagenase to explore the controlling factor in charge-charge interaction. The binding contribution of charged residues was accessed by mutating them to neutral ones. We found that none of the single residues, including Asp 189 of thrombin, can be said to contribute significantly to ion binding. The coulombic attraction and desolvation penalty are compatible to each other in most cases, and as a consequence, the net energetic differences are marginal.

This can be qualitatively understood by the interplay between desolvation and Coulombic interaction. A charged residue at an active site of a receptor has a strong coulombic attraction with a counter charge. However, the desolvation penalty offsets the interaction. The overall binding contribution of the residue may not be as great as it appears in an x-ray structure. This does not mean that electrostatic is not important for ionized ligand binding. In fact, we found that all charges and dipoles in a receptor need to be considered to predict the binding.

Figure 5 shows accumulated ΔG_a between test charges and two enzymes, thrombin and collagenase. In both cases, side chain contributions dominate the overall ΔG_a. For thrombin, the charged residues within 15 Å of the binding ion have a significant binding contribution while the remaining residues help maintain the binding. In contrast, the Coulombic binding of collagenase reaches a maximum at about 10 Å, then the trend reverses and ends up with a positive energy. In other words, collagenase will not bind to any negative charge at the S1' pocket. To compare with the experimental binding energies, we should exclude the first and perhaps second residues because they need to be desolvated to form a complex, which offsets their coulombic contribution significantly. These two examples show that the short-range electrostatic attraction may be necessary but not sufficient for the binding of a charged ligand. Charged residues that are distant from a receptor binding site may be more important than we had originally thought.

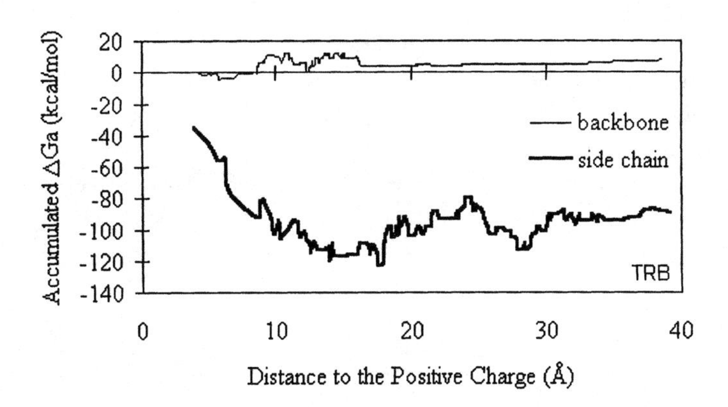

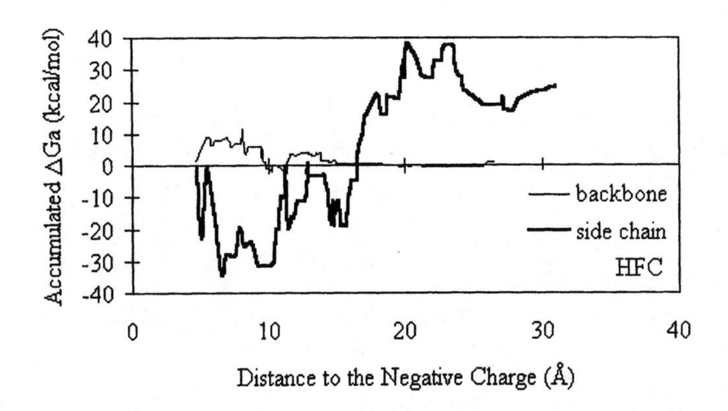

Figure 5. Accumulated ΔG_a for thrombin (TRB) and collagenase (HFC) with a counter charge at the active sites.

Currently, we are investigating the ionic binding of other binding systems. If the analyses for thrombin and collagenase are proven to be true in general, then we may just need a simple calculation to predict whether varying a charge on a ligand can enhance or reduce the binding affinity. To complete this study, more high-resolution structures of binding complexes are needed. The protonation states of residues also need to be validated. In short, one cannot predict ionic binding by simply looking at the 3D structure of a receptor complex.

Ligand-Based Design. Even without a receptor structure, the FDPB + SEC model can be useful for qualitative prediction. We can classify chemical modifications into two categories: electrostatic and hydrophobic. The former modification varies partial change in the ligand but has small or no size change. An example is the modification of ethane to ethanol. The latter modification involves a significant change in size but, a small partial charge change, such as the change of ethane to propane. If a ligand modification involves both partial charge and size changes, we can always treat them separately through a hypothetical intermediate.

Only $\Delta\Delta G_e$ needs to be considered in electrostatic modification. Among $\Delta\Delta G_e$ components, $\Delta\Delta G_a$ is very important to binding because it accounts for electrostatic interactions between a ligand and a receptor. For the solvation energies, the absolute value of $\Delta\Delta G_{s,f}$ is greater than that of $\Delta\Delta G_{s,b}$ due to less solvent exposure in the bound state. As an approximation to equation 4, we can have:

$$\Delta\Delta G \cong \Delta\Delta G_a - \Delta\Delta G_{s,f}. \tag{8}$$

This equation tells us that a modification of a nonpolar group to a polar group may enhance binding by lowering $\Delta\Delta G_a$. However, the net gain will be offset by increased $\Delta\Delta G_{s,f}$ and could result in a net loss of binding (positive $\Delta\Delta G_e$).

A hydrophobic modification will affect both $\Delta\Delta G_e$ and $\Delta\Delta G_n$. Because there is no charge change by the definition, $\Delta\Delta G_a$ can be neglected. Our analysis reveals that the absolute value of $\Delta\Delta G_{s,f}$ is always less than or equal to that of $\Delta\Delta G_{s,b}$ due to stronger protein-solvent interaction. In contrast, ΔA_f is always greater than or equal to ΔA_b in $\Delta\Delta G_n$ calculation (equation 7) because of less solvent exposure in the bound state. Substituting the leading terms of $\Delta\Delta G_e$ and $\Delta\Delta G_n$ into equation 2, we have another approximation:

$$\Delta\Delta G \cong \Delta\Delta G_{s,b} - 0.04\Delta A_f. \tag{9}$$

The first term in this equation favors a smaller hydrophobic modification because of a more solvated complex. The second term favors a bigger hydrophobic group due to the increase in the solvent accessible surface area.

We see that in each case, the overall binding energy difference depends on the balance of two energy components. In ligand-based design, we can only estimate one of these components (i.e., $\Delta\Delta G_{s,f}$ or ΔA_f), that explains why it is so difficult to predict the outcome of a ligand modification. However, the estimation of these values may give us informative limits of likelihood loss or gain in binding upon a molecular modification. A useful conclusion from this analysis is that a bigger hydrophobic modification tends to enhance binding by lowering the $\Delta\Delta G_{s,f}$ and increasing ΔA_f. This is consistent with numerous QSAR studies (*36*) of the relationship between binding and LogP, which is somewhat related to $\Delta G_{s,f}$ and ΔA_f.

Limitations. In addition to the omission of van der Waals interaction and conformational energies, there are certain limitations in the application of FDPB + SEC. First, successful applications depend on the quality of x-ray structures of binding complexes. Statistical mechanics also requires that the structure resembles both ensemble averages of a reference system and that of a modified system. This approximation certainly will not hold if the modified interaction is too large such as charging a ligand or substantial conformational change. In these cases, qualitative results may still be obtained.

Second, the protonation state of charged residues affects the quantitative results due to the effect of long range electrostatics. Fortunately, one can use FDPB to calculate the pKa shift of each residue. Because of the fast calculation speed, one can also calibrate protonation states against known binding activities.

Third, bound water has not been modeled explicitly in the FDPB + SEC model. The technical problem is the uncertainty of the position of two water hydrogens, which cannot be determined by the x-ray crystallography. Theoretically, a ligand, a receptor, and a bound water form a tri-molecular complex. If a modified ligand or receptor replaces the bound water, the system becomes bimolecular binding. We then need to address different degrees of freedom for the two systems. On the other hand, the high dielectric constant of ice supports the use of the continuum model for bound waters as an approximation.

It is interesting to compare the binding properties of explicit water with those of continuum model. An explicit bound water near a ligand and a receptor is usually regarded as a stabilizer of the complex showing multiple H-bonds to the ligand and the receptor. It is also said that replacing the water with a modified ligand will enhance binding entropically due to the increased degrees of freedom of that water. The continuum model of bound water also has dual functions. In electrostatic calculations, the water enhances receptor-ligand binding because of solvating the complex. In nonelectrostatic binding, it reduces the affinity due to the enlarged SASA of the complex (raises ΔA_b thereby $\Delta\Delta G_n$). More interestingly, both models agree that a bound water may enhance ligand-receptor binding due to enthalpy and weaken the binding due to entropy.

Final remarks. Most therapeutic targets do not have solved 3D-structures. This situation could soon change with the continuous increase of the 3D-structure database and advances in bioinformatics and protein homology modeling. With over 5,000 protein structures available today and no sign of slowing down, how to benefit from these resources has become one of the greatest challenges in pharmaceutical research. One can imagine that a reliable binding energy prediction method will significantly shorten the new drug discovery process and cut the cost of exploratory syntheses. Thus, FDPB + SEC and other binding energy calculation methods will play an increasingly important role in rational drug design.

Acknowledgment. The author is grateful to his colleagues and friends, M. E. Davis, M. K. Gilson, B. A. Luty, P. S. Vermersch, N. K. Vyas, M. Zacharias, F. A. Quiocho, J. A. McCammon, L. Minisci, D. Schneider, J. Wendoloski, T. Pelton, D. Kimonos and R. Vaz, for their helpful assistance and suggestions, and to an anonymous reviewer for the suggestion to use the term 'uniform dielectric' in the Electrostatic Calculation section.

Literature Cited.

1. McQuarrie, D. A. *Statistical Mechanics*; Harper and Row: New York, 1976.
2. Gilson, M. K. G., J. A.; Bush, B. L.; McCammon, J. A. *Biophysical J.* **1997,** *72*, 1047.
3. Wong, C. F.; McCammon, J. A. *J. Am. Chem. Soc.* **1986,** *108*, 3830.
4. Ajay; Murcko, M. A. *J. Med. Chem.* **1995,** *38*, 4953.
5. Shen, J.; Wendoloski, J. J. *Protein Science* **1995,** *4*, 373.
6. Harvey, S. C. *Proteins* **1989,** *5*, 78.
7. Shen, J.; Wendoloski, J. J. *J. Compt. Chem.* **1995,** *17*, 350.
8. Gilson, M. K.; Honig, B. *Proteins* **1988,** *4*, 7.
9. Karshikov, A.; Bode, W.; Tulinsky, A.; Stone, S. R. *Protein Science* **1992,** *1*, 727.
10. Shen, J.; Quiocho, F. A. *J Compt. Chem.* **1995,** *16*, 445.
11. Wendoloski, J. J.; Shen, J.; Oliva, M. R.; Weber, P. C. *Pharmac. Ther.* **1993,** *60*, 169.
12. Madura, J. D.; Makajima, Y.; Hamilton, R. M.; Wierzbick, A.; Warshel, A. *Structural Chemistry* **1996,** *7*, 131.
13. Zhang, T.; Daniel E. Koshland, J. *Protein Science* **1996,** *5*, 348.
14. Davis, M. E.; Madura, J. D.; Luty, B. A.; McCammon, J. A. *Comput. Phys. Comm.* **1991,** *62*, 187.
15. Nicholls, A.; Sharp, K. A.; Honig, B. *DelPhi* 3.0, Columbia University: New York, 1990.
16. Lazaridis, T.; Paulaitis, M. E. *J. Phys. Chem.* **1992.,** *96*, 3847.
17. Sitkoff, D.; Sharp, K. A.; Honig, B. *J. Phys. Chem.* **1994,** *98*, 1978.
18. Ben-Naim, A.; Marcus, Y. *J. Chem. Phys.* **1984,** *81*, 2016.
19. Rashin, A. A.; Bukatin, M. A. *J. Phys. Chem.* **1994,** *98*, 386.
20. Brunger, A. T. *X-PLOR Manual*; *2.0*; Yale University: New Haven, USA, 1991.
21. Brooks, B. R.; Bruccoleri, R. E.; Olafson, B. D.; States, D. J.; Swaminathan, S.; Karplus, M. *J. Compt. Chem.* **1983,** *4*, 187.
22. van Gunsteren, W. F.; Berendsen, H. J. C. Biomos: *Groningen Molecular Simulation (GROMOS) Library Manual,* Nijenborgh 4, The Netherlands, 1987
23. Vermersch, P. S.; Tesmer, J. J.; Quiocho, F. A. *J. Mol. Biol.* **1992,** *226*, 923.
24. Quiocho, F. A.; Wilson, D. K.; Vyas, N. K. *Nature* **1989,** *340*, 404.

25. Morgan, B. P.; Scholtz, J. M.; Ballinger, M. D.; Zipkin, I. D.; Bartlett, P. A. *J. Am. Chem. Soc.* **1991,** *113*, 297.
26. Shen, J. *J. Med. Chem.* **1997,** *40*, 2953.
27. Spurlino, J. C.; Smallwood, A. M.; Carlton, D. D.; Banks, T. M.; Vavra, K. J.; Johnson, J. S.; Cook, E. R.; Falvo, J.; Wahl, R. C.; Pulvino, T. A.; Wendoloski, J. J.; Smith, D. *Proteins* **1994,** *19*, 98.
28. Dhanaraj, V.; Ye, Q.-Z.; Johnson, L. L.; Hupe, D. J.; Ortwine, D. F.; J.B. Dunbar, J.; Rubin, J. R.; Pavlovsky, A.; Humblet, C.; Blundell, T. L. *Structure* **1996,** *4*, 375.
29. Holden, H. M.; Tronrud, D. E.; Monzingo, A. F.; Weaver, L. H.; Matthews, B. W. *Biochemistry* **1987,** *26*, 8542.
30. Bartlett, P. A.; Marlowe, C. K. *Biochemistry* **1987,** *26*, 8553.
31. Bamborough, P.; Cohen, F. E. *Current Opinion in Structural Biology* **1996,** *6*, 236.
32. Claeson, G. *Blood Coagulation and Fibrinolysis* **1994,** *5*, 411.
33. Bode, W.; Mayr, I.; Baumann, U.; Huber, R.; Stone, S. R.; Hofsteenge, J. *EMBO J.* **1989,** *8*, 3467.
34. Singh, J.; Conzentino, P.; Cundy, K.; Gainor, J. A.; Gilliam, C. L.; Gordon, T. D.; Johnson, J. A.; Morgan, B. A.; Schneider, E. D.; Wahl, R. C.; Whipple, D. A. *Bioorganic & Medicinal Chemistry Letters* **1995,** *5*, 337.
35. Wilson, K. P.; Black, J. A.; Thomson, J. A.; Kim, E. E.; Griffith, J. P.; Navia, M. A.; Murcko, M. A.; Chambers, S. P.; Aldape, R. A.; Raybuck, S. A.; Livingston, D. J. *Nature* **1994,** *370*, 270.
36. Hansch, C. L., A. *Exploring QSAR*; American Chemical Society: Washington, DC, 1995.

Chapter 5

SmoG: A Ligand Design Method Based on Knowledge-Based Parametrization of a Solvent Reorganization Model

Robert S. DeWitte and Eugene I. Shakhnovich

Department of Chemistry and Chemical Biology, Harvard University, 12 Oxford Street,Cambridge, MA 02138

In this chapter, we discuss SMoG (*1-2*) a *de novo* design program/methodology based on a knowledge-based approximation to the binding free energy. This approach is unique in that the representation of atomic interactions bears no relation to empirical force field terms, but rather is built solely on the consideration of the implicit behaviour of the solvent around various different solute atoms. The resulting scoring function evaluates instantaneously, and provides a smoothed potential energy surface of interaction. These features of the scoring function allow the space of potential ligands to be well searched by a metropolis monte carlo molecular growth algorithm that efficiently suggests lead candidates. We show how using SMoG to suggest compounds, followed by molecular modelling calculations and chemical intuition can be a fruitful approach to designing novel lead compounds with a high likelihood of binding.

Many programs have been written to try to design that bind to a particular protein with a known three dimensional structure (3-7). These methods have been built largely upon scoring functions that are either empirical (such as those used in molecular mechanics force-fields), rule-based approximations (for example, counting the numbers of hydrogen bonds, and computing buried surface areas), or some combination of the two of these. Unfortunately, these scoring functions provide only a poor estimate of the binding free energy of a ligand candidate, a limitation which minimizes the potential impact of these methods. The principal, and fundamental limitation of such evaluations of binding affinitiy is that they make no implicit reference to the change in the behaviour of the solvent atoms as the protein and ligand change from the unbound state to the bound state.

Water has a strong ability to order itself, as in ice, into rigid structures. However, at and around room temperature, the tendency is rather toward disorder than order, as the various hydrogen bonds between the water molecules fluctuate in the amorphous liquid. In this state, the individual molecules participate in hydrogen bonds to lower the overall energy of the bulk, but the participation is fleeting: a favourable entropic condition. A solute in water, though, can affect this balance. Hydrophilic and hydrogen bonding species participate in hydrogen bonds to a lesser or greater extent than the native water molecules, and thereby partially join into the dance, with a modest and favourable impact on the solvation free energy. Hydrophobic solutes, however, cannot participate in such interactions with the water molecules themselves. The result is that the water molecules near the solute orient themselves so as to participate in roughly the same number of hydrogen bonds as when they were in bulk. The result is a semi-rigid layer of water molecules encaging the solute, and this lower entropic state is manifest in poor solubility. Thus as in the problem of protein folding, hydrophobic effect supplies the driving force for protein-ligand binding (8).

Modelling the hydrophobic affect without explicit reference to an ensemble of water molecules is an open challenge to the community interested in applications of computing to biomolecular association problems. Our approach is to consider the length scale over which water can propagate the effect of a particular solute atom, and apply a model of this phenomenon to a database of protein-ligand crystal structures for parametrization. Bulk water itself has considerable correlative organizational behaviour, in which the correlation length, or the range of the order in the liquid extends about two solvation shells from the atom in question, or about 5 angstroms. Thus it is reasonable to expect that the solvation correlations between atoms also extend over this range. Put simply, atoms at this distance are connected to one another through the solvent.

It should also be born in mind that we are trying to address the question of lead discovery, and so we must be able to process a large number of candidate compounds, and evaluate them very efficiently in a broad classification of their prospective binding affinity. We are aided by the fact that we need not find the best solution to the problem, but rather should find a good solution to the problem. Thus, we can use an incomplete, optimization style algorithm to generate compounds. This, however, puts an additional strain on the evaluation methods, since it will need to be used several times in the optimization process. Thus, the scoring function and search method must be extremely fast. Moreover, the molecules that emerge from the method should be viewed as sketches, or prototypes, providing the seed ideas for the design of a true lead compound.

Methods

Coarse-graining and the knowledge-based potential. We implement here a coarse-grained model with a corresponding knowledge-based potential which treats both ligand and protein in an all-atom representation, but assumes a simplified form of their interaction. The reader is referred to Reference 1 as well as references therein for a description and justification of the the process whereby statistical observations p_{ij} can be converted into two-body parameters g_{ij} that sum to a notion of free energy. Here we simply state the result:

$$g_{ij} = -kT \log\left[\frac{p_{ij}}{<p>}\right]$$

where i and j refer to particular atom types in the protein and ligand respectively. <p> is a reference state which sets the zero for the energy scale, as the null hypothesis wherein all interaction types are equally likely.

It should be made clear that the choice of the interaction model is intrinsically a choice of length scales. We must determine the reasonable distances over which atoms project their chemical properties in order to accumulate the relevant statistics and apply the right model. Hence, knowledge-based potentials lend themselves naturally to coarse-graining techniques, where potential energy surfaces are smoothed by averaging all phenomena occuring below a cutoff length scale into properties describing the system at the specified length.

In our application, we chose a simple contact model: atoms within five angstroms are in contact, and those further apart are not in contact. With this definition, p_{ij} above refers to the probability that, say, a polar carbon in a ligand molecule and a hydrogen bond acceptor in the protein, will be found within five angstroms of each other. The parameters g_{ij} are used in the same way: each pair of atoms, one from the ligand (i) and one from the protein (j), that are within five angstroms contribute their corresponding g_{ij} to the free energy estimate.

One final aspect of the model is that the number of atom types is expanded to include some notion of the chemical personality of the various atoms. In other words, carbon atoms are broken into the categories of fatty carbons and polar carbons, oxygen atoms are either charged, hydrogen bond donors, or hydrogen bond acceptors. Similarly for nitrogen atoms, and some other atoms and ions are included, such as sulfur, phosphorus, fluorine, calcium and zinc. The model, together with the knowledge-based potential, is referred to as the design energy in this work.

Database. The database of protein-ligand complexes used to determine the g_{ij} parameters included the following complexes (all with RMSD (2.0 Å): 1art 1bcd 1bcx 1bic 1bit 1byb 1cah 1cam 1can 1cao 1caz 1chn 1cil 1cmp 1coy 1cra 1crq 1csc 1csh 1csi 1enc 1erb 1fel 1fem 1fen 1fkd 1fkf 1fkh 1gca 1gcd 1hcb 1hsl 1hvi 1hvk 1hvl 1hyt 1icm 1icn 1inc 1isc 1lcc 1lic 1lid 1lie 1lif 1lra 1lst 1mdq 1mfa 1mng 1olb 1pal 1pbe 1pbp 1ppf 1ppp 1ray 1raz 1snm 1sta 1sty 1swm 1syd 1thl 1tng 1tnh 1tni 1tnj 1tnk 1tnl 1tpp 1tro 2aae 2acq 2acr 2acs 2acu 2che 2csc 2ctc 2cut 2fke 2mbp 2pal 2rnt 2tbs 2xis 3cla 3cts 3dfr 3gch 3pat 3rnt 3sga 4csc 4gch 4pal 4sga 5cts 5sga 5tim 6rnt 7rnt 821p 8est 8tln 8xia 9est. Structural waters, where present, were considered as part of the protein.

The parameters that result are available as supplementary material to Reference 2.

Coarse-graining and the search algorithm. In principle, the combinatorial search space for molecular growth or docking algorithms is a rough energy landscape. Searching such a landscape requires careful algorithms, and long search times.

Fortunately, however, the identification of candidate lead molecules is not a search for the lowest free energy complex, but rather a low free energy complex (or several). Still, the search is a difficult process because of the multiple minimum problem. If the search space can be made more smooth by coarse graining, however, the searching method need not be as sophisticated. For this reason, SMoG employs a metropolis monte carlo growth algorithm. Such a search procedure quickly samples the configuration space and the molecular space under the bias of the interaction potential (knowledge-based energy in this case). In a coarse-grained ligand design search space, a simple, hasty, search algorithm such as the one presented here can do very well in finding low energy configurations.

Monte-Carlo Molecular Growth Algorithm. Directly in the binding region of the protein, simple organic molecules are generated by joining fragments with single bonds. Each step of the molecular growth proceeds as follows: two hydrogen atoms are selected: one from the fragment to be added, and one from the structure as generated so far. The new fragment is placed such that the hydrogen atoms are displaced and the atoms formerly bonded to those hydrogen atoms now form a single bond with a standard bond length. This procedure ensures that the new bond angles and bond lengths are reasonable approximations. Finally, the new functional group is oriented by torsional rotation about the new bond. Table I lists the fragments used in molecular growth.

Table I. Fragments used in the small molecule growth algorithm. Adapted with permission from reference 1.

Amide	Cyclohexene	Methane	Pyrimidine
Amine	1,2-Dithiane	n-Butane	Pyridine
Carbonyl	Ethane	Napthalene	Pyrrole
Carboxylic acid	Ethene	Nitrile	Sulfate
Chloride	Fluoride	Nitro	Sulfide
Cyanide	Furan	Phenyl	t-Butane
Cyclooctane	Glucose	Phosphate	Tetrahydrofurane
Cyclopentane	Hydroxyl	Propane	Tetrahydrothiene
Cycloheptane	Indole	Propene	Thiophene
Cyclohexane	Iodide	Purine	Trifluoromethane

In this manner, beginning with simple H_2 in the binding site, a molecule of any desired size can be generated, by continuing to add fragments. Notice that the growth is inherently branched because at each growth step, any hydrogen atom on the present structure is a potential site of growth.

Each fragment that is placed is oriented by torsional rotation about the new bond in fixed increments and all those orientations that are not sterically hindered are subject to energetic evaluation. That rotamer with the lowest energy is considered as a candidate for acceptance into the new molecule. This acceptance is determined by the Metropolis monte carlo criterion which compares the new energy per atom with that before this growth step. Any decrease is accepted, and any increase is accepted with probability $exp[-\Delta g/T]$ where $g=G/N$ is the free energy per atom, and T is an algorithmic temperature.

The preliminary selection of lowest allowed rotamer has two positive effects. First, it biases the molecule more quickly to low energy, since random selection of rotamers would lead to significantly more metropolis failures. Second, it is an indirect selection toward the tightest possible steric complementarity.

The Metropolis decision of acceptance or rejection of the new fragment is in place to allow the energy per atom to increase occasionally, as would need to be the case if the small molecule had grown into a tight steric region, and had no other recourse but to grow into the solvent or some other unoccupied region, where it would interact only marginally with the protein. This feature also provides the added benefit of generating molecules with bridging fragments where the bridge itself is not necessarily a strongly interacting part of the molecule.

The reader is referred to Reference 1 for a description of the analysis of the operating parameters used in this method. Under the operating conditions of sixty degree torsional increments, seventy percent Van derWaal contact radius and an algorithmic temperature of 10.0, each molecule of about twenty heavy atoms can be generated in a few seconds on a 100MHZ pentium computer running Linux.

Ligand Design Methodology. Figure one provides a general outline for ligand design using the SMoG approach.

At the first stage, it is helpful to get an appreciation for the binding site (for instance its shape and types of intermolecular interactions it may support) by allowing SMoG to generate a large number of molecules in the binding site. Our approach has been to generate one thousand molecules and record the structures of the best fifty. This computation is generally completed in a few hours. By viewing the molecules that fit into the space of the binding site, and form complementary chemical interactions, particularly paying attention to those molecular details that arise frequently, the chemist gains an immediate understanding of what molecular scaffolds are likely to be fruitful leads to follow. For example, one may observe that several of the high scoring molecules involve specific hydrogen bonds, attained through a specific functional group in a specific orientation. Alternatively, one may learn that the presence of lipophilic groups in a certain region are responsible for the high score of several other ligands. This 'consensus based' qualitative understanding forms the basis for further work with SMoG.

Whereas the high score of the molecules in stage one is usually due to the presence of one well placed molecular fragment, the desire at stage two is to build molecules that combine several of the positive features observed in stage one. This is done by selecting a few representative molecules from stage one, removing the parts of the molecule that are not important, and using the remaining strongly interacting, molecular fragment as a restart fragment. SMoG has the ability to continue growth from any molecule provided as input by selecting hydrogen atoms on it as points of further growth. Furthermore, the user may determine at which of these hydrogen atoms growth is allowed. For each moiety from stage one that is used as a restart fragment, a new line of molecules can be generated (generally another 1000 of which the best 50 are recorded). Each of these molecules will contain the tailored

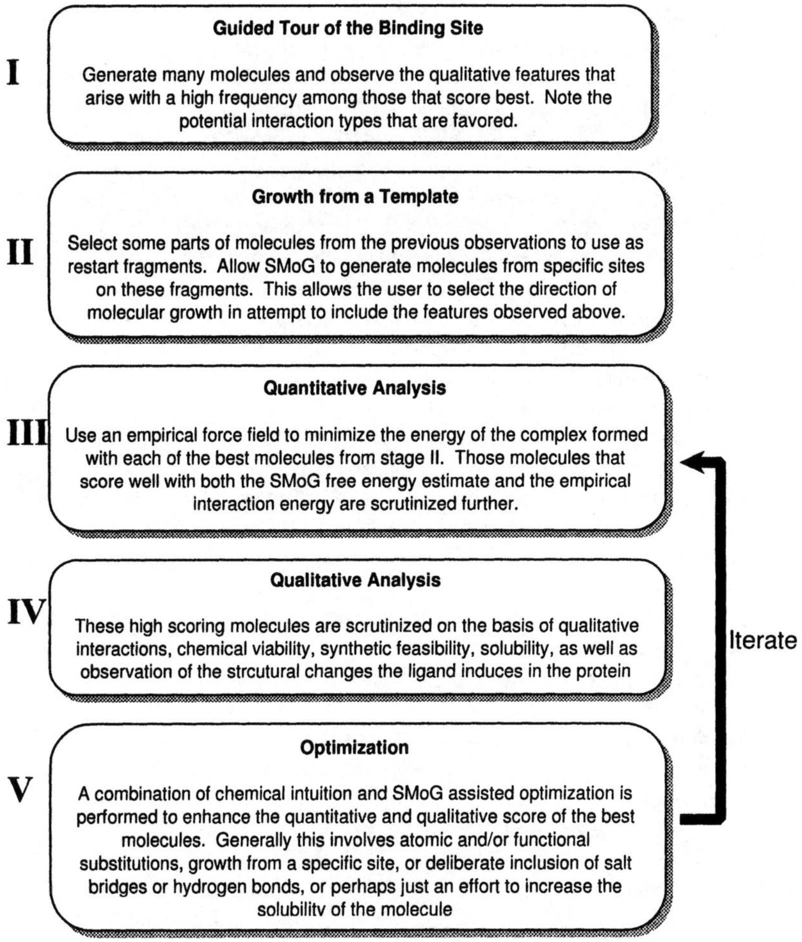

I

Guided Tour of the Binding Site

Generate many molecules and observe the qualitative features that arise with a high frequency among those that score best. Note the potential interaction types that are favored.

II

Growth from a Template

Select some parts of molecules from the previous observations to use as restart fragments. Allow SMoG to generate molecules from specific sites on these fragments. This allows the user to select the direction of molecular growth in attempt to include the features observed above.

III

Quantitative Analysis

Use an empirical force field to minimize the energy of the complex formed with each of the best molecules from stage II. Those molecules that score well with both the SMoG free energy estimate and the empirical interaction energy are scrutinized further.

IV

Qualitative Analysis

These high scoring molecules are scrutinized on the basis of qualitative interactions, chemical viability, synthetic feasibility, solubility, as well as observation of the strcutural changes the ligand induces in the protein

V

Optimization

A combination of chemical intuition and SMoG assisted optimization is performed to enhance the quantitative and qualitative score of the best molecules. Generally this involves atomic and/or functional substitutions, growth from a specific site, or deliberate inclusion of salt bridges or hydrogen bonds, or perhaps just an effort to increase the solubilitv of the molecule

Iterate

Figure 1. The stages of ligand design with SMoG.
(Reproduced from reference 2. Copyright 1997 American Chemical Society.)

interactions as well as a variety of other positive features. Thus, at the end of stage two, there will arise from each line of molecules a small number of candidates that incorporate several qualitative as well as quantitative interactions. At this point, a large part of the combinatorial problem of ligand generation has been overcome (i.e.: the architecture of the molecule has been decided). Depending on the situation at hand, it may be desirable to repeat stage two once more with some lines of molecules to optimize the collection of interacting functional groups even further.

Generally after stage two, one might have a few dozen structures to consider in the subsequent stages. At this point, it is important to determine which of these molecules to focus on in more detail during the subsequent design stages. It is important to realize that SMoG does not include an intramolecular interaction potential in its growth. Therefore, one should relax the slight strains that the molecules are carrying through minimization of the protein-ligand complex with an empirical potential (for example CHARMM). The empirical interaction energy of these relaxed complexes (especially the electrostatic component) is another useful measure of the quality of a molecule because the SMoG design potential does not explicitly account for electrostatic interactions between the molecule and the protein, and thus slightly undervalues hydrogen bond and salt bridge formation. Conversely, since hydrophobic interactions are largely solvent entropy effects, empirical calculations of vacuum interaction enthalpies undervalue the contribution of nonpolar interactions to binding free energy. Thus the two measures of interaction strength are somewhat complementary. Hence, the molecules that one should continue to focus on for the remainder of the design stages are those which have low CHARMM and SMoG energies. At present, we are moving to include explicit terms to handle electrostatic interaction events.

At stage four, the remaining molecules (perhaps a dozen) need to be scrutinized qualitatively with the goal of optimization in mind, rather than exclusion. The criteria with which to judge the molecules include chemical stability, ease of synthesis, internal strain energy, strain induced in the protein, and solubility. One should also determine if subsequent growth or manual optimization can introduce more hydrogen bonds, or capitalize on other features of the binding pocket, such as stacking with delocalized π-bonding systems. It is clear from our experience that a few molecules will emerge as having greater potential than the others because of the nature of the interactions they incorporate presently as well as features that suggest either simple manual changes leading to improvement, or directions in which automatic growth may enhance the binding interactions (using the whole current molecule as a restart fragment and allowing growth at only one or perhaps a few select hydrogen atoms).

In the fifth stage, the modifications suggested in stage four are introduced to the few select molecules that have the most potential, yielding yet another generation of structures which should be scrutinized quantitatively and qualitatively in stages three and four.

In the process of designing a molecule that is likely to be a strong binding ligand, stages three, four and five may need to be iterated several times until a candidate is found which is qualitatively sound and scores among the best molecules according to SMoG and CHARMM. As the process converges to a ligand, one may wish to use other modelling tools to analyze the molecules and enhance decision making. These may include conformational analysis to ensure that the binding mode of the molecule is not a highly strained conformer, and molecular dynamics simulation in solvent to observe the stability of the predicted complex

Results.

The Scoring Function Correlates Well with Experimental Binding Free Energies. In order to test the correlation between experimental binding free energies and the SMoG design procedure, SMoG was applied to the three protein-ligand complex systems for which structural and binding information has been published and is readily available. These examples include Purine Nucloside Phosphorylase (*9-14*), Src SH3 domain specificity pocket (*15-17*), and Human Immunodeficiency Virus-1 Protease (*18-19*). The molecular structures of each compound tested are presented in reference 1. Here, we merely summarize the results in figures 2 through 4.

Table II summarizes the overall correlation findings quantitatively.

Table II. Summary of Correlation Data. Adapted with permission from reference 1.

System	Correlation Coefficient	Number of Points	Probability
PNP	0.80	17	0.0020
SH3	0.81	8	0.1109
HIV	0.77	11	0.0501

The Monte Carlo Molecular Growth Algorithm Search is Sufficiently Exhaustive. Figure 5 demonstrates that the knowledge-based potential respects the native ligand (whose energy is marked as a dark stripe) as having extremely low energy. Moreover, molecules with a comparable energy are rare, but attainable in reasonable computation time since approximately five percent of generated molecules are comparable to the native ligand in each example.

The Methodology Produces Qualitatively and Quantitatively Interesting Ligands. The CD4 protein is an immunoglobin-family transmembrane receptor expressed in helper T-cells (Bour et. al.) It participates in contact between the T-cells and antigen-presenting cells by binding to the nonpolymorphic part of the class II major histocompatibility complex (MHC-II) protein, which is followed by the activation of the bound Lck kinase which leads to downstream activation events in T-cells. The Human Immunodeficiency Virus (HIV) gains entry into a T-cell by binding protein gp120 to the CD4 receptor. This gp120 binding site in the vicinity of Phe 43 of CD4 was the target for ligand design in this project (see Figure 8b).

Among the possible interactions that arose in stage one of the design process, it was apparent that π-π interaction with the phenyl ring of Phe 43 was important, as well as the formation of hydrogen bonds in the narrow pocket bounded by Lys 46 and Asp 56. After one pass through the five stages, the first generation of molecules was evident. These are shown in Figure 6, where one can see the common elements of a hydrogen-bonding core and a hydrophobic moiety in the same relative orientation in most molecules. Qualitative features, as well as the data in Table III led to the

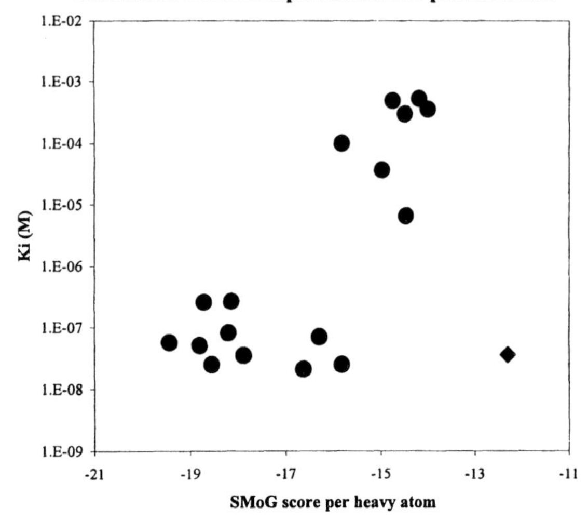

Figure 2. The correlation of measured binding constant (experimental result) with SMoG score (theoretical prediction) for a set of Purine Nucleoside Phosphorylase inhibitors that are not highly sensitive to the concentration of phosphate in the solution.
(Reproduced from reference 1. Copyright 1996 American Chemical Society.)

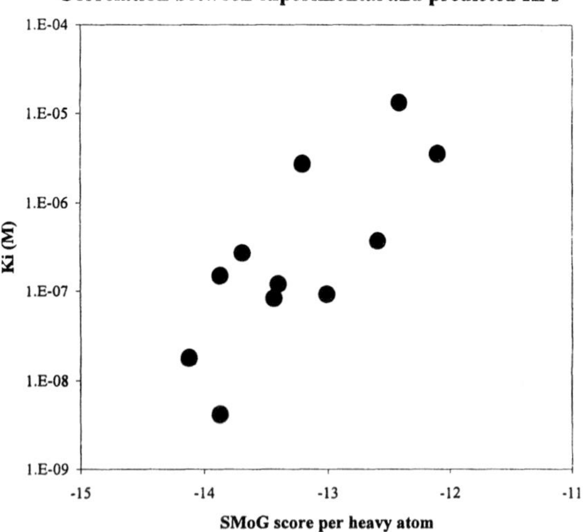

Figure 3. The correlation of measured binding constant (experimental result) with SMoG score (theoretical prediction) for a set of HIV-Protease inhibitors.
(Reproduced from reference 1. Copyright 1996 American Chemical Society.)

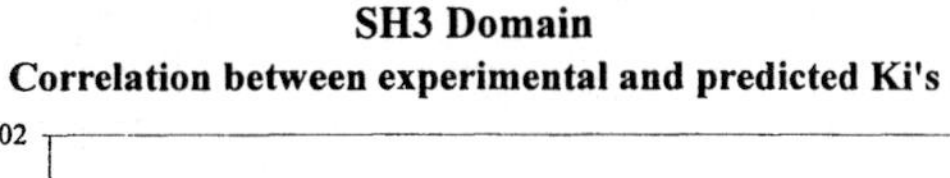

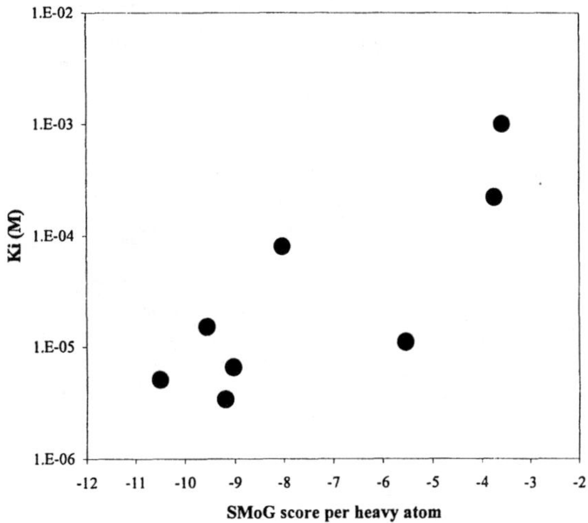

Figure 4. The correlation of measured binding constant (experimental result) with SMoG score (theoretical prediction) for a set of SH3 Domain ligands.
(Reproduced from reference 1. Copyright 1996 American Chemical Society.)

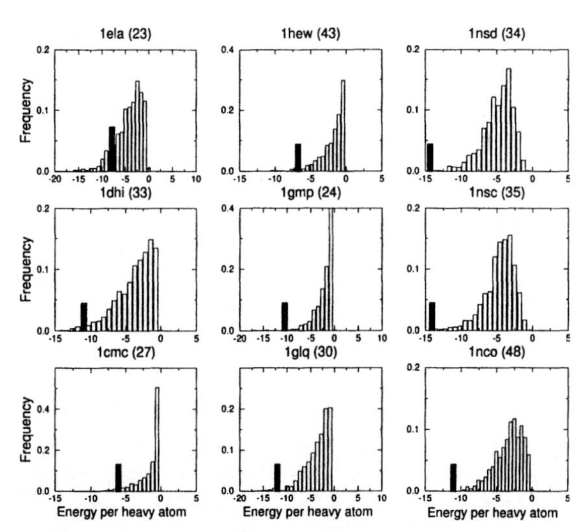

Figure 5. The distribution of SMoG score for the design of 1000 molecules of the same size as the native (*i.e.*: an extant) ligand whose number of atoms is shown in brackets. In each case, the SMoG score of the native ligand (shown in a heavy bar) lies in the tail of the distribution. Roughly 5% of molecules generated by SMoG score as strongly as the native ligand (or better).
(Reproduced from reference 1. Copyright 1996 American Chemical Society.)

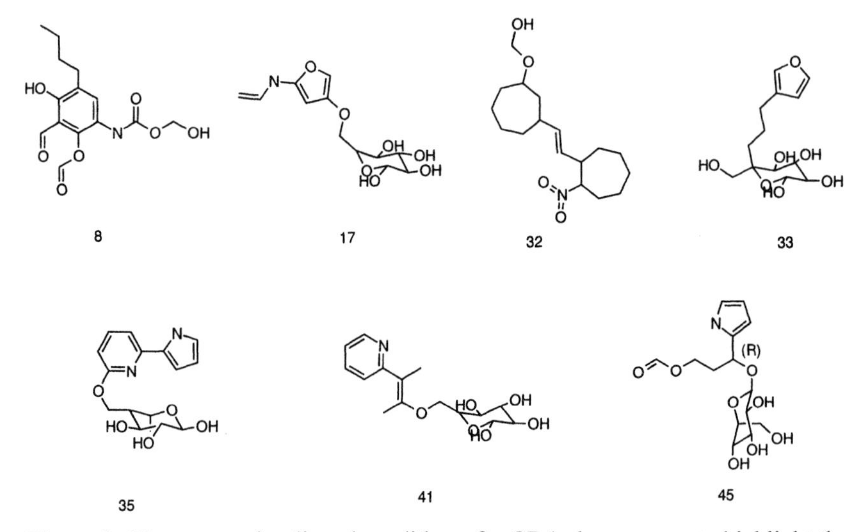

Figure 6. First generation ligand candidates for CD4, drawn so as to highlight the consensus in the selection and placement of similarly interacting fragments. (Reproduced from reference 2. Copyright 1997 American Chemical Society.)

selection of molecule 41 for further attention. Figure 7 and Table IV describe the evolution, through SMoG assisted optimization and manual editing suggested by chemical intuition, from Molecule 41 to the best candidate, molecule 41h. In particular, molecules 41b, 41e-g are the result of manual editing (guided by intuition) to improve the π-stacking interactions with Phe 43, and molecules 41c-d are the result of applying SMoG to extension of 41b into a small hydrophobic pocket. These extensions were not quantitatively advantageous (the pocket was slightly too small). Molecules 41h-i derive from an observation that the significant strain in 41e (a strong-scoring candidate) could be reduced by locking its bound conformation through a bridge that creates a fairly rigid fused ring system. Of these, 41h was the best candidate (the heteroatom allows the adoption of the interacting conformation of 41h). In this and all other cases, manual editing was performed the 3D Molecular Editor facility of Quanta, and subsequent minimization of energy with CHARMM. Evaluation of these results involves reference to the CHARMM interaction energy and recalculation of the SMoG score.

Figure 8 shows the three dimensional structure of Molecule 41h in the gp120 binding site of CD4. The interactions present include partial π-stacking with Phe 43, as well as four intermolecular hydrogen bonds with Lys 46 and Asp 56, and one intramolecular hydrogen bond which stabilizes the orientation of the pyridine group. The seven membered fused-ring bridge gives this molecule a great deal of rigidity in its bound conformation.

Table III. Quantitative analysis of the first generation CD4 candidates shown in Figure 6. Adapted with permission from reference 2.

Molecule	SMoG Score per heavy atom	CHARMM interaction Energy (kcal)
8	-26.2	-82.3
17	-30.0	-80.9
32	-28.5	-53.6
33	-36.3	-70.6
35	-26.8	-80.
41	-45.7	-99.
45	-26.9	-59.8

Discussion.

In this chapter we have shown that ligand design with SMoG as the first step and a continual resource has several advantages stemming principally from the speed with which it can provide an approximate prediction of relative binding free energy. In particular, the program can focus the chemist's search for novel lead compounds by suggesting novel scaffolds, and chemical architectures for which the desolvation driving force is expected to be strong. Within this framework, using a combination of modelling tools, intuition and SMoG calculations, the chemist can quickly move toward a set of compounds that are more highly optimized, even with respect to chemical synthesis (Reference 2 contains further discussion about revising compounds to achieve synthetic feasibility).

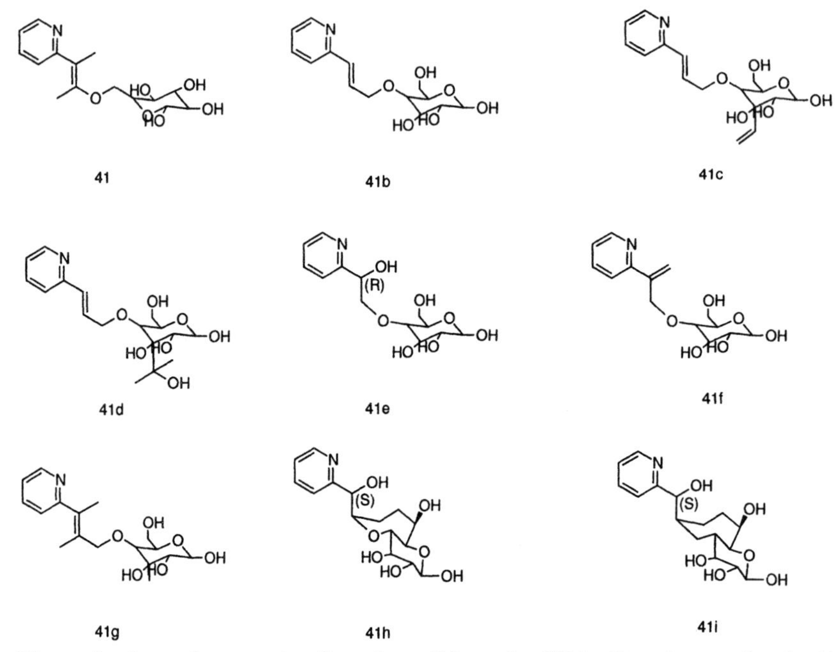

Figure 7. Second generation ligand candidates for CD4. Based on molecule 41 these candidates were derived by a combination of manual alteration and SMoG generated extension, always guided by both intuition and SMoG and CHARMM scores. (Reproduced from reference 2. Copyright 1997 American Chemical Society.)

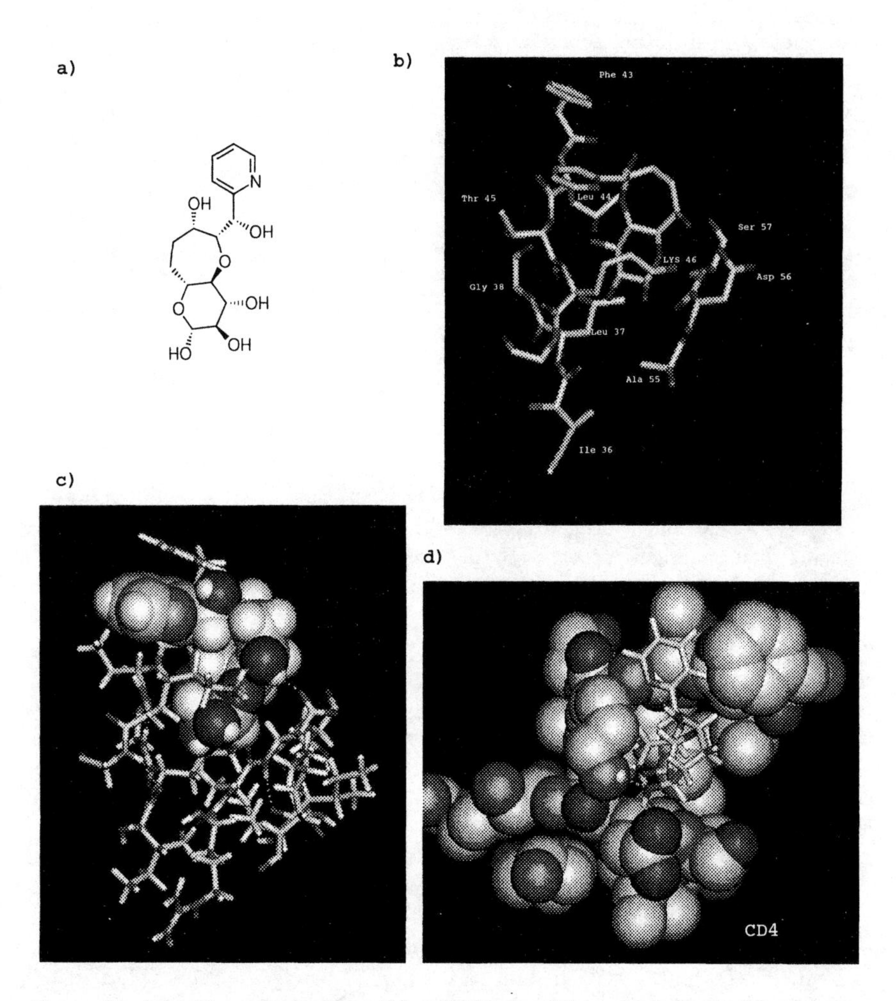

Figure 8. A highly optimized candidate (41h) for CD4, which forms 5 hydrogen bonds, as well as π-stacking interactions with Phe 43. (a) 2D molecular structure. (b) The gp120 binding site of CD4 with ligand candidate in place. (c) & (d) Space filling models of the predicted complex.

Table IV. Quantitative analysis of the second generation CD4 candidates shown in Figure 7. The strain energy is calculated as the difference in internal energy between the bound conformation and the conformation resulting from gas phase minimization with CHARMM. Adapted with permission from reference 2.

Molecule	SMoG Score per heavy atom	CHARMM Energies (kcal)		
		Interaction	Strain	Net
41	-45.7	-99	14.4	-84.3
41b	-47.9	-139	20.0	-119.0
41c	-45.1	-128	20.7	-107.3
41d	-46.1	-120	33.2	-86.8
41e	-50.6	-112	36.6	-75.4
41f	-50.5	-116	20.1	-95.9
41g	-49.4	-82	15.0	-67.0
41h	-49.9	-119	22.5	-96.5
41I	-48.0	-86	20.5	-65.6

One advantage of the design methodology proposed in this chapter is the ability to improve the qualitative features (size, shape, location, connectivity, synthetic feasibility, π-stacking, hydrogen bonds, salt bridges, etc.) of the molecules without reducing their quantitative scores. Because of this, the quality of the ligands that can be generated is simply a product of effort, insight and intuition on the part of the user. However, the insight and intuition are only needed as analysis tools, since SMoG continuously provides suggested alterations and extensions of molecules that form excellent chemical and spatial complimentarity with the protein binding site. In this sense, SMoG overcomes the otherwise intractable combinatorial task of generating optimal molecular scaffolds for scrutiny and optimization.

Careful examination of the Tables of SMoG scores and CHARMM interaction energies reveals that those subtle molecular features that were added manually to take advantage of a hydrogen bonding opportunity are not reflected very strongly in the SMoG score, but are reflected in the CHARMM interaction energy (particularly the electrostatic component). Conversely those attributes which correspond to increased hydrophobic interaction are reflected in SMoG's score, but not in the CHARMM interaction energy. This evidence supports using both measures of interaction energy, since their weaknesses and strengths are complimentary. It also implies that the accuracy of SMoG's prediction of binding free energy may be improved by adding specific terms to the form of the interaction potential that reflect electrostatic interactions such as hydrogen bonds and salt bridges.

SMoG's limitations include those implied in the simple methods with which chemical geometry is handled: inter-fragment bond lengths and angles are all assumed to be standard, and unvarying; the protein structure is considered fixed; and steric repulsions are either on or off, depending on a simple distance test. Other limitations are implementation dependent, and the program has been designed to allow flexibility in the choice of operating conditions. For example, smaller angle steps can be chosen to perform calculations more carefully, lower temperatures can be chosen, and the fragment library can be expanded.

Of course, as is the case with any design method, the crucial test of SMoG's merit will include the synthesis and measurement of the binding constant of a candidate ligand that was the direct result of SMoG design. We are currently pursuing this line of development vigorously in collaborative investigation. Preliminary experimental results with Carbonic Anhydrase are encouraging, but structural confirmation is still in progress at the time of writing.

We are confident that this approach, which is unique in many aspects, including the nature and source of the interaction potential and the growth algorithm, has much to offer the medicinal chemistry community because of its efficiency and the reliability of its scoring method. Moreover, as this brief account demonstrates, the approach to designing ligands is extremely flexible and fruitful.

Acknowledgements

The authors would like to reckognize the helpful assistance of S.L. Schreiber, S. Feng, J. Morken, J. Shimada, and A.V. Ishchenko, as well as funding support from the Packard Foundation and NSERC (Canada).

References

1. DeWitte, R. S.; Shakhnovich, E. I., *J. Am. Chem. Soc.* **1996,** *118,* 11733-11744.
2. DeWitte, R. S.; Ishchenko, A. V.; Shakhnovich, E. I., *J. Am. Chem. Soc.* **1997,** *119,* 4608-4617.
3. Goodford, P. J., *J. Med. Chem.* **1985,** *28,* 849-857.
4. Moon, J. B.; Howe, W. J., *Proteins.* **1991,** *11,* 314-328.
5. Bohm, H-J, *J. Comp-Aided Mol. Design.* **1992,** *6,* 593-606.
6. Lawrence, M. C.; Davis, P. C., *Proteins.* **1992,** *12,* 31-41.
7. Miranker, A.; Karplus, M., *Proteins.* **1991,** *11,* 29-41.
8. Dill, K. A. *Biochemistry.* **1990,** *29,* 7133-7155.
9. Tuttle, J. V.; Kernitzky, T.A., *J. Biol.Chem.* **1984**, *259*, 4065-4069.
10. Ealick, S. E.; Babu, Y. S.; Bugg, C. E.; Erion, M. D.; Guida, W. C.; Montgomery, J. A.; Secrist, J. A., *PNAS* **1991**, *88*, 11540-11544
11. Montgomery, J. A.; Niwas, S.; Rose, J. D.; Secrist, J.A.; Babu, Y. S.; Bugg, C. E.; Erion, M. D.; Guida, W. C.; Ealick, S. E., *J. Med. Chem.* **1993**, *36*, 55-69
12. Secrist, J. A.; Niwas, S.; Rose, J. D.; Babu, Y. S.; Bugg, C. E.; Erion, M. D.; Guida, W. C.; Ealick, S. E.; Montgomery, J. A., *J. Med. Chem.* **1993**, *36*, 1847-1854.
13. Erion, M. D.; Niwas, S.; Rose, J. D.; Subramanian, A.; Allen, M.; Secrist, J. A.; Babu, Y. S.; Bugg, C. E.; Guida, W. C.; Ealick, S. E.; Montgomery, J. A., *J. Med. Chem.* **1993**, *36*, 3771-3783.
14. Guida, W. C.; Elliot, R. D.; Thomas, H. J.; Secrist, J. A.; Babu, Y. S.; Bugg, C. E.; Erion, M. D.; Ealick, S. E.; Montgomery, J. A., *J. Med. Chem.* **1995**, *37*, 1108-1114.
15. Chen, J. K.; Lane, W. S.; Brauer, A. W.; Tanaka, A.; Schreiber, S. L., *J. Am. Chem. Soc.* **1993,** *115*, 12591-12592.
16. Feng, S.; Chen, J. K.; Yu, H.; Simon, J. A.; Schreiber, S. L., *Science* **1994**, *266*, 1241-1247.

17. Combs, A. P.; Kapoor, T. M.; Feng, S.; Chen, J. K.; Daude-Snow, L. F.; Schreiber, S. L., *J. Am. Chem. Soc.* **1996**, *118*, 287-288.
18. Abdel-Meguid, S. S.; Metcalf, B. W.;Carr, T. J.; Demarsh, P; DesJarlais, R. L.; Fisher, S.; Green, D. W.; Ivanoff, L.; Lambert, L.; Murthy, K. H. M.; Petteway, S. R., Jr.; Pitts, E. J.; Tomaszek, T. A., Jr.; Wonborne, E.; Zhao, B; Dreyer, G. B.; Meek, T. D., *Biochemistry* **1994**, *33*, 11671-11677.
19. Thompson, S. K.; Murthy, K. H. M.; Zhaong, B.; Wonborne, E.; Green, D. W.; Fisher, S. M.; DesJarlais, R. L.; Tomaszek, T. A., Jr.; Meek, T. D.; Gleason, J. G.; Abdel-Meguid, S. S., *J. Med. Chem.* **1994**, *37*, 3100-3107.

Chapter 6

The Evaluation of Multi-Body Dynamics for Studying Ligand–Protein Interactions: Using MBO(N)D to Probe the Unbinding Pathways of Cbz-Val-Phe-Phe-Val-Cbz from the Active Site of HIV-1 Protease

Donovan Chin [1], David N. Haney [2], Katya Delak [1], Hon M. Chun [1], and Carlos E. Padilla

[1] Moldyn Inc., 955 Massachusetts Avenue, Cambridge, MA 02139
[2] Haney Associates, 4212 93rd Avenue SE, Mercer Island, WA 98040

The speed and accuracy of the MBO(N)D multi-body dynamics program was compared to atomistic methods for studying the unbinding pathways of Cbz-Val-Phe-Phe-Val-Cbz (A74704) from HIV-1 protease using applied force simulations. The results from the applied force simulations using MBO(N)D show good comparison with the atomistic methods for the extraction forces, range of movement of the flaps of the protease, and residues encountered along the unbinding pathway; MBO(N)D, however, was *faster* than the atomistic method by a factor of eight. The applied force simulations were carried out as an example of MBO(N)D's ability to permit stable simulations with large time steps on systems that have large conformational changes. Applied force simulations provide information on the unbinding pathways between A74704 and HIV-1 that involve movement of the flaps of the protease—information that would be difficult or impossible to obtain through typical equilibration simulations. Both MBO(N)D and atomistic simulations suggest that the tips of the flaps of the protease may be important in the migration of the ligand into the active site. The implications of using MBO(N)D to study large conformational changes over long time scales for rational drug design are discussed.

This paper describes the use of a very fast multi-body dynamics method to simulate the unbinding pathways of the inhibitor, Cbz-Val-Phe-Phe-Val-Cbz (A74704), from the aspartyl protease of the human immunodeficiency virus (HIV-1). Simulating the unbinding pathways permits studies of protein-ligand adhesion forces, thermodynamic profiles of ligand binding into the active site, and large conformational changes that would not occur under equilibrium conditions. The two objectives are the following. First, to compare MBO(N)D and atomistic methods, and show that MBO(N)D is significantly faster than atomistic methods with comparable accuracy for the essential dynamics. Second, to present applied force simulations (AFS) methods as a potential new way of using molecular dynamics to study events that are inaccessible by current equilibrium methods.

The aspartyl protease has been the center of attention for drug discovery research for several years because its activity is central to the assembly and maturation of pathogenic HIV-1 by proteolytically cleaving the polyprotein products of the gag and gag-pol genes.(1) Several HIV protease crystal structures have been determined,(2,3) and their structures have led to the design of numerous HIV protease inhibitors—some of which have made it to market as treatments for the HIV infection.(4) This enzyme is homodimeric; it has a globular shape and consists of two separate polypeptide chains (99 residues each) bound together non-covalently to create the active form (Figure 1).

Early studies of the HIV protease structure showed that its conformation was substantially different depending on the presence or absence of a ligand bound at the active site.(5) More specifically, residues 43-56 on each of the monomers form an extended β hairpin structure that were either close together (bound ligand) or farther apart (unbound ligand). These flexible β hairpin structures are commonly referred to as "flaps." Mutations at the tips of these flaps—residues 45-53—have been implicated in conferring resistance to known inhibitors presumably because a particular mutation may reduce the mobility of the flaps thereby restricting the access of the inhibitor to the active site;(6) it is therefore of significant interest to understand the motions of these flaps.

The opening and closing of the flaps is likely to occur on time scales much longer than is currently accessible to typical molecular dynamics simulation (picoseconds to nanoseconds). Current simulation methods—where speed is derived from algorithmic approaches and not parallel processing—fall short of the computational speed needed for studying the opening and closing of the flaps.(7-19) We therefore use the following two strategies to address the problem. First, the use of a multi-body dynamics method that increases the computational speed by retaining only those variables that are associated with global motions. Second, the use of an applied force simulation (AFS) protocol that mimics the mechanism of atomic force microscopy (AFM) experiments to study the large conformational changes of the flaps. The use of the AFS protocol is particularly important because it allows the study of events that occur naturally on time scales much greater than is accessible from equilibrium simulations.(20-22) Each strategy is discussed in turn.

MBO(N)D or Multi-Body Order (N) Dynamics is a new molecular dynamics code developed by Moldyn that is designed around the concept of reduced variables and multigranularity.(23,24) The reduced variable approach of MBO(N)D is based on multi-body dynamics: that is, where groups of atoms are organized into interacting bodies in order to eliminate uninteresting high frequency events and permit much larger time steps than atomistic methods. Multigranularity is achieved in MBO(N)D by the simultaneous simulation of atoms, rigid bodies, and flexible bodies—flexible bodies are achieved by the addition of a few low frequency elastic modes. The mixing of body sizes and shapes can be customized for each system, property, or both. We have obtained increased computational speed by factors of up to 30 over traditional atomistic methods with a variety of molecular systems.(24) The computational speed of MBO(N)D is due primarily to the larger time steps.

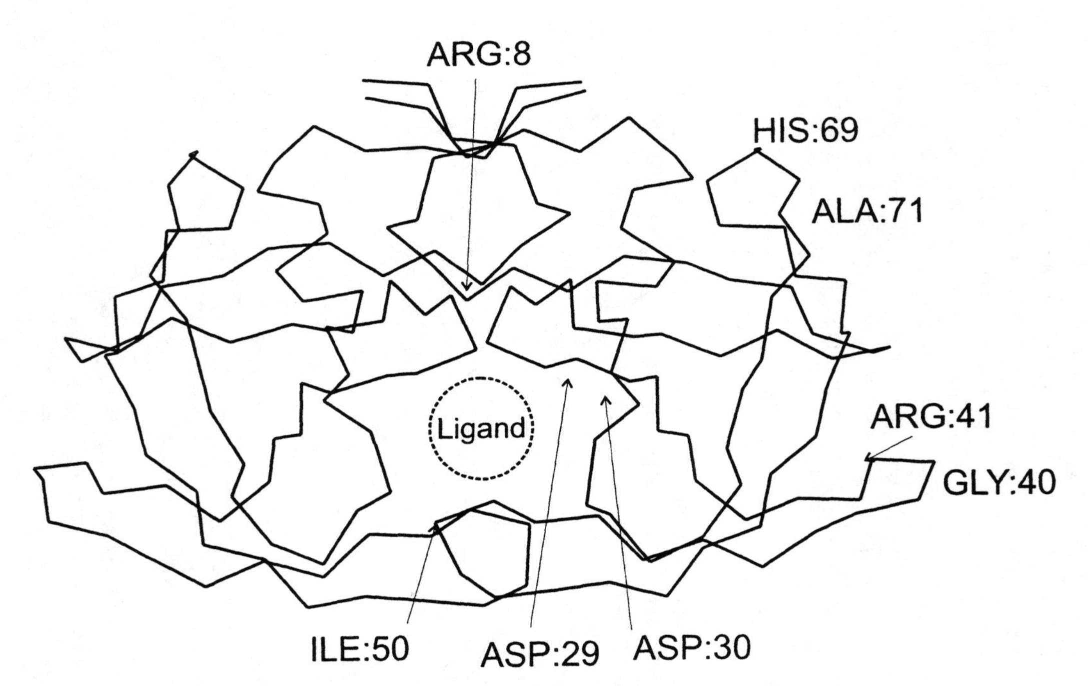

Figure 1. HIV protease (Cα trace), with the location of the A74704 ligand shown schematically. The locations of key residues—discussed in the substructuring section—are shown.

Despite the speed of MBO(N)D, it would not be possible to simulate the binding or unbinding dynamics of the A74704-HIV protease at this level of atomic detail (MBO(N)D currently uses an atomistic force field as discussed in the methodology section). We therefore chose to "force" the unbinding event by actively pulling the ligand from the binding site. That is, we effectively "compress" the experimental time scale down to that accessible though simulations. The strategy of AFS is an interesting and new approach that has resulted in new insight into the dynamics of a variety of different systems. The interest in AFS was inspired by AFM experiments where the most obvious experimental measure that can be compared to AFS is the extraction forces. Examples of AFS from the literature are the following. Grubmueller et al. carried out AFS on the biotin-streptavidin complex, and suggested that in the slow pulling regime there is a linear response of the extraction force to the rate of pulling. Their data in this slow region were extrapolated to the experimental pulling rate and extraction force, and the adhesion events associated with the simulated extraction forces studied in detail.(21) Schulten, et al. has recently argued that it may not be possible to accurately extrapolate extraction forces from the nanosecond time scale to microseconds (experimental scale),(25) but has nevertheless further defined the usefulness of AFS in studying events that occur on time scales several orders of magnitude greater than nanoseconds. Such events include studying the stick-slip motion of protein-ligand interactions during unbinding,(26) and in predicting a plausible binding pathway between bacteriorhodopsin and retinal.(25) Using a slightly different AFS protocol, Konrad et al. showed that the experimental extraction force (within the errors of the AFS and experimental data) of pulling apart the strands of DNA duplexes could be computed on the nanosecond time scale.(22) The conformations from their simulations of the DNA extension were helpful in understanding known experimental force-distance curves: that is, in the case where the DNA duplex apparently doubles in length while maintaining its original base parings.(22,27) It seems, therefore, that AFS can help in understanding the molecular motions, interactions, and deformations that occur during the aggregation of the biomolecules studied so far, but the ability of these methods to generate meaningful *absolute* values of the extraction force remains unclear.

It is important in AFS to apply the force slowly enough such that the system has time to relax energetically to the perturbation. These types of applied force simulations, therefore, can be slow (on the order of nanoseconds); multiple runs at different pulling rates (necessary to achieve convergence of the extraction force) only magnify the problem. Our solution to this problem is to carry out AFS using MBO(N)D.

The ability to efficiently simulate "rare" (on the time scale accessible to equilibrium methods) but important events in biomolecular recognition is powerful; the ability to validate these results with experiment is necessary. The usefulness of AFS in rational drug design, however, is not yet well defined. To this end, we have begun an extensive study here at Moldyn to evaluate AFS methods, and to test the extent to which MBO(N)D can increase the efficiency of these calculations: the work described in this paper represents our preliminary results.

Methods

MBO(N)D is currently interfaced to the CHARMM molecular modeling program.(28) Most of the simulation conditions used with MBO(N)D are similar to those used with atomistic methods; differences between the simulation conditions of MBO(N)D and CHARMM are noted below.

Starting Structure. The initial structure of the HIV-1/A74704 complex was obtained from the Brookhaven databank (pdb code: 9HVP); only the coordinates of the protein and the ligand were used subsequently. The placement and orientation of polar hydrogens were added to the complex using HBUILD.(29) We used the PARAM19 parameter set from Harvard University,(28) which involves the united atom representation for non-polar atoms. The non-bond conditions were those from PARAM19: interactions were updated heuristically and cut off beyond 8 Å; energies were shifted to 0 at 7.5 Å. Solvent was implicitly modeled through the use of a distance-dependant dielectric in the Coulombic interactions. These conditions are not the most optimal for realistic studies of protein–ligan unbinding, but are adequate for comparative studies between MBO(N)D and atomistic methods.

High-energy deformations and stresses in the crystallographic complex were removed from the complex by minimizing the energy of the system in the following way. We first constrained the main chain atoms (-CONCα-) using a harmonic force constant of 1000 kcal/mol•Å, and applied Adopted-Basis Newton Raphson (ABNR) for 500 steps; we then performed another 500 steps of ABNR without constraints. The root-mean-square deviation (RMSD) between the non-hydrogen atoms of the minimized structure and the crystal structure was 0.7 Å for the main-chain atoms, and 1.5 Å for the side chain atoms. This energy-minimized structure was used as the starting point for both the atomistic and MBO(N)D simulations.

Applied Force Simulations. We extracted the ligand from the protein by applying a protocol similar to that of Konrad (Scheme I).(22) In this method, a constant force is applied between a fixed point far away from the complex (45 Å) and a proximate atom on the ligand; the center of mass (COM) of the protein was held in place by a harmonic force constant. An important point is that the protein was allowed to rotate about its COM, which allows for some stresses incurred during the pulling to be relieved. Clearly the initial orientation of the pulling vector (Scheme I) is important to the results of the system. We have hypothesized that the current orientation of the pulling vector relative to the complex in Scheme I is plausible (although we know of no data that involves AFM and this HIV complex). Our primary goal in this paper, nevertheless, is to highlight the capabilities of MBO(N)D when compared to atomistic methods, and therefore the orientation of the pulling vector need not be the optimal one.

Temperature was maintained throughout the simulation by periodically scaling the velocities every 0.5 ps such that the average total temperature was 298 K. The total simulation length consisted of many short 10 ps segments. After each 10 ps segment, the value of the applied tension was increased by a constant amount, and the simulation repeated. The values of the applied pulling force increments between the ligand and a

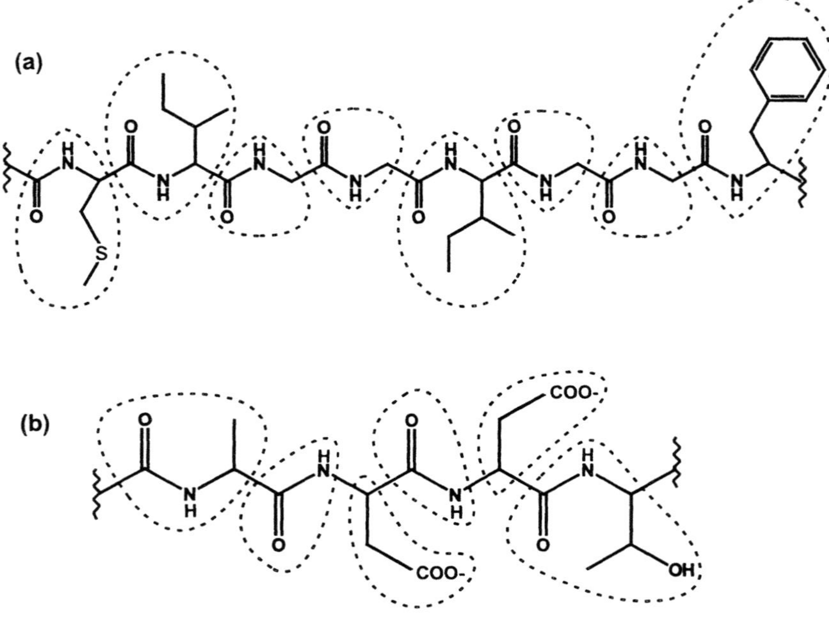

Scheme I. A schematic of the pulling experiment for HVP. HVP is drawn as a Cα trace; the ligand is drawn as ball and stick. The orientation of the pulling vector is shown; a constant force is applied along this vector between a fixed point 45 Å away in space and the nearest atom on the ligand.

point far from the protein ligand complex were 556, 278, 139, 69.5, 34.8, 27.8, 20.9, 13.9, and 6.95 piconewtons (pN) per 10 ps. The lengths of time for these runs ranged from 40 ps (for the fastest applied pulling rate) to 1030 ps. (for the slowest pulling rate). The cycle of incrementing tension and equilibrating was repeated until the minimum distance between the ligand and protein was > 8 Å (that is, beyond the non-bond cut off).

Atomistic Simulation. The constant temperature simulation was carried out at 298 K using the Leapfrog integrator in CHARMM for each of the 10 ps cycles mentioned in the pulling protocol. A time step of 1 fs was used along with SHAKE constraints applied to covalent bonds that involved hydrogens. We did not use a higher time step for the atomistic method because a 1 fs time step with SHAKE represents a typical use for the Leapfrog integrator and the CHARMM force field; in fact, this combination may be the most meaningful.(30)

Substructuring of Atoms into Bodies. An integral part of using MBO(N)D is grouping the atoms in a molecule into an appropriate combination of bodies, effectively subdividing a macromolecule into smaller chemically and physically meaningful components.(24) The bodies can either be rigid or flexible. The dynamics of flexible bodies can be modeled by a reduced set of elastic modes, emphasizing the lowest frequency modes that correspond to the overall motions of the body. High-frequency modes (local vibrations) that are not important to the event of interest are eliminated.(8,31,32) The bodies are allowed to undergo large motions relative to each other, but within each body, the relative motions among the grouped atoms is small (flexible body) or zero (rigid body).

Substructuring of the 9HVP Complex. Reproducing the proper global motions with MBO(N)D requires that the important hinges are identified correctly. Determining these hinge points can be difficult for a biomolecule, however, from a single static or quasi-static crystal or NMR structure.(33) We therefore used two crystal structures—open (3HVP) and closed (4HVP)—to provide us with more pseudo-dynamical information on the hinges for HIV protease. Using these two structures, we characterized the hinge motion of HIV protease by analyzing the ϕ and ψ angles (Figure 2a), the temperature factors (Figure 2b), and the pseudo-dihedral angle (Figure 2c; defined as the dihedral between four consecutive Cα atoms). These analyses all have similarities in the positions of "peaks" of the motion. That is, high motion is seen near residues 10, 17, 37-39, 49-52, 56, 60, 68-69, 73 and 81 (Figure 2a-c). We therefore expect to define smaller bodies in these regions to allow more mobility.

Our approach to substructuring HIV protease—based on extensive experience on other protein complexes—involved the following steps. First, we created bodies roughly 2-3 residues in size with "hinges" at the ϕ dihedral for the entire protein. Second, we inserted smaller bodies (one residue in size) for those regions of higher motions as noted in the above analyses. For example, there are many small bodies in the flaps where more motion is expected. We will refer to this substructuring strategy as “h1” (Scheme II and IIIa), which resulted in 62 bodies for each HIV protease

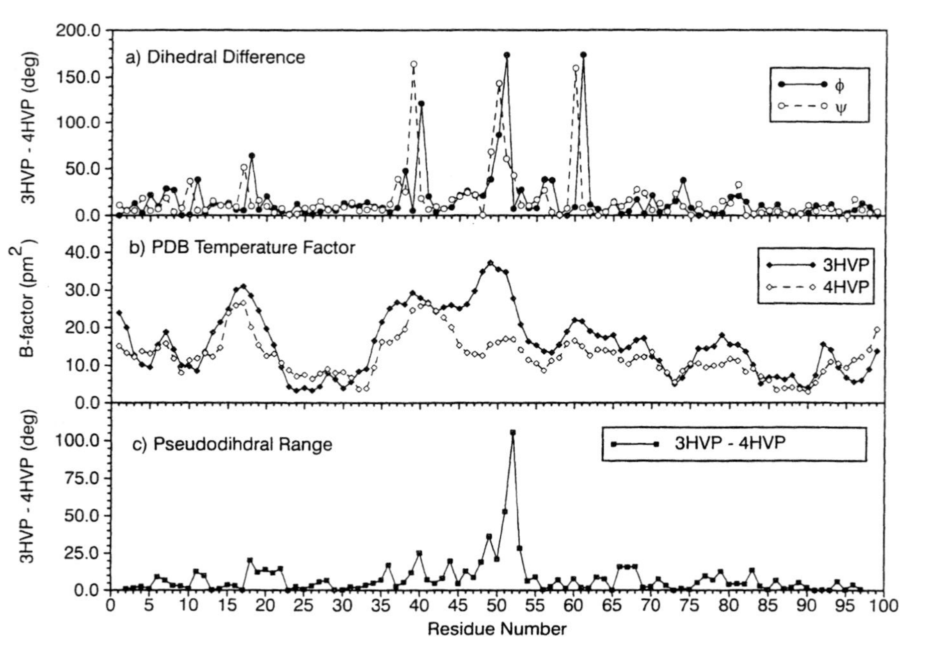

Figure 2. (a) The difference between Φ and Ψ angles for the open (3HVP) and closed (4HVP) structures of HIV-1. (b) Temperature factor analysis (c) Pseudodihedral analysis—defined as the dihedral between four consecutive Cα atoms.

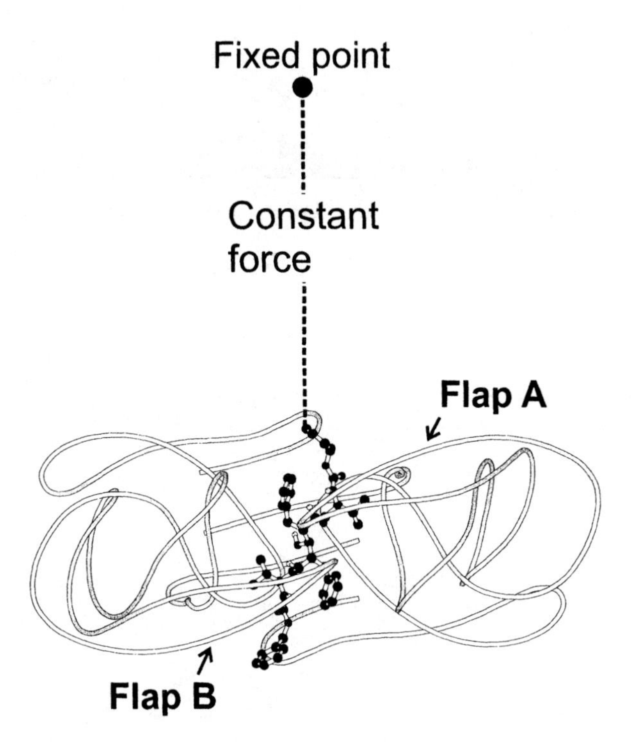

Scheme II. The substructuring schemes for HIV protease and A74704. Each block for h1 and h2 represents a body and the residues contained within. For example, in the h1 the first body contains residues 1-3 (up to ψ of residue 3). Only one of the two HIV protease polypeptide chains is shown, but both are substructured identically. The substructuring of A74704 is shown (the dashed lines delineate the bodies).

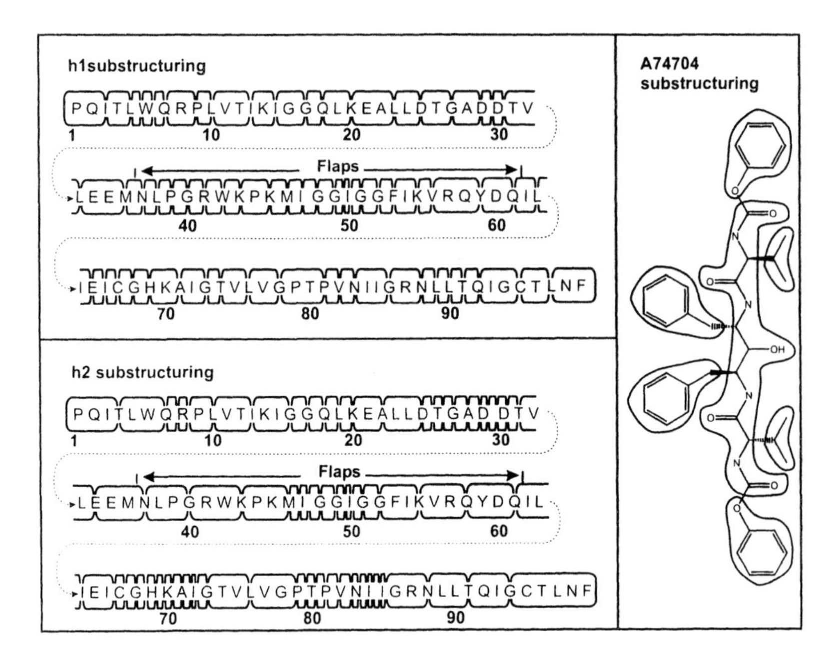

Scheme III. A close up view of the body definitions (dashed lines) for (a) residues 46-53 (-MIGGIGGF-) used in h1 and h2, and (b) residues 28-31 (-ADDT-) used in h2. Glycines can present special problems when it is desirable to have small bodies associated with them: that is, while the substructuring shown in (b)—Cα plus side-chain atoms as one body; adjacent peptide planes as other bodies—can result in more motion for Gly, it would unfortunately leave the Cα as a single body, which would severely limited the time step and negate much of the advantage of MBO(N)D. The substructuring of Gly shown in (a) represents a compromise between speed and accuracy.

polypeptide chain (for a total of 124). The smallest body size was for a single residue body containing Gly, which had 5 atoms in the extended atom representation. There were many bodies that contained two residues, and the largest body contained four residues.

We wanted to see if allowing movement in the side chains of residues proximate to the ligand would affect the dynamics of extraction—an analysis of the side-chain motion of the residues suggested significant movement in this region (not shown). We therefore inserted bodies into h1 that were defined by hinges about the ϕ and ψ angles of a given residue to the residues that were within 3.5 Å of the ligand (the number of atoms in the previous bodies were adjusted accordingly). This substructuring resulted in bodies that contained the Cα and side chain atoms, with the adjacent peptide planes as separate bodies. We will refer to this substructuring strategy as "h2" (Scheme II and IIIb); it is slightly different than h1 in that it has additional bodies in the four regions 25-30, 45-50, 65-75, and 80-85 for a total of 138 bodies. The inhibitor was divided into seven bodies, with hinges at the χ_1 angles of the side chains (Scheme II).

MBO(N)D Simulation. Extensive details of the MBO(N)D methodology can be found elsewhere,(24) and only a concise overview of the equations of motion are described here. The equations of motion in MBO(N)D use a body-based description of the system. In this description, the force vector, G, of each body in the system is described in equation 1.

$$G = G_{ff} + \Omega M U + \frac{1}{2} U M_{,j} U - \dot{M} U \tag{1}$$

where the first term, G_{ff}, accounts for chemical interactions embedded in the force field (CHARMM); Ω contains three skew-symmetric matrices of linear and angular velocities, and accounts for gyroscopic and coriolis effects; U contains the same linear and angular velocities as in Ω but in vector form; M is the generalized inertia matrix. The last two terms in eq. 1 account for the change in the body's inertia matrix due to deformation of the body. The subscript, j, represents the derivative with respect to the jth modal coordinate—that is, if the bodies are flexible; if they are rigid then the last two terms in eq. 1 are not used. The force field evaluations are first calculated in the atomistic model, and the resulting force vector is processed to obtain the generalized forces in terms of body torques, linear forces, and deformational forces for MBO(N)D:

$$G_{ff} = \begin{pmatrix} \sum_j r_j \times f_j \\ \sum_j f_j \\ \sum_j \phi_j^T f_j \end{pmatrix} \tag{2}$$

where the summation over j includes all atoms within the body being considered; r_j is the vector from the body reference origin to atom j; f_j is the total force applied to atom j and ϕ_j represents the jth partition of the body-based mode vectors for the flexible body.

Results and Discussion

Comparison and validation of the dynamics of ligand extraction between MBO(N)D and atomistic methods are the two main goals of this study. Nine different pulling rates where studied; the longest simulation (slowest pull) was ~1.4 ns. The key properties analyzed were the extraction forces for the ligand, the angle between the flaps, and the distances between the ligand and the protein during the extraction. Each result is discussed in turn.

Extraction Forces. There is good agreement between the extraction forces from MBO(N)D and atomistic simulations in the slow pulling region (the area of most interest) as shown in Figure 3. Here, h1 converges to a constant extraction force faster than h2; the faster convergence is because h1 has fewer degrees of freedom (DOF) than h2. MBO(N)D resulted in a factor of 8 increase in speed over atomistic runs. The range of values for the extraction forces in Figure 3 underscores the need for carrying out multiple simulations with different rates of pulling. Only when the extraction forces have converged within a reasonable range can the contribution to the dynamics from inertial effects be understood.

The orientation of the pulling vector used in this study resulted in an asymmetric movement of the flaps in HIV (Figure 4). The flaps have a certain "handedness" to them, which is reflected in its movement during extraction. During the applied force simulation, the ligand interacts most predominantly with one of these flaps resulting in the asymmetric movement.

Range of Motion of the Angle Between the Flaps. We used a definition by Collins et al. for the flap angles in HIV protease.(6) The flap angle is defined between two vectors, and each vector is defined between the Cα's of residues 40 and 50 in the flaps of each monomer. The comparisons between the atomistic and MBO(N)D simulations were variable over the different applied pulling forces (Figure 5). The results for the slower, more dynamically meaningful runs, suggest that the values of the inter-flap angles of h1 follow more closely the values from atomistic than h2; but h2 results in a greater range (maximum value – minimum value) than h1. The h2 scheme has more DOF than h1, which results in greater range of motion for the flaps. These data suggest that the additional movement allowed by the smaller bodies in the active site (h2) may not contribute significantly to the motion of the flaps and to the extraction forces. While h1 results in less range of motion than h2, it is apparently enough motion to allow stable extraction forces (Figures 3, and 4).

Close Residues Along the Unbinding Pathway. We investigated those residues in the HIV protease that were closest to the ligand during the extraction. In other words, by computing the distances of residues that were within 3.5 Å of the ligand during the pulling simulations, we are able to get a rough spatial description of the unbinding pathway. Figure 6 shows a representative probability distribution function (pdf) of these distances (taken from the slowest run 1.4 ns; applied force increment of 6.95 pN), and demonstrates the overall consistency of the unbinding pathways between MBO(N)D and atomistic simulations. Residues 7-9, and 24-31 line

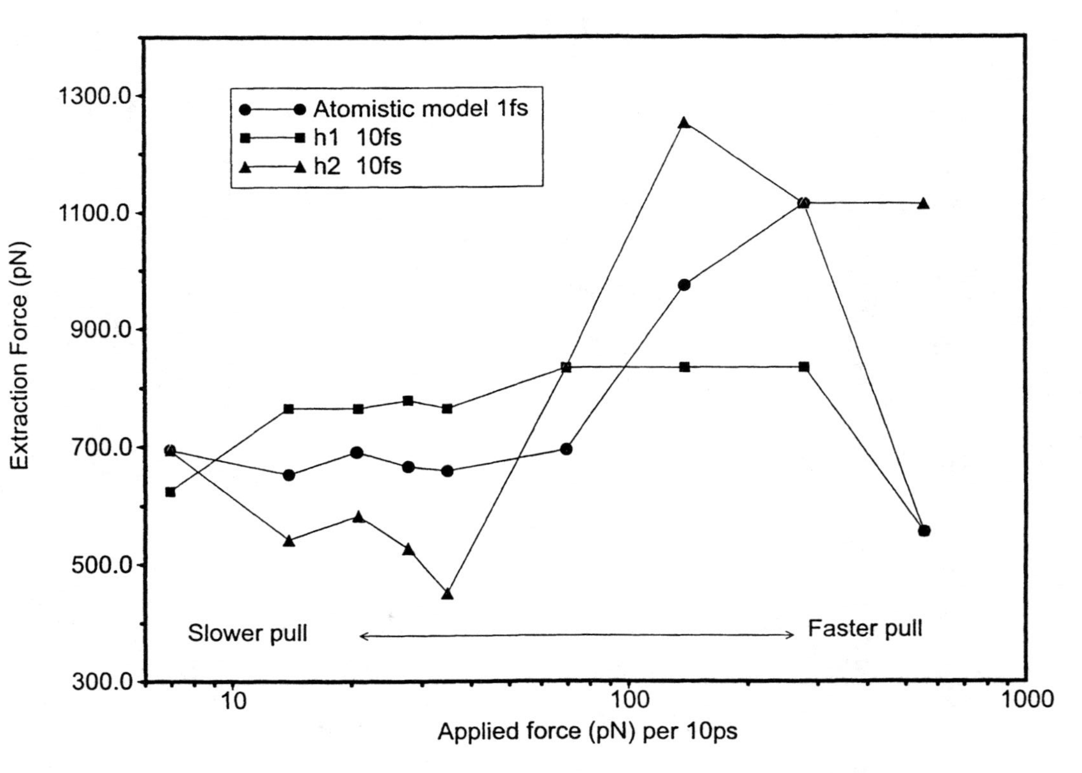

Figure 3. Extraction forces from atomistic and MBO(N)D simulations.

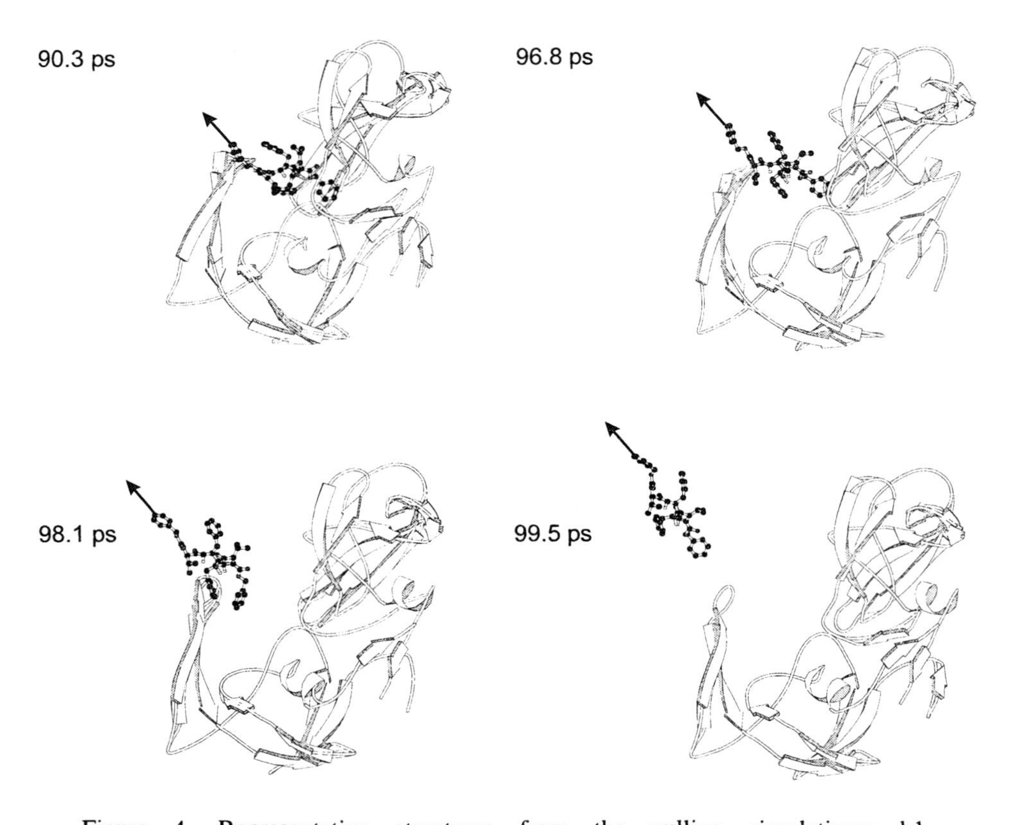

Figure 4. Representative structures from the pulling simulations; h1 substructuring, applied force of 6.95 pN per 10ps.

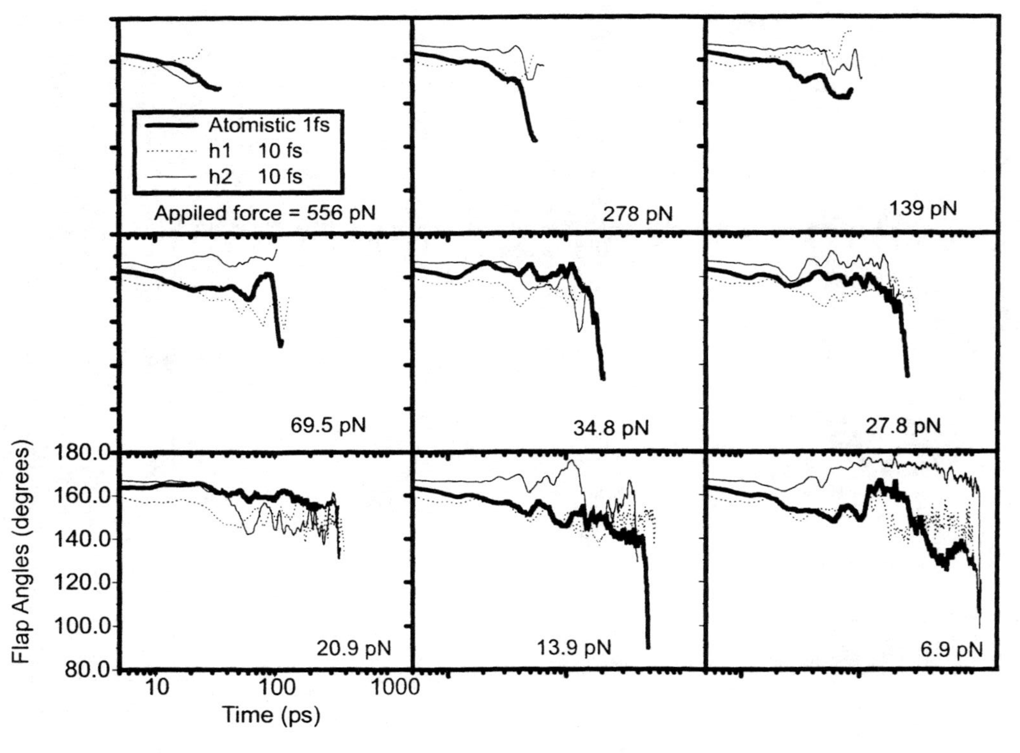

Figure 5. Inter-flap angles from atomistic and MBO(N)D simulations.

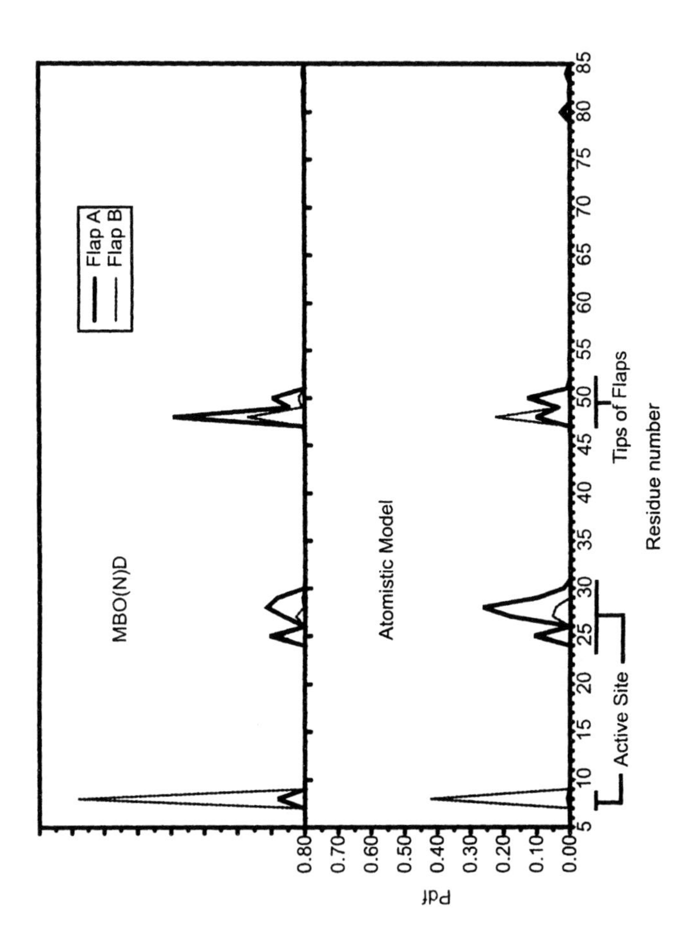

Figure 6. The close residues from atomistic and MBO(N)D simulations (h1 substructuring, applied force of 6.95 pN per 10ps).

the active site, but are not in the flaps; residues 48-50 (Gly-Gly-Ile) are associated with the tips of the flaps (Figures 1 and 6). There is some discrepancy, however, between the relative heights of the pdfs for these residues in each of the flaps from the MBO(N)D and atomistic simulations. While MBO(N)D correctly identifies the range of residues at the tips that interact with A74704, the pdf from MBO(N)D is higher for flap A at residue 48 than for flap B because at this level of substructuring we expect to lose some of the details in the dynamics. Not all of the discrepancies will be due to the nature of MBO(N)D since there is an inherent amount of randomness between different simulation conditions that can contribute to the discrepancies (for example, Figure 5).

The fact that MBO(N)D seems to compare well to atomistic simulations in many, but not all, properties underscores the trade off between accuracy and speed. Said differently, an even finer substructuring—smaller bodies—at these tips may reproduce the correct relative heights of the pdfs, but would probably result in less computational speed due to the need for smaller time steps with smaller bodies.

The contacts observed between the ligand and the tips of the flaps, nevertheless, are particularly interesting because these residues are proximate to locations where mutations have been suggested to restrict the motion of these tips (which presumably restricts access to the active site) and therefore impart resistance to inhibitors.(6) MBO(N)D with AFS could be used to see if the mutated form of the protease would result in higher extraction forces with A74704—harder to pull out—than the wild-type form as was concluded in a similar study by Collins et al.(6)

Conclusions

We have reported preliminary results of an extensive set of nanosecond simulations involving pulling the A74704 ligand from the active site of HIV-1. MBO(N)D results in an 8-fold improvement in speed over atomistic methods with good agreement in accuracy using several metrics: the extraction forces converged to similar values; the ranges of motion for the flaps are similar; the general binding pathways are consistent. The h1 substructuring seems to provide reasonable results when compared to the atomistic results; the results from the h2 substructuring suggest that the additional motion of the side chains in the active site do not significantly affect the over all behavior of unbinding. Our simulations suggest, therefore, that the key motions for the binding of A74704 occur in the flaps and not in the active site. More specifically, the results from both atomistic and MBO(N)D simulations suggest that the tips of the flaps (residues 45-53) may play an important role during the binding and unbinding of a ligand to HIV-1. The tradeoff between MBO(N)D and atomistic methods is that most of the finer details of motion are sacrificed for speed with retention of the global motions.

One area of ongoing research is the use of more realistic solvent models with MBO(N)D. To this end, there are two main strategies that are being explored by us, and each is discussed in turn.

Explicit solvent. We have carried out MBO(N)D simulations on other protein-ligand systems here at Moldyn where explicit molecules of water were used to fill the cavity of the active site, and to coat the surface of the protein. Each water in this system was treated as a single rigid body, and their inclusion in the MBO(N)D simulation did not limit the use of a high time step (unpublished). Our motive for using this strategy was based on the work by Steinbach and Brooks, where the authors showed that a thin shell of water was sufficient to reproduce the motional behavior of myoglobin in bulk solvent.(34) Our experience with this strategy is that these molecules of water are relatively immobile with respect to bulk waters due to interactions with the protein; it is this reduced mobility that permits the high time step despite the small size of these bodies. The substructuring of these waters—grouping of two of more molecules of water into bodies, which could be made flexible with modes—is a related strategy that we are exploring extensively. Substructured waters may permit even higher time steps. The use of single rigid waters that are far from the protein does indeed severely limit the time step, and would not afford any speed advantages with MBO(N)D.

Reduced variable solvent approaches. There are several methods that can be used to model solvent in a more meaningful way than the simple (and crude) distance-dependent dielectric method. For example, continuum electrostatic methods, which include the spatial electrostatic potential of the system;(35,36) solvent potential methods, which are based on the exposed solvent accessible surface area and empirical parameterization;(37,38) Langevin dynamics, which represents the viscous effects of solvent through the appropriate frictional dissipation and fluctuation terms.(39,40) These methods are consistent with the reduced variable approach of MBO(N)D, and we are currently exploring their use.

Applied force simulations on problems in rational drug design are attractive for two reasons. First, they may be compared to appropriate AFM experiments. Second, they provide new structural and dynamical information not seen in typical equilibrium simulation strategies. That is, applied force simulations provide information about events at and away from the primary binding site, and about the strength of the interaction along this pathway. Current equilibrium simulation methods and computational resources cannot be expected to explore events in protein-ligand complexes that have energetic barriers above a few kcal/mol. Transition state theory helps qualitatively to define the time scales associated with various barrier heights: for example, transition barriers that are 5, 10, or 15 kcal/mol will require approximately nanoseconds, microseconds, and milliseconds respectively in time to traverse.(41) Non-equilibrium methods, therefore, are one way to study long time scale events in protein-ligand complexes.

Determining the contribution of inertial effects to the dynamics of a single AFS that result from extracting the ligand faster than the system can relax energetically and structurally can be difficult. We have shown, however, that there exists a region below which inertial effects seem to be negligible, but this requires multiple simulations at different rates of applied force. Strategies to estimate the amount of inertial effects during a given simulation, and therefore aid in reducing the number of simulations on a particular protein-ligand complex are being explored by us.

One possible use of AFS in rational drug design is to identify alternative binding sites away from the primary binding site, and then use this information to connect these sites through a suitable linker.(42) An important advantage of AFS over other similar but static methods that involve minimization, we believe, is that it includes contributions from conformational entropy: the compensatory effects between enthalpy and entropy is one of the most challenging problems facing the use of computations in rational drug design.(43) The usefulness of AFS methods, however, is related not only to the rate of the applied force, but to the orientation of the pulling vector. While a single or multiple pulling vectors can be easily used when there is sufficient knowledge of the system, a very large number of pulling vectors may be needed for systems where the knowledge of plausible pathways is vague or non-existent. Incomplete sampling of all possible pathways is a weakness of similar search methods. A large number of AFS simulations can be computationally expensive, and therefore MBO(N)D stands to contribute significantly to this problem. Another important limitation in AFS method for rational drug design is that it does not give free energies. Data from AFS methods, nevertheless, are related to k_{off} ($K_d = k_{off} / k_{on}$), and this information would be useful for a series of compounds with similar values of k_{on}

MBO(N)D permits larger time steps than atomistic methods; the increased speed will be important in the study of very long time scale events (> tens of nanoseconds). One criteria for MBO(N)D to reproduce the atomistic behavior of a system is that high frequency events such as van der Waals clashes are not strongly coupled to the dominant low frequency motions. In fact, systems that exhibit large motions about hinges—for example, DNA bending and stretching,(22) and hinge bending domains in proteins(44)—are good candidates for MBO(N)D.(24)

Acknowledgments

This work was supported in part by a grant (70NANB5H1078) from the National Institute of Standards and Technology (NIST) Advanced Technology Program (ATP).

Literature Cited

1. Debouck, C.; Gorniak, J. G.; Strickler, J. E.; Meek, T. D.; Metcalf, B. W.; Rosenberg, M. *Proc Natl Acad Sci U S A* **1987**, *84*, 8903-8906.
2. Greer, J.; Erickson, J. W.; Baldwin, J. J.; Varney, M. D. *J. Med. Chem.* **1994**, *37*, 1035-1054.
3. Appelt, K. *Perspectives in Drug Discovery and Design* **1993**, *1*, 23-48.
4. Deeks, S. G.; Smith, M.; Holodniy, M.; Kahn, J. O. *JAMA* **1997**, *277*, 145-153.
5. Miller, M.; Schneider, J.; Sathyanarayana, B. K.; Toth, M. V.; Marshall, G. R.; Clawson, L.; Selk, L.; Kent, S. B.; Wlodawer, A. *Science* **1989**, *246*, 1149-1152.
6. Collins, J.; Burt, S. K.; Erickson, J. W. *Nature. Struc. Biol.* **1995**, *2*, 334.
7. McCammon, J. A.; Harvey, S. *Dynamics of Proteins and Nucleic Acids*; Cambridge University Press: Cambridge, 1987.
8. Amadei, A. B.; Linssen, M.; Berendsen, H. J. C. *Proteins: Struct. Funct. Gen.* **1993**, *17*, 412.
9. deGroot, B. L.; Amedi, A.; D.M.F., v.; Berendsen, H. J. C. *J. Biomol. Str. Dyn.*, **1996**, *13*, 741.
10. Gibson, K. D.; Scheraga, H. *J. Comp. Chem* **1990**, *1*, 468.
11. Hao, M. H.; Pincus, M. R.; Rackovsky, S.; Scheraga, H. *Biochemistry* **1993**, *32*.
12. Mazur, K.; Abagyan, R. A. *J. Comp. Phys* **1989**, *6*, 118.
13. Procacci, P.; Berne, B. J. *J. Chem. Phys* **1992**, 101, 2421.
14. Mazur, K.; Dorofeev, V. E.; Abagyan, R. A. *J. Comp. Phys.* **1991**, *92*, 261.
15. Ryckaert, J. P.; Ciccotti, G.; Berendsen, H. J. C. *J. Comp. Phys* **1977**, *23*, 327.
16. Schlick, T.; Olson, W. K. *J. Mol. Biol.* **1992**, *223*.

17. Space, B.; Rabitz, H.; Askar, A. *J. Chem. Phys.* **1993**, *99*, 9070.
18. Tuckerman, M. E.; B.J., B.; Martyna, G. J. *J. Chem. Phys.* **1992**, *97*.
19. Zhang, G.; Schlick, T. *J. Comp. Chem.* **1993**, *14*, 1212.
20. Lebrun, A.; Lavery, R. *Nucleic Acids Res.* **1996**, *24*, 2260.
21. Grubmuller, H.; Heymann, P. T. *Science* **1996**, *271*, 997.
22. Konrad, M. W.; Bolonick, J. I. *J. Am. Chem. Soc.* **1996**, *118*, 10989.
23. Chun, H. M.; Turner, J. D.; Frisch, H. P. *Paper AAS 89-457, AAS/AIAA Conf., Stowe, VT* **1987**, 89-457.
24. Chun, H. M.; Padilla, C.; Alper, H. A.; Chin, D. N.; Watanabe, M.; Soosar, K.; Blair, K.; Nagle, R.; Karplus, M.; Haney, D. N. *J. Comp. Chem.* **1998**, *in preparation.*
25. Izrailev, S.; Stepaniants, S.; Balsera, M.; Oono, Y.; Schulten, K. *Biophys. J.* **1997**, *72*, 1568-1581.
26. Isralewitz, B.; Izrailev, S.; Schulten, K. *Biophys. J.* **1997**, *73*, 2972-2979.
27. Noy, A.; Vezenov, D.; Kayyem, J.; Meade, T.; Lieber, C. *Chem. Biol.* **1996**, *4*, 519-527.
28. Brooks, B. R.; Bruccoleri, R. E.; Olafson, B. D.; States, D. J.; Swaminathan, S.; Karplus, M. *J. Comp. Chem.* **1983**, *4*, 187.
29. Brunger, A.; Karplus, M. *Proteins: Struct. Funct. Genet.* **1988**, *4*, 148.
30. Watanabe, M.; Karplus, M. *J. Phys. Chem.* **1995**, *99*, 5680.
31. Turner, J. D.; Weiner, P. K.; Chun, H. M.; Lupi, V.; Gallion, S.; Singh, U. C. *Computer Simulation of Biomolecular Systems: Theoretical and Experimental Applications,*; Turner, J. D.; Weiner, P. K.; Chun, H. M.; Lupi, V.; Gallion, S.; Singh, U. C., Ed.; ESCOM, Leiden, 1993; Vol. Chapter 2.
32. vanAalten, D. M. F.; Amadei, A.; Linssen, A. B. M.; Eijsink, V. G. H.; Vriend, G.; Berendsen, H. J. C. *Proteins* **1995**, *22*, 45-54.
33. Wriggers, W.; Schulten, K. *Proteins: Struc. Func. Gen.* **1997**, *29*, 1-14.
34. Steinbach, P. J.; Brooks, B. R. *Proc. Natl. Acad. Sci. USA* **1993**, *90*, 9135.
35. Still, W. C.; Tempczyk, A.; Hawley, R. C.; Hendrickson, T. *J. Am. Chem. Soc.* **1990**, *112*, 6127.
36. Scheafer, M.; Karplus, M. *J. Phys. Chem.* **1996**, *100*, 1578.
37. Fraternali, F.; van Gunsteren, W. F. *J. Mol. Biol.* **1996**, *256*, 939.
38. Wesson, L.; Eisenberg, D. *Protein Sci.* **1992**, *1*, 227.
39. Pastor, R. W. *Mol. Phys.* **1988**, *65*, 1409-1419.
40. Allison, S. A. *Macromolecules* **1991**, *24*, 530-536.
41. Warhsell, A. *Computer Modeling of Chemical Reactions in Enzymes and Solutions*; John Wiley & Sons: New York, 1997.
42. Jain, A.; Whitesides, G. M.; Alexander, R. S.; Christianson, D. W. *J. Med. Chem.* **1994**, *37*, 2100.
43. Chin, D. N.; Lau, A. Y.; Whitesides, G. M. *J. Org. Chem.* **1998**, *63*, 938-945.
44. Gerstein, M.; Lesk, A.; Chothia, C. *Biochemistry* **1994**, *33*, 6739-6749.

Chapter 7

Calculation of Relative Hydration Free Energy Differences for Heteroaromatic Compounds: Use in the Design of AMP Deaminase Inhibitors

Mark D. Erion and M. Rami Reddy

Metabasis Therapeutics Inc., 9360 Towne Centre Drive, San Diego, CA 92121

A strategy for designing potent AMP deaminase inhibitors is described which entails the identification of compounds that undergo covalent hydration in aqueous solution to generate hydrated analogs exhibiting close structural resemblance to the transition state structure. Using a combination of quantum mechanical calculations and the free energy perturbation methodology, relative hydration free energy differences ($\Delta\Delta G_{hyd}$) were calculated for a variety of pteridine, purine, quinazoline and pyrimidine analogs. Calculated results were in good agreement with experimental data. Differences in the extent of hydration were attributed to electronic and steric effects and to differences in aromaticity based on calculated bond separation energies. The potential value of hydration free energy calculations to drug design was demonstrated by showing that the sum of $\Delta\Delta G_{hyd}$ and the relative binding free energy ($\Delta\Delta G_{bind}$) of the hydrated molecule complexed to adenosine deaminase (ADA) accurately accounted for the 400-fold difference in inhibitory potency of two ADA inhibitors.

Adenosine is a naturally occurring nucleoside that elicits a vast array of pharmacological effects used to preserve cellular and organ function during times of ischemia (*1, 2*). Adenosine is produced by cells in response to oxygen deprivation through net breakdown of intracellular stores of ATP. Transport of adenosine out of the cell leads to activation of adenosine receptors located on adjacent cells. Since adenosine undergoes rapid metabolism, only receptors on cells near the site of adenosine production are activated and therefore only the pharmacology associated with adenosine receptor activation in that tissue is observed. For example, adenosine produced in the heart is associated with effects on blood flow, heart rate and myocardial protection from ischemic damage, whereas adenosine in the CNS is associated with neuroprotection, antiseizure activity and analgesia. Each activity is usually a result of activation of one or more of the four adenosine receptor subtypes, namely A1, A2a, A2b and A3.

Not surprisingly, the profound pharmacological activity associated with adenosine receptor activation has attracted enormous interest within the pharmaceutical industry over the past two decades. Through these efforts, receptor subtype specific agonists have been discovered and shown to exhibit good activity in animal models of cardiovascular, CNS, inflammatory and metabolic diseases. Unfortunately, no adenosine receptor agonist has been successfully developed due, in nearly every case, to an unacceptable therapeutic window resulting from simultaneous activation of receptors at sites unrelated to the disease.

An alternative strategy that was envisioned to have greater potential for producing compounds with a wider therapeutic window has been under intense study at Metabasis Therapeutics. The strategy takes advantage of the site- and event-specific nature of adenosine production (oxygen-limited cells) by using compounds, known as Adenosine Regulating Agents (ARAs), that further enhance the levels of extracellular adenosine at these sites through modulation of enzyme activities and biochemical pathways involved in adenosine metabolism and production (Figure 1) (*3-5*).

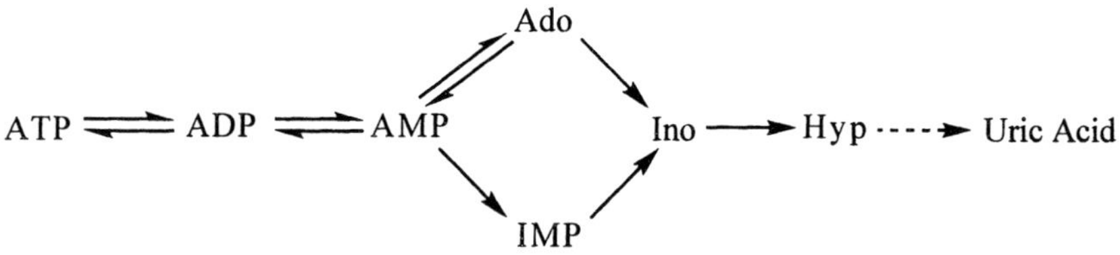

Figure 1: Purine Catabolic Pathway

One target that is especially noteworthy based on its role in the purine salvage and catabolic pathways is AMP deaminase (AMPDA) (*6*). AMPDA catalyzes the deamination of AMP to IMP (Figure 2) and consequently can

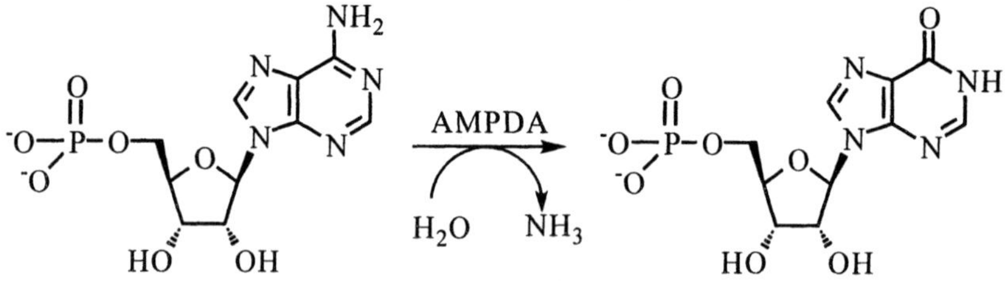

Figure 2: AMP Deaminase-Catalyzed Reaction

indirectly control adenosine production by diverting the ATP breakdown product, AMP, away from adenosine. Inhibition of AMPDA is anticipated to result in elevated extracellular adenosine levels only at sites undergoing ischemia, since flux through AMPDA is nearly undetectable under normal conditions due to the low basal AMP concentrations (≈ 1 μM) coupled with the high AMPDA K_M (≈ 1 mM). In contrast, flux through AMPDA increases dramatically during ischemia since net ATP breakdown produces AMP levels that rapidly approach the K_M of AMPDA.

Inhibitors of AMPDA, like all AMP-binding proteins, represent a considerable design challenge. First, AMP sites are typically very hydrophilic with a multitude of positively-charged amino acid residues in direct contact with the phosphate portion of the molecule. These electrostatic interactions comprise a large proportion of the AMP binding affinity as readily observed by the typical 10^6-fold loss in binding affinity that characterizes non-phosphorylated analogs. Herein lies the dilemma for the medicinal chemist, since retention of charge on the ligand in order to maintain binding affinity prevents passive diffusion of the inhibitor into cells and inhibition of intracellular enzymes such as AMPDA. An additional challenge in the design of AMPDA inhibitors stems from the large number of AMP binding proteins and therefore the necessity for the design of highly specific ligands to reduce the risk of producing additional drug toxicities.

Design of AMPDA Inhibitors

Transition state (TS) mimics often represent the most potent and specific enzyme inhibitors (*7*). High affinity is achieved because these inhibitors engage in the full complement of enzyme interactions that are made between the enzyme and the substrate in the TS. Many of these interactions are either absent or less favorable in the ground state structure, which of course is the characteristic required for efficient catalysis, i.e. net stabilization of the TS structure and a lowering of the energetic barrier to reaction. High specificity is also achieved through TS mimicry because usually only one enzyme can both catalyze a specific reaction and recognize a unique set of small molecule substrates. Both properties were particularly attractive for our purposes especially since a large enhancement of binding affinity could enable the discovery of AMPDA inhibitors that retain sufficient binding affinity in the absence of substantial molecular charge and therefore are able to enter cells and inhibit AMPDA.

The TS structure that is recognized and stabilized by AMPDA is not known (*8*) but is expected to be similar to the TS structure for the related enzyme, adenosine deaminase (ADA), on the basis of the reaction (deamination of adenine) and their high sequence homology in the purine binding site. The TS of ADA is postulated from the high resolution X-ray structure of ADA complexed with 6-hydroxy-1,6-dihydropurine riboside (*9*). Accordingly, deamination is postulated to occur by an initial rate-limiting zinc-assisted hydration of the 1,6 double bond of adenine to produce the tetrahedral intermediate hydrate **1**. Irreversible breakdown of the intermediate produces the 6-oxo purine product and ammonia.

Three compounds are thought to inhibit ADA by TS mimicry (Figure 3). In each case, a single hydroxyl, located in a position analogous to the hydroxyl on the hydrated intermediate, is known to be absolutely essential for inhibitory activity. Coformycin (**3**), a microbial metabolite, and its 5'-monophosphate reversibly inhibit ADA and AMPDA ($K_i \approx 10^{-10}$ M), respectively, with binding affinities that represent some of the most potent ever discovered for a small molecule inhibitor (*10*). Despite the absence of the 5'-phosphate, coformycin also is a modest inhibitor of AMPDA ($K_i \approx 10^{-6}$ M). The fact that it is an extremely potent inhibitor of ADA, however, limits its use for some chronic indications, since potent inhibition of ADA produces severe immunosuppression. Recently, we reported a series of N3-substituted coformycin aglycone analogs that are 1000-fold more potent AMPDA inhibitors ($K_i \approx 1$ nM) than coformycin (*11-13*) and are highly selective for AMPDA relative to ADA ($K_i >> 10^{-5}$ M). These compounds are currently undergoing pharmacological evaluation in a variety of animal models of disease to explore the potential of AMPDA as an adenosine regulating agent target.

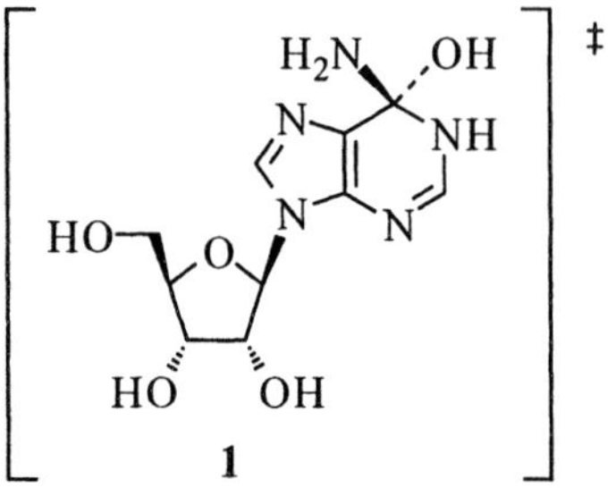

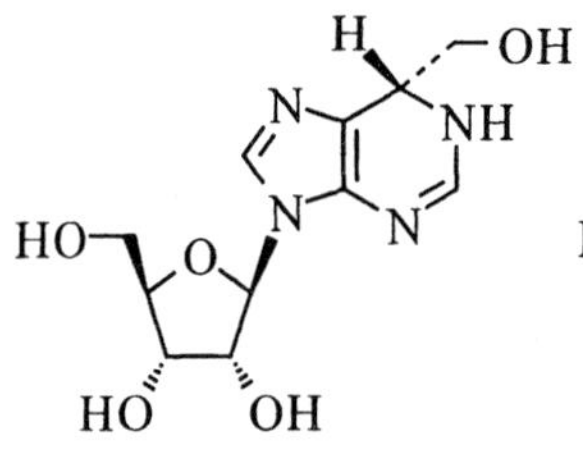

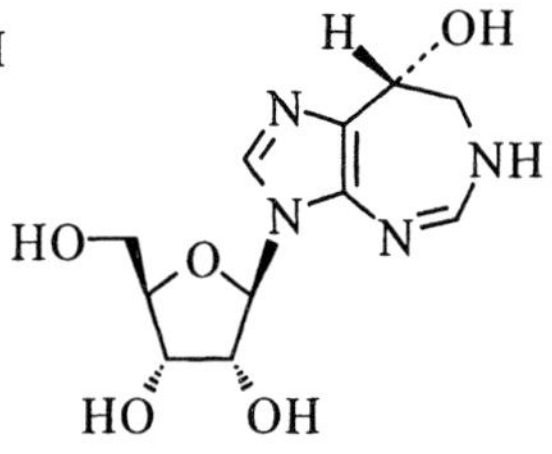

Figure 3: ADA and AMPDA TS Inhibitors

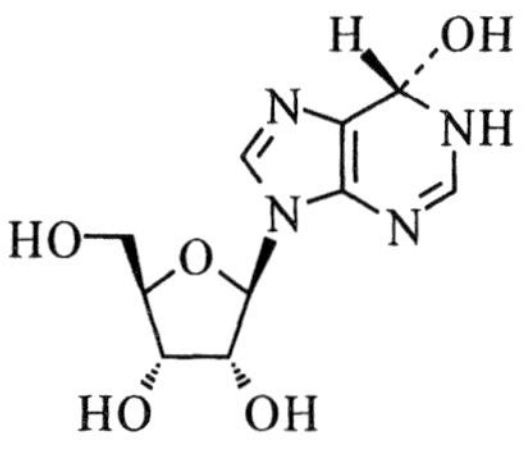

The TS mimic with the highest reported affinity for ADA is the 1,6-hydrate of purine riboside, i.e. **4,** which is not particularly surprising given its close structural resemblance to the ADA TS structure (*14*). The molecular basis for the high affinity is apparent from the X-ray structure of the ADA complex, which shows multiple interactions between the active-site residues and the zinc with the 6-hydroxyl group (*9*). Unlike coformycin-based TS mimics, no previous efforts have been reported designed to exploit the high affinity of **4**. The most likely reason stems from the instability of **4**, which exists nearly exclusively as purine riboside in solution as evident from the highly unfavorable equilibrium constant ($K_{eq} = 10^{-7}$)

that characterizes the hydration reaction (Figure 4). Purine riboside exhibits an apparent inhibition constant of ADA of approximately 10^{-5} M. Since the species responsible for inhibition of ADA is the hydrated molecule, and since the apparent equilibrium constant is related to the hydration equilibrium constant and the inhibitory constant for the hydrated molecule (K_i^*) by $K_i\,(app) = K_i^*(1 + 1/K_{eq}) \approx K_i^* K_{eq}^{-1}$, K_i^* has been calculated to be a remarkable 10^{-12} M (*15*).

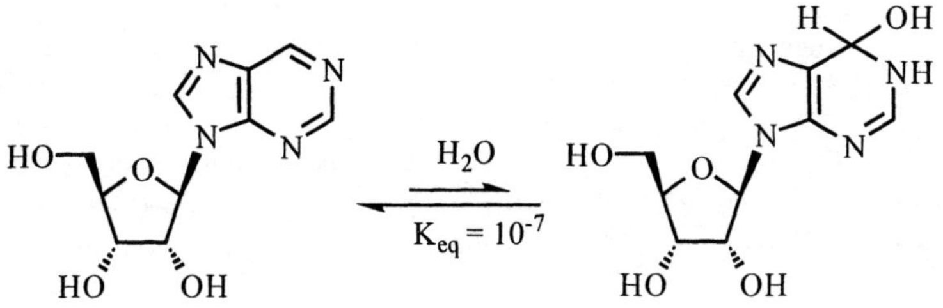

$$\xrightarrow[K_{eq} = 10^{-7}]{H_2O}$$

Figure 4: Hydration of Purine Riboside (**5**)

Our strategy for the design of potent, cell-penetrable AMPDA inhibitors was to modify the purine base in a manner that enhanced hydration without impairing the binding of the hydrated species to the AMP binding site (Figure 5). To test this strategy we analyzed potential modifications of purine riboside for their ability to enhance hydration. Modifications that enhanced hydration from 10^{-7} to 10^{-4} without diminishing K_i^* would enhance the apparent inhibition by greater than 1000-fold, i.e. from 1 μM to 1 nM. Modifications identified in this work were anticipated to be transferable to our discovery efforts on AMPDA inhibitors based on the high homology between AMPDA and ADA in the purine binding site.

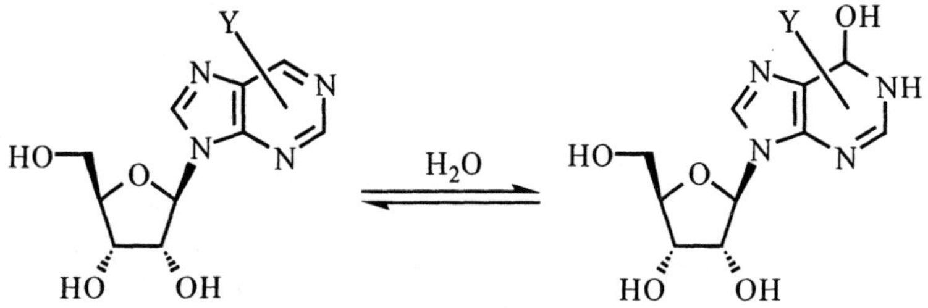

		5	Analog
K_{eq}	=	10^{-7}	10^{-4}
K_i^*	=	10^{-13} M	10^{-13} M
K_i (app)	=	10^{-6} M	10^{-9} M

Figure 5: Drug Design Strategy

Computer-Aided Drug Design

Design of potent deaminase inhibitors depends on our ability to accurately predict the effect of the modifications on both the hydration equilibrium and the affinity of the hydrated molecule for the deaminase binding site. Accordingly, accurate calculations of both were considered essential for prediction of the overall inhibitory potential of various purine riboside analogs. Analysis of the hydrate binding affinity was envisioned to entail calculation of relative binding affinities using the well-described free energy perturbation method and a computer model derived from the X-ray coordinates of the ADA:6-hydroxy-1,6-dihydropurine riboside complex. Accurate calculation of the hydration equilibrium constant required calculation of the free energy difference for the hydration reaction which is related to the differences in free energy between the hydrated product and the reactants in the gas phase and in solvent (eq 1).

$$\Delta G_{hyd} = -RT\ln K_{eq} = \Delta G_{gas} + \Delta\Delta G_{sol} \tag{1}$$

Hydration Free Energy Calculations. The gas phase free energies (ΔG_{gas}) were calculated using energies obtained from ab initio quantum mechanical calculations at the 6-31G** basis set level on fully-geometry optimized anhydrous and hydrated compounds. As detailed in an earlier publication (*16*), efforts were made to enhance the accuracy of these calculations by including zero point and vibrational energies as well as by including electron correlation energy contributions using second, third and fourth order Moller-Plesset perturbation theory and QCISD(T) correlation methods at the 6-31G** basis set level.

The solvation free energy differences were calculated using molecular dynamics (MD) simulations in conjunction with the thermodynamic cycle perturbation (TCP) approach (*17*). The TCP cycle therefore entailed a computational transformation of the unhydrated molecule (R) to the hydrated molecule (P) in the gas phase and in the presence of SPC/E waters (*16*) (Figure 6).

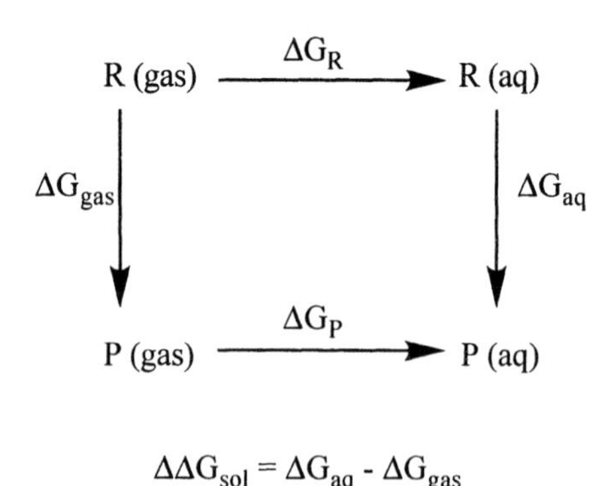

Figure 6: TCP Cycle for Calculation of $\Delta\Delta G_{sol}$

To test the accuracy of the approach, a small set of carbonyl-containing compounds with experimentally-determined hydration equilibrium constants were evaluated. Results showed that although the correct trends were observed, significant differences were observed between the calculated and experimental results. Inaccuracies in the hydration free energy difference were attributed to inaccuracies in both ΔG_{gas} and $\Delta\Delta G_{sol}$. Gas phase quantum mechanical free energies varied significantly depending on the level of theory. Errors in the solvation free energy differences were attributed to the difference in hybridization between the reactant (*sp*^{2}) and product molecules (*sp*^{3}) and the slow convergence of the calculations as a result of this large structural perturbation.

To improve the accuracy of the results, the relative hydration free energy difference for two similar hydration reactions was calculated in order to minimize systematic errors. Relative hydration free energies were considered suitable for our purposes, since the primary aim was to determine whether a structural modification produced a compound with enhanced or diminished hydration. Several significant advantages are associated with calculation of relative hydration free energies compared to absolute free energies. First, the free energy contribution of the water molecule completely cancels since it is common to both reactions. Second, the solvation free energy calculation entails the difference of two thermodynamic cycles, one for the reactants and one for the products. Consequently, the calculation converges more rapidly because the structural perturbation is markedly less pronounced. Last, accurate results are possible at lower levels of quantum mechanical theory. The major disadvantage of the approach, however, is that it requires evaluation of two hydration reactions that bear relatively close structural resemblance.

Calculation of relative hydration free energies for a set of carbonyl compounds gave results similar to experimental findings (*18*). Good agreement was obtained regardless of the molecular factor influencing the extent of hydration. Factors included differences in sterics and electronics near the carbonyl as well as differences in angle strain, which is the factor that likely accounts for the high propensity of cyclopropanone to hydrate in aqueous solutions.

Table I: Relative Hydration of Carbonyl-Containing Compounds (1-2)

R1	R2	$\Delta\Delta G_{hyd}$ (calc)[a] (kcal/mol)	$\Delta\Delta G_{hyd}$ (expt) (kcal/mol)
HCHO	CH_3CHO	-4.7	-4.3
CH_3CHO	CH_3COCH_3	-4.2	-4.0
CH_3COCH_2Cl	CH_3COCH_3	-3.1	-3.4
$CH_3COCOOH$	CH_3COCH_3	-3.7	-4.0
CH_3COCF_3	CH_3COCH_3	-8.2	–
(cyclopropanone)=O	CH_3COCH_3	-14	–

[a] Standard errors ranged from ± 0.4 - 0.5 kcal/mol

Hydration of Heteroaromatic Compounds

Heteroaromatic compounds represent another class of organic compounds known to undergo covalent hydration (*19-20*). Hydration equilibria for numerous azanaphthalenes were experimentally determined in the 1960s and 1970s using a variety of spectroscopic techniques. A subset of these compounds with equilibrium constants ranging from 10^{-7} to greater than 10^{3} were chosen to assess the accuracy of our method for calculating relative hydration free energies (*21*). In one study, pteridine, which hydrates approximately 20% in aqueous solutions, was compared to several analogs in which the only difference was the absence or the presence of nitrogens in the aromatic ring fused to the pyrimidine ring. Although the structural differences were relatively small and distant from the hydration site, the difference in the extent of hydration was large. As shown in Table II, the calculated and experimental results were very similar. Accurate results were also obtained for a triazanaphthalene compound that fails to hydrate in the neutral form but exists almost exclusively in the hydrated form as the cationic species.

Table II: Relative Hydration of Heteroaromatic Compounds ($\Delta\Delta G_{hyd}$) (1-2)

P1	P2	Calc[a] (kcal/mol)	Expt (kcal/mol)
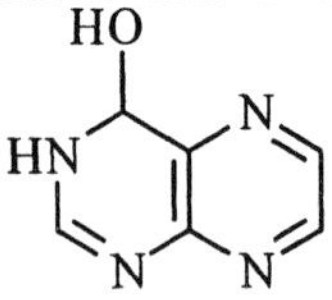 HO HN N N N	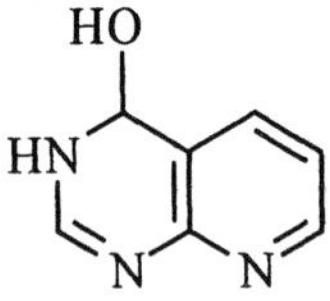 HO HN N N	-2.0	-2.5
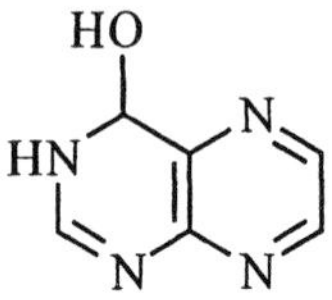 HO HN N N N	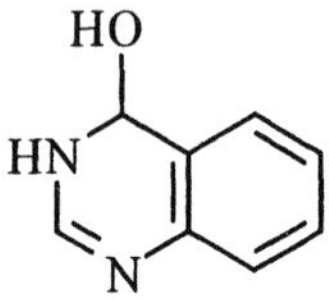 HO HN N	-4.0	-5.1
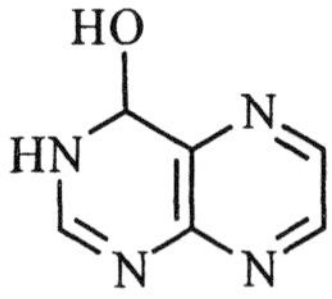 HO HN N N N	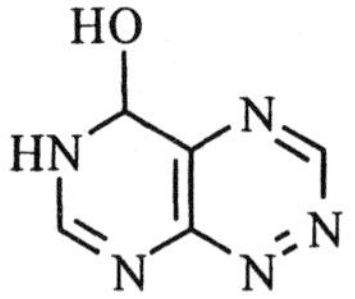 HO HN N N N N	6.1	6.2
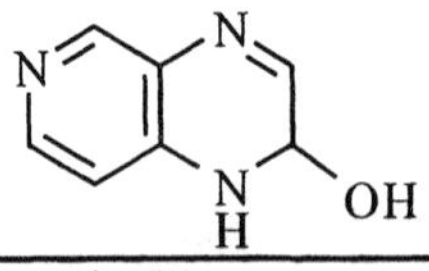 N N N H OH	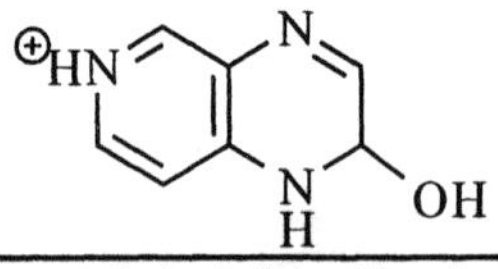 ⊕HN N N H OH	11.0	8.2

[a] Standard errors ranged from ± 0.4 - 0.5 kcal/mol

Good agreement was also achieved for azanaphthalene compounds containing a substituent that shifts the hydration equilibrium. For example, a study of 4-methylpteridine (**6**) and 4-trifluoromethyl pteridine (**7**) showed that substituents can, in some instances, not only change the extent of hydration but also the preferred site of hydration (*22*) (Figure 7). Experimental results indicate that the 4-trifluoromethyl analog hydrates initially across the 5,6- and 7,8-double bonds to form the dihydrate. Over time the dihydrate is converted to the 3,4-double bond hydrate. The anhydrous form is not detected. In contrast, the 4-methyl analog is poorly hydrated and the only hydrated species detected in solution is the dihydrate and not the 3,4-double bond monohydrate. The molecular factors attributed to the difference is reported to be a mixture of the methyl group sterically destabilizing the 3,4-hydrate and the trifluoromethyl group stabilizing the hydrate through inductive effects. The calculated results are consistent with these findings showing a 9.1 kcal/mol difference between the methyl and trifluoromethyl analogs for hydration across the 3,4-double bond and a 5.9 and -1.9 kcal/mol difference between the 3,4-double bond hydrate and the 7,8-double bond hydrate for the methyl and trifluoromethyl substituted compounds, respectively. Thus, the method can be used to accurately predict the site of hydration as well as the extent of hydration.

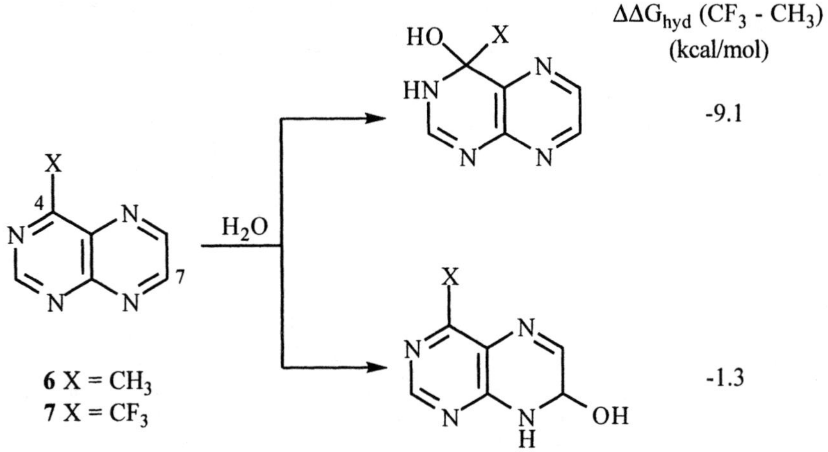

Figure 7: Calculated Relative Hydration Free Energies of 4-Substituted Pteridines

Differences in Purine and Pteridine Hydration. In contrast to pteridine, 9-methylpurine exists in the hydrated form to an extremely limited extent. Experimental estimations of the hydration equilibrium constant for purine riboside suggest that it is about 10^{-7} or nearly 7 orders of magnitude less favorable than K_{eq} for pteridine. This finding is remarkable considering that both compounds hydrate across the 1,6-double bond of the pyrimidine ring and both contain an heteroaromatic ring fused at the 4- and 5-positions of the pyrimidine ring by aromatic nitrogen atoms. The only difference is the position of one carbon atom, which is found in the aromatic ring in pteridine and attached to N9 of purine. Since this difference is unlikely to result in large differences in steric or inductive effects at the 1,6-double bond, we speculated that the differences in hydration could be due to differences in resonance energy lost upon hydration of the pyrimidine ring. Clearly, hydration of an heteroaromatic ring results in a loss of aromaticity in the hydrated product. Heteroaromatic groups that are less aromatic will thereby suffer less loss in resonance energy and therefore be more likely to hydrate.

Calculation of bond separation energies at the 6-31G** basis set level for the hydrated and anhydrous forms of both 9-methylpurine and pteridine supported this hypothesis (Figure 8), since pteridine lost 0.58 kcal/mol resonance energy upon hydration, whereas the purine analog lost 9.43 kcal/mol. The difference in resonance energy lost between the two heterocycles is therefore approximately 8.9 kcal/mol or a value very similar to the their relative hydration free energy difference. These results suggest that the major factor responsible for the large difference in hydration free energy between pteridine and 9-methylpurine is a difference in aromaticity.

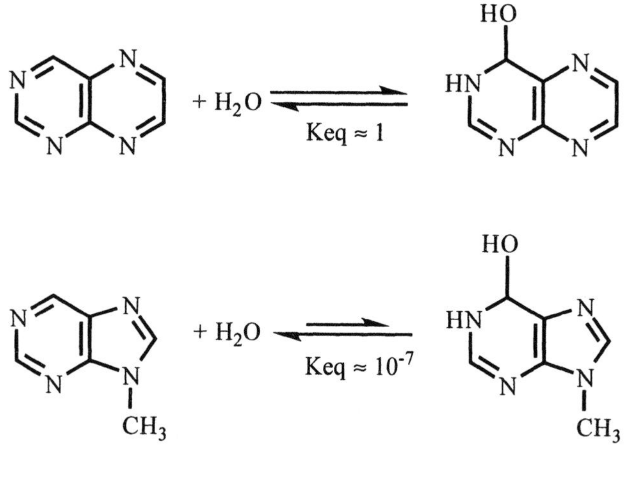

$+ H_2O$ ⇌ Keq ≈ 1

$+ H_2O$ ⇌ Keq ≈ 10^{-7}

Purine - Pteridine (kcal/mol)		
ΔΔΔE (BSE)	=	-8.9
$\Delta\Delta G_{hyd}$ (calc)	=	9.3
$\Delta\Delta G_{hyd}$ (expt)	≈	8.8

Figure 8: Purine vs. Pteridine Hydration

Calculation of Relative Inhibitor Potencies. Results from our studies of heteroaromatic hydration suggested that analogs of purine riboside (PR) with enhanced hydration could be designed by incorporation of electron withdrawing substituents near the hydration site or by replacement of the purine base with a less aromatic base. To achieve our goal of identifying analogs with improved ADA inhibitory potency , we calculated, as reported in detail elsewhere (*21*), the relative binding affinity for the hydrated species complexed to the ADA binding site cavity in order to determine which modifications enhance hydration and are not detrimental to binding affinity (Figure 9).

$$\mathrm{PR} \underset{K_{eq}}{\overset{H_2O}{\rightleftharpoons}} \mathrm{PR - H_2O} \underset{K_i^*}{\overset{\mathrm{ADA}}{\rightleftharpoons}} \mathrm{ADA{:}\ [PR - H_2O]}$$

$$\mathrm{PR'} \underset{K_{eq'}}{\overset{H_2O}{\rightleftharpoons}} \mathrm{PR' - H_2O} \underset{K_i^{*'}}{\overset{\mathrm{ADA}}{\rightleftharpoons}} \mathrm{ADA{:}\ [PR' - H_2O]}$$

Figure 9: Relative Inhibitor Potency

The apparent inhibitory potency ($K_{i\ (app)}$) as measured experimentally is related to both the hydration equilibrium constant and the intrinsic binding affinity of the hydrated molecule (K_i^*) by eq 2. Accordingly, the relative inhibitory potency between two analogs is related to the relative free energy difference by eq 3. Computationally, ΔG_{rel} is determined simply by summing the relative free energies for hydration and hydrate binding affinity as shown in eq 4.

$$K_{i\ (app)} = K_i^* (1 + 1/K_{eq}) \approx K_i^* K_{eq}^{-1} \quad (2)$$

$$\Delta G_{rel} = -RT\ln[K_i(app)/K_i'(app)] \quad (3)$$

$$\Delta G_{rel} = \Delta\Delta G_{hyd} + \Delta\Delta G_{bind} \quad (4)$$

To test whether we could accurately calculate the relative inhibitor potency for a pair of ADA inhibitors, we studied purine riboside (**5**) and 8-azapurine riboside (**8**). Previously, Townsend et al., showed that the 8-aza analog was a 400-fold more potent ADA inhibitor (*23*). The molecular reason for this enhancement in potency was not determined but could be either due to enhanced hydration or due to enhanced ADA binding affinity of the hydrated species. To determine the reason we calculated the relative difference in free energies for both hydration and binding.

The relative binding free energy difference was calculated using a computer model of the murine adenosine deaminase (ADA)-6-hydroxy-1,6-dihydropurine riboside (**4**) complex generated from the X-ray structure (pdb file name: 2ADA). All molecular dynamics, molecular mechanics and TCP calculations were carried out with the AMBER program (*24*) using an all atom force field and SPC/E potentials (*25*) to describe water interactions. The aqueous phase and complex molecular dynamics simulations were conducted as reported elsewhere (*16*). The free energy difference for binding was calculated using the second cycle (Figure 10) and represents the difference of the free energy between purine riboside hydrate and its 8-aza analog (**9**) in the complex and in solvent. The structural perturbation was carried out over 51 windows with each window comprising 2.5 ps of equilibration and 5 ps of data collection. The free energy difference is the average of four calculations, i.e. forward and reverse mutations starting with purine riboside hydrate and 8-azapurine riboside hydrate. The results indicated that the 8-aza analog loses 3.1 ± 0.7 kcal/mol of binding energy thereby eliminating this potential explanation for the 400-fold improvement in ADA inhibitor potency exhibited by 8-azapurine riboside.

The factors that account for the loss in binding affinity of 8-azapurine riboside hydrate were delineated in subsequent studies. Using the first TCP cycle (Figure 10), the difference in desolvation free energy was calculated and shown to favor purine riboside hydrate by 1.1 ± 0.5 kcal/mol. The remaining portion of the lost binding energy, i.e. 2 kcal/mol, associated with the 8-azapurine riboside hydrate binding is likely due to a loss in intrinsic binding affinity which may arise from an unfavorable electrostatic interaction between the 8-nitrogen and Asp296 as observed in the energy minimized, MD-equilibrated ADA complex.

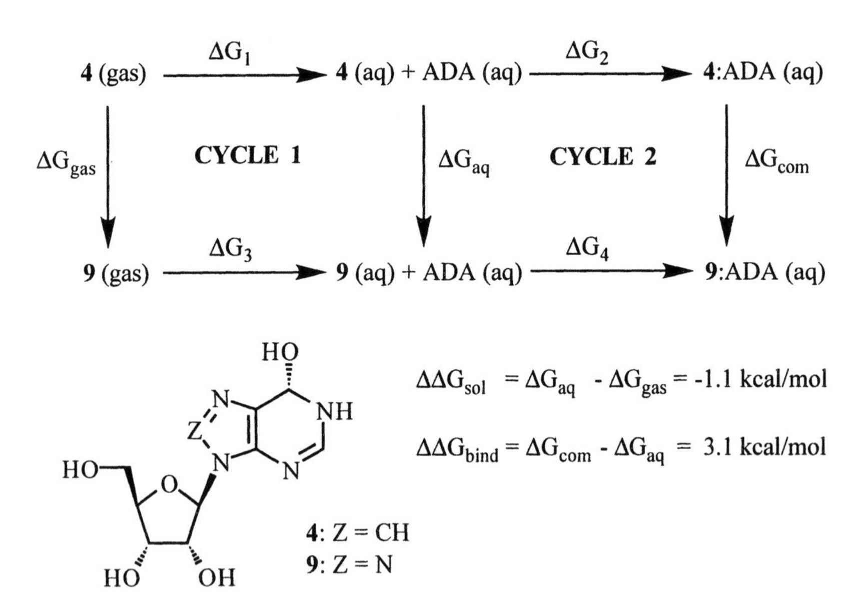

Figure 10: TCP Cycles for Relative Solvation and Binding Free Energies

The other possible explanation for the 400-fold improvement in inhibitory potency exhibited by 8-azapurine riboside (**8**) is that the 8-aza analog hydrates to a much larger extent than purine riboside (**5**). Calculation of the relative hydration free energy difference between 9-methylpurine and 8-aza-9-methylpurine strongly supported this possibility, since the difference was 7.1 kcal/mol or approximately a 5-order of magnitude rightward shift in the equilibrium constant for the 8-aza analog. Calculation of bond separation energies for the hydrated and unhydrated molecules suggested that the large difference in hydration arises from a large relative loss in resonance energy incurred by purine compared to 8-azapurine during the hydration reaction.

The calculated results provide a clear explanation for the difference in inhibitory potency between purine riboside and its 8-aza analog (Table III). The relative hydration free energy difference indicates that the 8-azapurine analog hydrates about 160,000-fold greater than the corresponding purine analog ($\Delta\Delta G_{hyd}$ = -7.1 kcal/mol) whereas the 8-aza analog suffers approximately a 200-fold decrease in binding affinity ($\Delta\Delta G_{bind}$ = 3.1 kcal/mol). The net effect, however, is a 4.0 kcal/mol enhancement in inhibitory potency in favor of the 8-aza analog which translates to a predicted K_i (app) for 8-azapurine riboside of 2×10^{-8} M; a value very close to the experimental result of 4×10^{-8} M (*23*).

Table III: Inhibitory Potential of Purine Riboside and its 8-aza Analog

Purine Riboside (**5**)		**8 - 5** (calc) (kcal/mol)		8-Azapurine Riboside (**8**)	
K_{eq}	$= 1.1 \times 10^{-7}$	$\Delta\Delta G_{hyd}$	= -7.1	K_{eq}	$= 1.8 \times 10^{-2}$
K_i^*	$= 1.8 \times 10^{-12}$ M	$\Delta\Delta G_{bind}$	= 3.1	K_i^*	$= 3.4 \times 10^{-10}$ M
K_i (app)	$= 1.6 \times 10^{-5}$ M	ΔG_{rel}	= -4.0	K_i (app)	$= 1.9 \times 10^{-8}$ M
				K_i (app)	$= 4.0 \times 10^{-8}$ M (expt)

Summary

Rational drug design using computational methods is useful in prioritizing potential target compounds if the calculated results accurately predict the experimental findings. Our study analyzing the difference in ADA inhibitory potency between purine riboside and 8-azapurine riboside illustrates the importance of calculating both the relative hydration free energy and the relative binding free energy for molecules that act as enzyme inhibitors only after undergoing covalent hydration. In addition, our studies of heteroaromatic hydration suggest that various ring substituents and ring modifications can significantly enhance purine riboside hydration and that this effect could be useful in the design of potent ADA and AMPDA inhibitors.

Acknowledgments

We thank Ms. Lisa Weston for her assistance in preparing this manuscript.

Literature Cited

1. Erion, M. D. *Ann. Rep. Med. Chem.* **1993**, *28*, 295-304.
2. Jacobson, K. A.; van Galen, P. J. M.; Williams, M. *J. Med. Chem.* **1992**, *35*, 407.
3. Mullane, K. M. *Cardiovasc. Res.* **1993**, *27*, 43-47.
4. Rosengren, S.; Bong, G. W.; Firestein, G. S. *J. Immunol.* **1995**, *154*, 5444-5451.
5. Erion, M. D.; Ugarkar, B. G.; DaRe, J.; Castellino, A. J.; Fujitaki, J. M.; Dixon, R.; Appleman, J. R.; Wiesner, J. B. *Nuc. Nucleotides* **1997**, *16*, 1013-1022.
6. Smolenski, R. T.; Suitters, A.; Yacoub, M. H. *J. Mol. Cell Cardiol.* **1992**, *24*, 91-96.
7. Wolfenden, R. *Acc. Chem. Res.* **1972**, *5*, 10-18.
8. Bagdassarian, C. K.; Schramm, V. L.; Schwartz, S. D. *J. Am. Chem. Soc.* **1996**, *118*, 8825-8836.
9. Wilson, D. K.; Rudolph, F. B.; Quiocho, F. A. *Science* **1991**, *252*, 1278-1284.
10. Frieden, C.; Kurz, L. C.; Gilbert, H. R. *Biochemistry* **1980**, *19*, 5303.
11. Erion, M. D.; Bookser, B. C.; Kasibhatla, S. R. *PCT Patent Appl.* **1994**, WO 94/18200.
12. Bookser, B. C.; Kasibhatla, S. R.; Cottrell, M. B.; Probst, G.; Appleman, J. R.; Erion, M. D. *Book of Abstracts;* 214th ACS National Meeting Las Vegas, NV; American Chemical Soc.: Washington, D.C., **1997**; MEDI 077.
13. Kasibhatla, S. R.; Bookser, B. C.; Probst, G.; Appleman, J. R.; Erion, M. D. *Book of Abstracts;* 214th ACS National Meeting Las Vegas, NV; American Chemical Soc.: Washington, D.C., **1997**; MEDI 078.
14. Jones, W.; Kurz, L. C.; Wolfenden, R. *Biochemistry* **1989**, *28*, 1242-1247.
15. Jones, W.; Wolfenden, R. *J. Am. Chem. Soc.* **1986**, *108*, 7444-7445.
16. Erion, M. D.; Reddy, M. R. *J. Comp. Chem.* **1995**, *16*, 1513-1521.
17. Tembe, B. L.; McCammon, J. A. *Comput. Chem.* **1982**, *8*, 281-283.
18. Bell, R. P. *Adv. Phys. Org. Chem.,* **1966**, *4*, 1-29.
19. Albert, A. *Adv. Heterocyc. Chem.* **1976**, *20*, 117-143.
20. Perrin, D. D. *Adv. Heterocyc. Chem.* **1965**, *4*, 43-73.
21. Erion, M. D.; Reddy, M. R. *J. Am. Chem. Soc.* (in press)
22. Clark, J.; Pendergast, W. *J. Chem. Soc. C.* **1969**, 1751-1754.
23. Shewach, D. S.; Krawczyk, S. H.; Acevedo, O. L.; Townsend, L. B. *Biochem. Pharmac.*, **1992**, *44,* 1697-1700.
24. Weiner, S.J.; Kollman, P.A.; Case, D.A.; Singh, U.C.; Ghio,C.; Alagoha,G.; Profeta, Jr., S.; Weiner, P.K. *J. Am. Chem. Soc.* **1984**, *106*, 765-784.
25. Berendsen, H.J.C.; Grigera, J.R.; Straatsma, T.P. *J. Phys.Chem.* **1987**, *91*, 6269.

Chapter 8

New Tools for Rational Drug Design

Gregory D. Hawkins,[1] Jiabo Li,[1] Tianhai (Tony) Zhu,[1] Candee C. Chambers,[1,3] David J. Giesen,[1,4] Daniel A. Liotard,[2] Christopher J. Cramer,[1] and Donald G. Truhlar[1]

[1] Department of Chemistry and Supercomputer Institute, University of Minnesota, Minneapolis, MN 55455
[2] Laboratoire de Physico-Chimie Theorique, Université de Bordeaux, 351 Cours de la Liberation, 33405 Talence Cedex, France

We have developed two new tools for molecular modeling that can be very useful for computer-aided drug design, namely class IV charges and the SM*x* series of solvation models. This contribution overviews the current status of our efforts in these areas, including the CM2 charge model and the SM5 series of solvation models. The solvation models may be used to estimate partition coefficients for phase transfer equilibria of organic solutes between water and 1-octanol, the most widely used mimic of cellular biophases, and also between water and other solvents that have been used for this purpose, e.g., hexadecane and chloroform.

1. Introduction

The partitioning of an organic solute between an aqueous phase (*aq*) and a nonpolar medium (*np*) is critical for many phenomena in biological and medicinal chemistry. In particular this partitioning can be critical for drug delivery, binding, and clearance. Predictions of the relative free energy of organic molecules in aqueous and nonpolar media can be very useful for predicting the bioavailability of potential drugs. Lipid-like nonpolar media are especially important because they mimic cell membranes, and the lipophilic character of organic compounds is one of the most widely used predictors of their bioactivity. The lipid solubility of a molecule correlates with its ability to enter the brain (i.e., pass the blood-brain barrier) or other parts of the central nervous system and is generally believed to have a large influence on pharmacological properties.

[3] Current address: Departments of Physics and Chemistry, Mercyhurst College, Erie, PA 16504
[4] Current address: Eastman Kodak Company, Rochester, NY 14650

The lipophilic character of a molecule is typically measured quantitatively by its partitioning between an organic phase and water. 1-Octanol is the most widely used solvent for mimicking biophases in this respect, and Hansch and Dunn[1] have attempted to rationalize the success of correlations based on 1-octanol by noting that proteins (with their amide groups) and lipid phases (with their ester and phosphate functionalities) both present accessible hydrogen-bonding opportunities to drug molecules, and the OH functional group of octanol can serve as a hydrogen bond acceptor or donor to mimic such effects, while the molecule is large enough to remain overall hydrophobic. The partitioning coefficient P of organic solutes between water and 1-octanol is widely used in property-activity relationships in rational drug design, and a very large amount of work concerned with the measurement and/or prediction of such partition coefficients has been reported. The reader is referred to representative articles for further references.[1–15]

Hexadecane is another important example of a nonpolar solvent because solute–hexadecane interactions, like solute–1-octanol interactions, are recognized as a surrogate for hydrophobic interactions of molecules with lipid bilayers or other cellular material[16–20] or with the nonpolar active site of an enzyme or receptor. In such models, the partition coefficient of a solute between an alkane solvent and water provides some indication of how likely it is to penetrate the bilayer, skin, brain, central nervous system, or other biophase or to bind to the nonpolar site in (or on) the protein. The difference between log P for an amphiphilic solvent like 1-octanol or 1-hexanol and apolar, aprotic inert solvents like straight-chain alkanes or cyclohexane is generally interpreted as a measure of the hydrogen-bond donor capacity of solutes.[21–24] Furthermore this difference has been used in rational drug design because it correlates with brain/blood and cerebrospinal/blood partitioning equilibria.[25]

Another solvent that has been used for similar purposes as 1-octanol and hexadecane is chloroform. Reynolds[26] has discussed the utility of water/chloroform partition coefficients for correlating membrane permeability and bioactivity properties that depend on such permeability.

The ability to understand the solvation of organic solutes in nonpolar media is also important for conformational analysis of bioactive compounds. A recent example of the importance of solvent effects on conformation is the interpretation of octanol/water and heptane/water partition coefficients for the immunosuppressant cyclosporin A in terms of solvent-dependent conformational changes and of the relationship of these changes to solvent-dependent inhibitory activity.[24]

Historically, most attempts to develop predictive models for solvation free energies or partitioning coefficients have involved multivariate quantitative structure-property relationships (QSPRs).[27–33] More recently, methods for including solvent electrostatic effects self consistently in quantum mechanical solute descriptions have advanced vigorously,[34–45] and such models are preferred for making predictions on molecules outside the QSPR training sets or for transition states. Accurate quantitative predictions must include nonelectrostatic effects as well, and we have developed successful models for quantum mechanical self-consistent electrostatics in both aqueous solutions[46–59] and organic solvents.[55–65]

An especially important aspect of the framework of our model is that only solute atoms are treated explicitly; the solvent is treated as a continuous fluid. There are three kinds of terms in the solvation free energy: long-range electrostatic contributions (labeled ENP, to denote that they include self-consistent solute electronic and nuclear contributions and solute-solvent electric polarization effects), intermediate-range cavity-structural (CS) contributions, and short-range cavity-dispersion (CD) effects. Hydrogen bonding affects all three terms, ENP, CD, and CS.

The functional forms and parameters of the electrostatic model for organic solvents are identical to those for water except that the dielectric constant, ε, of the organic solvent replaces the dielectric constant of water. The electrostatic treatment involves a three-dimensional integration over the free energy density due to electric polarization of the solvent in the regions of space not occupied by the solute,[38,42,51,66,67] and therefore it reflects the solute shape realistically. The solute electronic wave functions and solute internal energies are calculated with semiempirical molecular orbital theory,[68] *ab initio* Hartree-Fock theory,[69] or density functional theory.[70] The competition between solvent polarization and solute distortion is accounted for by placing solvation terms inside the effective one-electron Hamiltonians for the molecular orbitals.[42,71–73]

The atomic partial charges needed for the electrostatic solvation terms may be calculated by conventional Mulliken analysis or by class IV[50,74,75] charge models. The latter capability is a particular strength of our solvation model since these charges, according to previous validation,[74,75] yield remarkably accurate electrostatic properties, and in addition they are very inexpensive to calculate. Accurate atomic partial charges are of great interest for molecular modeling in general and their usefulness extends beyond solvation modeling.[76] Thus we shall review our recent progress in this area as a separate topic.

In addition to electrostatics, our solvation models also include non-electrostatic effects in the first solvation shell. These effects are modeled in terms of solvent-accessible surface areas[77,78] and semiempirical atomic surface tensions.[79] The solvent dependence of our predicted free energies of solvation comes from two sources: (i) The electrostatic term contains the factor $(1 - \varepsilon^{-1})$, where ε is the dielectric constant of the solvent. (ii) The atomic surface tensions are determined separately for water and organic solvents, and in the latter case they depend on one or more of the following solvent descriptors: n, the index of refraction; α and β, Abraham's[80–83] hydrogen bond acidity and basicity parameters (converting our notation to his, α is $\Sigma\alpha_2^H$ and β is $\Sigma\beta_2^H$); γ, the macroscopic surface tension of the solvent; and two descriptors which depend upon the fraction of non-hydrogenic atoms within the solvent which are aromatic carbon or electronegative halogen atoms (we define "electronegative halogen atoms" as F, Cl, and Br since these are the halogen atoms that are more electronegative than carbon[84]). A major advantage of using these parameters is that they are available for almost all possible solvents. Should one desire to treat an unusual solvent for which a and b are not known, three possibilities present themselves. First, they could be determined by generating the kind of partition coefficient data and fits used originally by Abraham.[80–83] Second, they could be determined by correlating them against other acidity or basicity scales[85,86] that *are* known for the solvent of interest. Third, Murray and Politzer[86] have shown that Abraham's single-site hydrogen-bond acidity and basicity parameters (α_2^H and β_2^H) correlate well with maxima and minima of calculated electrostatic potentials on the molecular surface, and these single-site parameters can be used to estimate $\Sigma\alpha_2^H$ and $\Sigma\beta_2^H$ in most cases.

Section 2 summarizes the current status of class IV charges. Section 3 presents a level chart of SM5 models. Section 4 summarizes the performance of several SM5 models for free energies of solvation in water, 1-octanol, hexadecane, and chloroform.

2. Class IV charges

Partial atomic charges may be classified as follows:[74]

Class I: non-quantum-mechanical charges, for example, the empirical charges in a molecular mechanics force field;

Class II: charges obtained directly from wave functions without calculating physical observables, for example, charges obtained by Mulliken[87] or Löwdin[88] population analysis;

Class III: charges obtained by fitting to electrostatic potentials or multipole moments computed from wave functions, for example, ChElPG[89] charges;

Class IV: charges mapped from class II or class III charges with semiempirical parameters designed to make the mapped charges better reproduce experimental multipole moments or converged quantum mechanical electrostatic potentials or multipole moments.[50,74,75]

We have presented two models for class IV charges: Charge Model 1[50,74] (CM1) and Charge Model 2[75] (CM2).

In the CM1 model, we computed zero-order charges by Mulliken analysis and mapped them as nonlinear functions of calculated bond orders with 15–19 parameters based on data (experimental dipole moments and calculated electrostatic potentials) for compounds containing H, C, N, O, F, Si, S, Cl, Br, and I. Parameters were determined for AM1[91–93] and PM3[94] semiempirical molecular orbital wave functions. We achieved root-mean-square errors in the dipole moments of 0.27 D for maps based on AM1 wave functions and 0.20 D for maps based on PM3 wave functions.[74]

In the CM2 model we computed zero-order charges by Löwdin analysis and mapped them as quadratic functions of calculated bond orders with 20 parameters based on 198 experimental dipole moments for compounds containing H, C, N, O, F, Si, P, S, Cl, Br, and I. Parameters were determined for AM1, for four different basis sets for ab initio Hartree-Fock wave functions (MIDI!,[95] MIDI!(6D), 6-31G*,[69] and 6-31+G*[69]), and for four combinations of basis set (MIDI!, MIDI!(6D), or 6-31G*) with density

Table 1. Partial atomic charges and dipole moments for β-propiolactone[a]

	HF/MIDI!			BPW91/MIDI!		
	Mulliken	Löwdin	CM2	Mulliken	Löwdin	CM2
partial charges:						
O-1	-0.82	-0.46	-0.36	-0.57	-0.33	-0.31
C-2	1.09	0.58	0.56	0.76	0.41	0.51
O (carbonyl on C-2)	-0.69	-0.41	-0.44	-0.50	-0.31	-0.41
C-3	0.09	0.10	0.10	-0.00	0.04	0.02
H-1,2 on C-3	0.20	0.09	0.07	0.22	0.10	0.10
C-4	-0.58	-0.25	-0.20	-0.51	-0.26	-0.26
H-3,4 on C-4	0.25	0.13	0.10	0.19	0.13	0.13
dipole moment (D)	7.69	4.71	4.31	6.02	3.81	4.21

[a]dipole moment from HF/MIDI! density: 4.18 D; from BPW91 density: 3.41 D; from experiment: 4.18 D

functional (BPW91[96,97] or B3LYP[98–100]). We achieved root-mean-square errors in dipole moments in the range 0.17–0.19 D for HF/6-31G*, B3LYP/MIDI!, BPW91/6-31G*, HF/MIDI!, and BPW91/MIDI!, 0.20–0.21 D for two cases with the MIDI!(6D) basis, 0.25 D with AM1, and 0.41 D with HF/6-31+G*, the latter value reflecting the difficulty of obtaining accurate charges from wave functions with diffuse basis functions. On the average, errors in the dipoles computed as expectation values from the full wave functions were about 1.8 times larger than those computed from the CM2 charges.[75]

As an example of the predictions of the CM2 charge model, consider β-propiolactone. The experimental dipole moment is 4.18 D, and the use of BPW91/MIDI! wave functions yields 3.41 D, whereas the CM2 model based on this same BPW91/MIDI! wave function for β-propiolactone yields 4.21 D. The partial charges on the oxygen atoms differ by as much as 0.25 when obtained by Mulliken analysis of HF/MIDI! and BPW91/MIDI! wave functions and by as much as 0.13 for Löwdin analysis. But the mapped charges from these two quite different wave functions agree within 0.05. Full results are given in Table 1.

3. Summary of SM5 models

Aqueous/nonpolar partitioning is usually quantified by the partition coefficient P or its, logarithm ("log P"), where

$$P = \frac{[solute]_{np}}{[solute]_{aq}} \tag{1}$$

Another (equivalent) definition of P is

$$\log P = \Delta\Delta G_S^0/(-2.303RT) \quad (2)$$

where

$$\Delta\Delta G_S^0 = \Delta G_S^0(np) - \Delta G_S^0(aq), \quad (3)$$

$G_S^0(solv)$ is the standard-state free energy of solvation of the solute in solvent *solv*, R is the gas constant, and T is temperature.

The standard-state free energy of solvation in water is written as

$$\Delta G_S^0 = \Delta G_{ENP} + \sum_\alpha G_{CDS,\alpha} \quad (4)$$

where α denotes one of the atoms of the solute and

$$G_{CDS,\alpha} = A_\alpha \sum_i f_{\alpha i}(R)\tilde{\sigma}_{\alpha i} \quad (5)$$

where $f_{\alpha i}$ is a function of the geometry R of the solute (actually it depends only on selected bond distances, and it has no dependence on bond angles or dihedral angles) and $\tilde{\sigma}_{\alpha i}$ is a surface tension coefficient. The standard-state free energy of solvation in an organic solvent has the same form as for water except that $\tilde{\sigma}_{\alpha i}$ is not a constant but rather depends on solvent descriptors. The solvent descriptors are generally n, α, β, and γ. In some cases (SM5.4 parameterizations) special parameters are used for chloroform, benzene, and toluene; in other cases (SM5.42R, SM5.2R, and SM5.0R parameterizations) two special solvent descriptors are added to the four mentioned in the previous sentence, in particular descriptors computed from the fraction of nonhydrogenic solvent atoms that are aromatic carbons or electronegative halogens. Some $\tilde{\sigma}_{\alpha i}$ values are independent of α and have $f_{\alpha i} = 1$; these are sometimes called the CS terms. The other terms are sometimes called CD terms; however, one should be cautious about physical interpretations of the individual terms.

The actual parameterizaiton is carried out as follows: First the nonlinear parameters are fixed based on a variety of considerations, including trends over solutes and solvents for solvation free energies of neutrals and ions. Then the surface tension coefficients are fit to a large set of data taken chiefly from the tabulation of Cabani *et al.*[101] for ΔG_S^0 of neutrals in water and mostly computed from log P values from the MedChem data base[102] for organic solvents.

In the present paper we consider solutes containing H, C, N, O, F, S, Cl, Br, and I. (Some, but not all, models are also parameterized for solutes containing P, but P-containing solutes are not discussed in this chapter.) As an example of the size of the training set, we consider the training set used for solutes with H, C, N, O, F, S, Cl, Br, and I in the SM5.2R model. This training set has data for 43 ions and 248 neutrals in water. It also has 1836 data points for 227 neutrals in 90 organic solvents. The SM5.2R parameterizations have 46 surface tension coefficients for organic solvents and 25 for water.

Table 2. Mean Unsigned Error (kcal/mol) in the Aqueous Solvation Free Energies Predicted by Selected SMx Solvation Models.

Solute Class	Data Points	SM5.4/		SM5.2R/				SM5.0R
		AM1	PM3	MNDO(d)	MNDO	AM1	PM3	
Unbranched Alkanes	8	0.6	0.6	0.7	0.7	0.6	0.6	0.5
Branched Alkanes	5	0.7	0.7	0.5	0.5	0.5	0.4	0.3
Cycloalkanes	5	0.2	0.1	0.4	0.4	0.4	0.4	0.5
Alkenes	9	0.5	0.3	0.3	0.3	0.2	0.2	0.2
Alkynes	5	0.2	0.2	0.1	0.1	0.2	0.2	0.1
Arenes	8	0.2	0.2	0.2	0.2	0.2	0.2	0.5
Alcohols	16	0.5	0.4	0.2	0.2	0.2	0.2	0.3
Ethers	12	0.8	0.9	0.6	0.6	0.5	0.6	0.6
Aldehydes	6	0.3	0.4	0.3	0.3	0.3	0.3	0.3
Ketones	12	0.4	0.4	0.5	0.5	0.3	0.4	0.4
Carboxylic Acids	5	0.8	0.8	0.4	0.4	0.4	0.4	0.5
Esters	13	0.5	0.5	0.3	0.3	0.3	0.3	0.3
Bifunctional CHO	5	0.4	0.4	0.5	0.5	0.4	0.4	0.5
Water, Dihydrogen	2	1.6	1.2	0.0	0.0	0.0	0.0	0.9
Aliphatic Amines	15	0.8	0.8	0.5	0.5	0.6	0.5	0.5
Aromatic Amines	10	0.7	0.7	0.8	0.8	0.6	0.7	1.0
Nitriles	4	0.5	0.5	0.5	0.5	0.4	0.3	0.7
Nitrohydrocarbons	6	0.5	0.1	0.1	0.1	0.5	0.4	0.4
Amides & Ureas	4	2.6	1.2	1.4	1.4	1.1	1.1	2.2
Bifunctional HCN and HCNO	5	0.9	1.1	0.9	0.9	0.8	0.9	1.1
Ammonia & Hydrazine	2	2.8	3.1	0.2	0.2	0.4	0.2	1.1
Thiols	4	0.3	0.2	0.6	0.6	0.5	0.6	0.3
Sulfides	6	0.6	0.5	1.2	1.0	1.1	1.0	0.5
Disulfides	2	0.2	0.2	0.1	0.1	0.0	0.0	0.1
Fluorinated Hydrocarbons	6	0.6	0.4	0.7	0.7	0.4	0.5	1.1
Chloroalkanes	13	0.3	0.3	0.3	0.4	0.3	0.8	0.4
Chloroalkenes	5	0.7	0.5	0.6	0.4	0.7	1.0	1.0
Chloroarenes	8	0.2	0.3	0.8	0.9	0.3	0.5	0.3
Brominated Hydrocarbons	14	0.3	0.2	0.2	0.2	0.4	0.2	0.4
Iodinated Hydrocarbons	8	0.3	0.2	0.3	0.3	0.3	0.6	0.3
Other Halo Compounds	25	0.6	0.8	0.7	0.7	0.6	1.0	0.8
All solutes:	248	0.6	0.5	0.5	0.5	0.4	0.5	0.5

Table 3. Mean Unsigned Error (kcal/mol) in the 1-Octanol Solvation Free Energies Predicted by Selected SMx Solvation Models.

Solute Class	Data Points	SM5.4/		SM5.2R/				SM5.0R
		AM1	PM3	MNDO(d)	MNDO	AM1	PM3	
Unbranched Alkanes	8	0.4	0.3	0.1	0.1	0.4	0.3	0.1
Branched Alkanes	2	0.1	0.1	0.1	0.1	0.2	0.2	0.1
Cycloalkanes	4	0.6	0.6	0.4	0.4	0.3	0.3	0.4
Alkenes	6	0.6	0.4	0.2	0.2	0.6	0.4	0.2
Alkynes	4	0.3	0.3	0.2	0.2	0.2	0.2	0.1
Arenes	8	0.2	0.2	0.3	0.3	0.3	0.3	0.3
Alcohols	16	0.2	0.3	0.5	0.5	0.5	0.5	0.4
Ethers	11	0.6	0.5	0.5	0.5	0.6	0.5	0.5
Aldehydes	4	0.5	0.5	0.4	0.4	0.4	0.4	0.5
Ketones	10	1.0	1.0	0.8	0.8	0.8	0.8	0.9
Carboxylic Acids	5	0.7	0.7	0.3	0.3	0.4	0.4	0.1
Esters	9	1.2	1.1	0.3	0.3	0.4	0.3	0.6
Bifunctional CHO	4	1.1	1.0	0.8	0.8	0.9	0.8	0.6
Water, Dihydrogen	2	1.2	1.1	0.7	0.7	0.8	0.7	0.5
Aliphatic Amines	9	0.6	0.5	0.4	0.4	0.6	0.5	0.4
Aromatic Amines	7	0.8	0.5	0.5	0.5	0.5	0.5	0.6
Nitriles	4	0.7	0.6	0.2	0.2	0.4	0.2	0.5
Nitrohydrocarbons	6	0.7	0.1	0.2	0.2	0.7	0.4	0.1
Amides & Ureas	1	1.7	0.2	2.6	2.6	3.1	2.1	2.5
Bifunctional HCN and HCNO	3	2.0	1.6	0.7	0.7	0.7	0.6	0.9
Hydrazine	1	2.0	3.3	1.7	1.7	1.7	1.7	1.8
Thiols	2	0.3	0.2	0.4	0.4	0.5	0.4	0.3
Sulfides	3	0.8	0.7	1.0	0.6	0.8	0.6	0.3
Disulfides	1	0.0	0.0	0.1	0.2	0.2	0.3	0.3
Fluorinated Hydrocarbons	2	0.4	0.2	1.2	1.2	0.6	0.3	0.5
Chloroalkanes	7	0.3	0.3	0.3	0.6	0.4	0.5	0.5
Chloroalkenes	3	0.5	0.4	0.8	0.5	0.8	1.1	1.0
Chloroarenes	6	0.7	0.5	0.7	0.9	0.3	0.3	0.3
Brominated Hydrocarbons	12	0.3	0.4	0.3	0.4	0.3	0.3	0.2
Iodinated Hydrocarbons	5	0.2	0.2	0.5	0.6	0.7	0.5	0.6
Other Halo Compounds	15	0.7	0.7	0.7	0.7	0.6	0.9	0.8
All solutes:	180	0.6	0.5	0.5	0.5	0.5	0.5	0.5

Table 4. Mean Unsigned Error (kcal/mol) in the Hexadecane Solvation Free Energies Predicted by Selected SMx Solvation Models.

Solute Class	Data Points	SM5.4/		SM5.2R/				SM5.0R
		AM1	PM3	MNDO(d)	MNDO	AM1	PM3	
Unbranched Alkanes	9	0.5	0.5	0.4	0.4	0.2	0.3	0.4
Branched Alkanes	5	0.5	0.5	0.6	0.6	0.4	0.5	0.7
Cycloalkanes	4	1.2	1.2	0.4	0.4	0.2	0.3	0.5
Alkenes	6	0.4	0.2	0.2	0.2	0.4	0.3	0.2
Alkynes	5	0.2	0.1	0.1	0.1	0.1	0.2	0.1
Arenes	9	0.4	0.3	0.4	0.4	0.5	0.5	0.5
Alcohols	17	0.2	0.2	0.2	0.2	0.2	0.3	0.3
Ethers	9	0.3	0.3	0.3	0.3	0.3	0.3	0.2
Aldehydes	7	0.1	0.2	0.3	0.3	0.2	0.3	0.4
Ketones	12	0.3	0.3	0.2	0.2	0.2	0.2	0.3
Carboxylic Acids	5	0.2	0.2	0.4	0.4	0.5	0.4	0.5
Esters	13	0.5	0.4	0.3	0.3	0.3	0.3	0.3
Bifunctional CHO	1	0.2	0.3	0.1	0.1	0.2	0.0	0.4
Water, Dihydrogen	2	0.5	0.5	0.1	0.1	0.1	0.1	0.1
Aliphatic Amines	8	0.3	0.3	0.1	0.1	0.1	0.1	0.1
Aromatic Amines	9	0.4	0.3	0.2	0.2	0.2	0.2	0.2
Nitriles	4	0.2	0.1	0.2	0.2	0.2	0.1	0.2
Nitrohydrocarbons	6	0.5	0.2	0.2	0.2	0.6	0.3	0.1
Amides & Ureas	1	1.2	1.0	0.1	0.1	0.0	0.2	0.4
Bifunctional HCN and HCNO	0	...	...	...	...	...	...	...
Ammonia	1	0.7	0.5	0.4	0.4	0.5	0.3	0.4
Thiols	3	0.5	0.5	0.5	0.5	0.4	0.4	0.4
Sulfides	5	0.2	0.2	0.8	0.6	0.7	0.6	0.5
Disulfides	2	0.1	0.1	0.3	0.3	0.3	0.3	0.3
Fluorinated Hydrocarbons	4	0.7	0.6	0.4	0.4	0.4	0.4	0.6
Chloroalkanes	7	0.2	0.1	0.2	0.2	0.3	0.2	0.2
Chloroalkenes	4	0.4	0.3	0.6	0.5	0.7	0.7	0.5
Chloroarenes	3	0.3	0.2	0.2	0.2	0.4	0.2	0.3
Brominated Hydrocarbons	12	0.1	0.1	0.2	0.2	0.2	0.2	0.4
Iodinated Hydrocarbons	8	0.1	0.2	0.3	0.3	0.3	0.2	0.2
Other Halo Compounds	8	0.4	0.4	0.6	0.7	0.6	0.4	0.3
All solutes:	189	0.3	0.3	0.3	0.3	0.3	0.3	0.3

Table 5. Mean Unsigned Error (kcal/mol) in the Chloroform Solvation Free Energies Predicted by Selected SMx Solvation Models.

Solute Class	Data Points	SM5.4/		SM5.2R/				SM5.0R
		AM1	PM3	MNDO(d)	MNDO	AM1	PM3	
Unbranched Alkanes	1	0.2	0.3	0.2	0.2	0.1	0.1	0.4
Branched Alkanes	0	...	...	...	...	...	...	...
Cycloalkanes	1	1.1	1.2	0.0	0.0	0.2	0.0	0.1
Alkenes	0	...	...	...	...	...	...	...
Alkynes	0	...	...	...	...	...	...	...
Arenes	6	0.1	0.2	0.2	0.2	0.7	0.5	0.2
Alcohols	14	0.4	0.3	0.3	0.3	0.3	0.3	0.4
Ethers	6	0.4	0.4	0.6	0.6	0.5	0.5	0.5
Aldehydes	3	0.6	0.5	0.9	0.9	0.9	1.0	0.8
Ketones	3	0.2	0.2	0.4	0.4	0.3	0.3	0.4
Carboxylic Acids	5	0.2	0.2	0.1	0.1	0.2	0.2	0.4
Esters	9	0.2	0.2	0.5	0.5	0.4	0.5	0.7
Bifunctional CHO	2	0.8	0.8	0.7	0.7	0.8	0.8	0.8
Water, Dihydrogen	1	1.5	1.2	0.3	0.3	0.1	0.2	0.7
Aliphatic Amines	8	0.4	0.2	0.5	0.5	0.6	0.6	0.5
Aromatic Amines	8	0.3	0.4	0.8	0.8	0.4	0.8	1.0
Nitriles	2	0.2	0.2	0.7	0.7	0.7	0.4	1.0
Nitrohydrocarbons	2	0.1	0.2	0.2	0.2	0.3	0.2	0.5
Amides & Ureas	2	1.5	0.5	2.3	2.3	2.2	2.1	2.5
Bifunctional HCN and HCNO	3	3.0	2.8	1.3	1.3	1.3	1.3	1.0
Ammonia & Hydrazine	2	2.4	2.9	0.8	0.8	0.6	0.8	0.7
Thiols	1	0.7	0.7	0.7	0.6	0.5	0.6	0.8
Sulfides	4	1.0	1.0	1.2	1.2	1.2	1.2	1.2
Disulfides	0	...	...	...	...	...	...	...
Fluorinated Hydrocarbons	1	0.3	0.1	1.0	1.0	0.8	0.8	0.3
Chloroalkanes	0	...	...	...	...	...	...	...
Chloroalkenes	0	...	...	...	....	...	...	...
Chloroarenes	2	0.2	0.2	0.6	0.7	0.6	0.6	0.3
Brominated Hydrocarbons	1	0.0	0.2	0.7	0.7	1.0	0.8	0.4
Iodinated Hydrocarbons	1	0.1	0.2	0.9	1.0	1.4	0.5	0.9
Other Halo Compounds	4	0.6	0.5	0.6	0.5	0.6	0.7	0.7
All solutes:	92	0.5	0.5	0.6	0.6	0.6	0.6	0.6

Table 6. Free Energy of Solvation and Partition Coefficient Results for 1,2 Ethanediol.

Model	ΔG_{ENP}	G_{CDS}	ΔG_S^o	log $P_{org/water}$	
				theory	experiment
	water				
SM5.4/AM1	-6.5	-2.3	-8.8		
SM5.4/PM3	-6.3	-2.9	-9.2		
SM5.2R/MNDO(d)	-1.9	-7.0	-8.9		
SM5.2R/MNDO	-1.9	-7.0	-8.9		
SM5.2R/AM1	-2.8	-6.4	-9.2		
SM5.2R/PM3	-2.2	-7.1	-9.3		
SM5.0R			-8.7		
	1-octanol				
SM5.4/AM1	-5.9	-1.4	-7.2	-1.1	-1.4
SM5.4/PM3	-5.8	-1.9	-7.7	-1.1	
SM5.2R/MNDO(d)	-1.7	-6.3	-8.0	-0.7	
SM5.2R/MNDO	-1.7	-6.3	-8.0	-0.7	
SM5.2R/AM1	-2.5	-5.6	-8.2	-0.7	
SM5.2R/PM3	-2.0	-6.3	-8.2	-0.8	
SM5.0R			-8.1	-0.4	
	hexadecane				
SM5.4/AM1	-3.2	0.1	-3.1	-4.2	-4.8
SM5.4/PM3	-3.2	-0.2	-3.4	-4.3	
SM5.2R/MNDO(d)	-0.9	-2.6	-3.5	-4.0	
SM5.2R/MNDO	-0.9	-2.6	-3.5	-4.0	
SM5.2R/AM1	-1.3	-2.2	-3.5	-4.2	
SM5.2R/PM3	-1.1	-2.6	-3.7	-4.1	
SM5.0R			-3.8	-3.6	
	chloroform				
SM5.4/AM1	-5.0	-0.2	-5.2	-2.6	-2.4
SM5.4/PM3	-5.0	-0.5	-5.5	-2.7	
SM5.2R/MNDO(d)	-1.5	-3.7	-5.2	-2.7	
SM5.2R/MNDO	-1.5	-3.7	-5.2	-2.7	
SM5.2R/AM1	-2.1	-3.3	-5.4	-2.8	
SM5.2R/PM3	-1.7	-3.7	-5.4	-2.9	
SM5.0R			-5.1	-2.6	

Tables 2–5 show the mean unsigned deviations in standard-state free energies of solvation for various classes of solutes in water and the three organic solvents singled out in the introduction. In each table we show the application of several models to the same set of data, namely our latest and largest training set, excluding phosphorus-containing compounds, except that in Tables 2, 3, and 5, the SM5.4/PM3 results are based on one less data point because hydrazine is excluded when PM3 is used to optimize geometries. Tables 2–5 show that we have uniformly small

Table 7. Free Energy of Solvation and Partition Coefficient Results for Thioanisole.

Model	ΔG_{ENP}	G_{CDS}	ΔG^o_S	log $P_{org/water}$	
				theory	experiment
	water				
SM5.4/AM1	-3.9	0.7	-3.3		
SM5.4/PM3	-3.0	-0.3	-3.3		
SM5.2R/MNDO(d)	-1.0	-1.7	-2.7		
SM5.2R/MNDO	-1.0	-1.9	-2.9		
SM5.2R/AM1	-3.8	0.8	-3.0		
SM5.2R/PM3	-2.7	-0.3	-3.0		
SM5.0R			-3.4		
	1-octanol				
SM5.4/AM1	-3.5	-3.9	-7.4	3.0	2.7
SM5.4/PM3	-2.7	-4.6	-7.3	2.9	
SM5.2R/MNDO(d)	-0.9	-5.2	-6.1	2.5	
SM5.2R/MNDO	-0.9	-5.5	-6.4	2.6	
SM5.2R/AM1	-3.4	-3.0	-6.4	2.5	
SM5.2R/PM3	-2.4	-4.1	-6.5	2.6	
SM5.0R			-6.4	2.2	
	hexadecane				
SM5.4/AM1	-1.8	-5.1	-6.9	2.6	
SM5.4/PM3	-1.4	-5.7	-7.0	2.7	
SM5.2R/MNDO(d)	-0.5	-5.6	-6.1	2.5	
SM5.2R/MNDO	-0.5	-5.8	-6.3	2.5	
SM5.2R/AM1	-1.8	-4.3	-6.1	2.3	
SM5.2R/PM3	-1.2	-5.0	-6.3	2.4	
SM5.0R			-6.3	2.1	
	chloroform				
SM5.4/AM1	-2.9	-4.8	-7.8	3.3	2.4
SM5.4/PM3	-2.3	-5.7	-8.0	3.4	
SM5.2R/MNDO(d)	-0.8	-6.5	-7.3	3.3	
SM5.2R/MNDO	-0.8	-6.7	-7.4	3.3	
SM5.2R/AM1	-2.9	-4.8	-7.7	3.4	
SM5.2R/PM3	-2.0	-5.7	-7.7	3.4	
SM5.0R			-7.4	2.9	

mean errors. Notice that some solute classes are not well represented in the data sets for specific solvents, and in fact some solute classes are not represented at all in some solvents. The SM5 solvation models are able to treat such cases because all the data for free energies of solvation in organic solvents are fit simultaneously, and the number of solvent descriptors is much smaller than the total number (90) of organic solvents. We believe in this way we have captured all the major physical effects.

Tables 6-8 were included to examine a couple of individual examples, namely, 1,2-ethanediol, thioanisole, and *p*-dichlorobenzene. These tables show the partitioning of the predicted solvation free energy

Table 8. Free Energy of Solvation and Partition Coefficient Results for *p*-Dichlorobenzene.

Model	ΔG_{ENP}	G_{CDS}	ΔG_S^o	log $P_{org/water}$	
				theory	experiment
	water				
SM5.4/AM1	-2.1	1.1	-1.0		
SM5.4/PM3	-1.3	0.2	-1.2		
SM5.2R/MNDO(d)	-1.0	-0.6	-1.6		
SM5.2R/MNDO	-1.4	-0.3	-1.7		
SM5.2R/AM1	-2.5	1.5	-1.0		
SM5.2R/PM3	-1.8	0.2	-1.6		
SM5.0R			-1.0		
	1-octanol				
SM5.4/AM1	-1.9	-3.7	-5.6	3.3	3.4
SM5.4/PM3	-1.2	-4.5	-5.7	3.3	
SM5.2R/MNDO(d)	-0.9	-5.0	-5.9	3.2	
SM5.2R/MNDO	-1.3	-4.8	-6.1	3.2	
SM5.2R/AM1	-2.2	-3.2	-5.5	3.3	
SM5.2R/PM3	-1.6	-4.2	-5.8	3.1	
SM5.0R			-5.6	3.4	
	hexadecane				
SM5.4/AM1	-1.0	-4.7	-5.7	3.4	3.7
SM5.4/PM3	-0.6	-5.1	-5.7	3.3	
SM5.2R/MNDO(d)	-0.5	-5.4	-5.9	3.1	
SM5.2R/MNDO	-0.7	-5.2	-5.9	3.1	
SM5.2R/AM1	-1.2	-4.3	-5.5	3.3	
SM5.2R/PM3	-0.9	-4.9	-5.8	3.1	
SM5.0R			-5.8	3.5	
	chloroform				
SM5.4/AM1	-1.6	-4.3	-5.9	3.6	3.9
SM5.4/PM3	-1.0	-5.1	-6.1	3.6	
SM5.2R/MNDO(d)	-0.8	-6.1	-6.8	3.8	
SM5.2R/MNDO	-1.1	-5.9	-6.9	3.8	
SM5.2R/AM1	-1.9	-4.7	-6.6	4.1	
SM5.2R/PM3	-1.4	-5.5	-6.9	3.9	
SM5.0R			-6.6	4.2	

between the electrostatic (ΔG_{ENP}) and non-electrostatic (G_{CDS}) components as well as the logarithm of the partition coefficient between selected organic solvents and water. The SM5.4 models utilize class IV charges and are designed to optimize solute geometry in the presence of the solvent reaction field. Note that the absolute value of the ΔG_{ENP} term is generally much larger for the SM5.4 parameterizations than for the SM5.2R models which incorporate the less-accurate class II charges. In general, class IV charges lead to greater charge separation within a solute molecule, which results in a larger $|\Delta G_{ENP}|$. Our method of parameterizing the

Table 9. Absolute Value (kcal/mol) of the ΔG_{ENP} and G_{CDS} Terms in Selected SMx Models[a]

Model	$\langle\lvert\Delta G_{ENP}\rvert\rangle$	$\langle\lvert G_{CDS}\rvert\rangle$
	water	
SM5.4/AM1	4.5	1.5
SM5.4/PM3	3.7	1.3
SM5.2R/MNDO(d)	1.7	2.3
SM5.2R/MNDO	1.7	2.3
SM5.2R/AM1	3.2	1.9
SM5.2R/PM3	2.4	2.0
SM5.0R	0.0	3.7
	1-octanol	
SM5.4/AM1	4.0	2.0
SM5.4/PM3	3.4	2.5
SM5.2R/MNDO(d)	1.5	4.2
SM5.2R/MNDO	1.5	4.2
SM5.2R/AM1	2.8	2.9
SM5.2R/PM3	2.1	3.6
SM5.0R	0.0	5.7
	hexadecane	
SM5.4/AM1	2.1	2.4
SM5.4/PM3	1.8	2.8
SM5.2R/MNDO(d)	0.8	3.6
SM5.2R/MNDO	0.8	3.6
SM5.2R/AM1	1.5	2.9
SM5.2R/PM3	1.1	3.3
SM5.0R	0.0	4.5
	chloroform	
SM5.4/AM1	3.4	2.6
SM5.4/PM3	2.9	3.1
SM5.2R/MNDO(d)	1.3	4.4
SM5.2R/MNDO	1.3	4.4
SM5.2R/AM1	2.4	3.5
SM5.2R/PM3	1.8	4.0
SM5.0R	0.0	5.5

[a]Reported averages are for 67 organic solutes for which experimental solvation free energies are available in water, hexadecane, octanol, and chloroform. (A total of 268 data points.)

remaining non-electrostatic term (G_{CDS}) to the experimental solvation free energies allows the diminished electrostatics obtained with the less expensive SM5.2R models to be compensated for by the G_{CDS} term, resulting in fairly accurate absolute solvation free energies and partition coefficients. This approach was taken to the limit in the SM5.0R model

Table 10. Mean Unsigned Errors in Predicted Solvation Free Energies, Organic/Water Partition Coefficients, and Free Energy of Transfer for Selected SMx methods[a]

Model	MUE ΔG_S^o	MUE $\Delta\Delta G^o_{org/water}$	MUE $\log P_{org/water}$
	water		
SM5.4/AM1	0.54		
SM5.4/PM3	0.47		
SM5.2R/MNDO(d)	0.45		
SM5.2R/MNDO	0.44		
SM5.2R/AM1	0.43		
SM5.2R/PM3	0.39		
SM5.0R	0.46		
	1-octanol		
SM5.4/AM1	0.56	0.63	0.46
SM5.4/PM3	0.51	0.54	0.40
SM5.2R/MNDO(d)	0.43	0.38	0.28
SM5.2R/MNDO	0.42	0.38	0.28
SM5.2R/AM1	0.47	0.42	0.31
SM5.2R/PM3	0.41	0.37	0.27
SM5.0R	0.40	0.40	0.38
	hexadecane		
SM5.4/AM1	0.29	0.49	0.36
SM5.4/PM3	0.29	0.49	0.36
SM5.2R/MNDO(d)	0.27	0.45	0.33
SM5.2R/MNDO	0.26	0.45	0.33
SM5.2R/AM1	0.27	0.51	0.38
SM5.2R/PM3	0.26	0.47	0.35
SM5.0R	0.30	0.30	0.43
	chloroform		
SM5.4/AM1	0.32	0.46	0.33
SM5.4/PM3	0.27	0.39	0.28
SM5.2R/MNDO(d)	0.45	0.45	0.33
SM5.2R/MNDO	0.46	0.45	0.33
SM5.2R/AM1	0.47	0.45	0.33
SM5.2R/PM3	0.44	0.41	0.30
SM5.0R	0.50	0.50	0.41

[a]Reported averages are for 67 organic solutes from our training set for which experimental solvation free energies are available in water, hexadecane, octanol, and chloroform. (A total of 268 data points.)

which contains no explicit electrostatic or SCF treatment. Although it is likely that the very inexpensive SM5.0R approach will have difficulty predicting solvation free energies in cases where the charge distribution within a given solute differs significantly from the implicit distributions parameterized into the model, SM5.0's predictions for the example molecules and overall training set are reasonably similar to those predicted by SM5 models with more rigorous electrostatic treatments.

To compare the performance of the SMx models in water, 1-octanol, hexadecane, and chloroform, we selected the subset of organic solutes from the training set for these solvation models for which the experimental free energy of solvation is known for all four solvents. This subset contains 68 molecules. Table 9 compares the average absolute value of the ΔG_{ENP} and G_{CDS} terms for various SMx models in 4 solvents. As mentioned earlier, the SM5.4 methods typically have the largest $\langle|\Delta G_{\mathrm{ENP}}|\rangle$, while the SM5.2 methods with the usually smaller class II charges have $\langle|\Delta G_{\mathrm{ENP}}|\rangle$ that are only one half to one-third as large as the SM5.4 counterpart. It is interesting to note that the class II charges produced by the AM1 Hamiltonian appear to be the most similar to the class IV charges, while the MNDO and MNDO(d) Hamiltonians produce much less charge separation. The $\langle|G_{\mathrm{CDS}}|\rangle$ terms have opposite trends to make up for the differences in the electrostatics.

Table 10 contains the mean unsigned errors in the absolute free energies of solvation from the gas phase into each of our four considered solvents as well as the unsigned error in the free energy of transfer from water to an organic solvent and the resulting error in the log of the estimated partition coefficient. Note that in general the SM5.4 and SM5.2 models perform similarly in both the mean unsigned error of the absolute free energies and the mean unsigned error of the log of the partition coefficient. The results in chloroform are an exception to this trend. The SM5.4 models were especially reparameterized for chloroform and hence they do achieve a significantly improved mean unsigned error in the absolute free energies of solvation. (The SM5.2R and SM5.0R models are parameterized for chloroform solvent at the same time as 89 other organic solvents, although a solvent descriptor is included which helps distinguish electronegative-halogen-containing solvents.) However, both the SM5.4 and SM5.2R models are shown to perform similarly in their ability to predict chloroform/water partition coefficients. SM5.0R generally is shown to have slightly larger errors than the SM5.4 and SM5.2R parameterizations, but still produces answers that are within reason considering the simplicity of the model.

4. Concluding remarks

We have developed a number of universal solvation models based on quantum mechanical treatment of the solutes, with solute polarizability included self-consistently. Both electrostatics and first-solvation-shell effects are treated by 3-D modeling. Hydrogen bonding of solute with solvent and solute disruption of solvent-solvent hydrogen bonding are both included. Solute functionality is recognized on the basis of atomic numbers and geometry only; thus the inconvenience (and occasional ambiguity) of assigning atomic types is avoided.

The solvation models are parameterized directly in terms of free energies, which are the critical thermodynamic quantities for predicting equilibria. One possible application is the prediction of log $P_{\text{octanol-water}}$, which is a widely used measure of lipophilicity, the movement of organic compounds through cells, and drug activity. We hope the models will be useful for a variety of purposes in the humanistic endeavor of designing better drugs.

5. Acknowledgments

This work was supported in part by the National Science Foundation.

References

1. Hansch, C.; Dunn, W. J. III *J. Pharm. Sci.* **1972**, *61*, 1.
2. Fujita, T.; Iwasa, J.; Hansch, C. *J. Am. Chem. Soc.* **1964**, *86*, 5175.
3. Leo, A.; Hansch, C.; Elkins, D. *Chem. Rev.* **1971**, *71*, 525.
4. Leo, A.; Jow, P. Y. C.; Silipino, C.; Hansch, C. *J. Med. Chem.* **1975**, *18*, 865.
5. Franke, R.; Dove, S.; Kuhne, B. *J. Med. Chem.* **1977**, **14**, 363.
6. Funisaki, N.; Hada, S.; Neya, S. *J. Phys. Chem.* **1985**, *89*, 3046.
7. Camilleri, P.; Watts, S. A.; Boraston, J. A. *J. Chem. Soc. Perkin Trans. II* **1988**, *188*, 1699.
8. Lewis, D. F. V. *J. Comput. Chem.* **1989**, *10*, 145.
9. Dunn, W. J. III: Nagy, P. I.; Collantes, E. R. *J. Am. Chem. Soc.* **1991**, *113*, 7898.
10. Suzuki, T. *J. Comput.-Aided Molec. Des* **1991**, *5*, 149.
11. Richards, N. G. J.; Williams, P. B.; Tute, M. S. *Int. J. Quantum Chem. Quantum. Biol. Symp.* **1991**, *18*, 299.
12. Dallas, A. J.; Carr, P. W. *J. Chem. Soc. Perkin Trans. 2,* **1992**, 2155.
13. Leo, A. *J. Chem. Rev.* **1993**, *93*, 1281.
14. Viswanadhan, V. N.; Reddy, M. R.; Bacquet, R. J.; Erion, M. D. *J. Comp. Chem.* **1993**, *14*, 1019.
15. Slater, B.; McCormack, A.; Avdeef, A.; Comer, J. E. A. *J. Pharm. Sci.* **1994**, *83*, 1280.
16. Collander, R. *Physiol. Plant.* **1954**, *7*, 420
17. Mayer, S.; Maickel, R. P.; Brodie, B. B. *J. Pharmacol. Exptl. Therap.* **1959**, *127*, 205.
18. Tute, M. S. *Adv. Drug Res.* **1971**, *6*, 1.
19. Venable, R. M.; Zhang, Y.; Hardy, B. J.; Pastor, R. W. *Science* **1993**, *262*, 223.
20. Solé-Violan, L.; Devallez, B.; Postel, M.; Riess, J. G. *New J. Chem.* **1993**, *17*, 581.
21. Sieler, P. *Eur. J. Med. Chem.* **1974**, *9*, 473.
22. Abraham, M. H.; Chadha, H. S.; Mitchell, R. C. *J. Pharm. Sci.* **1994**, *83*, 1257.
23. Tayar, N. E.; Tsai, R.-S.; Testa, B.; Carrupt, P.-A.; Leo, A. *J. Pharm. Sci.* **1991**, *80*, 590.
24. Tayar, N. A.; Mark, A. G.; Vallat, P.; Brunne, R. M.; Testa, B.; van Gunsteren, W. F. *J. Med. Chem.* **1993**, *36*, 3757.
25. Young, R.-C.; Mitchell, R. C.; Brown, T. H.; Ganellin, C. R.; Griffiths, R.; Jones, M.; Rana, K. K.; Saunders, D.; Smith, I. R.; Sore, N. E.; and Wilks, T. J. *J. Med. Chem.* **1988**, *31*, 656.

26. Reynolds, C. H. *J. Chem. Inf. Comput. Sci.* **1995**, *35*, 738.
27. Hine, J.; Mookerjee, P. K. *J. Org. Chem.* **1975**, *40*, 287.
28. Cabani, S.; Gianni, P.; Mollica, V.; Lepori, L. *J. Solution Chem.* **1981**, *50*, 563.
29. Abraham, M. H.; Whiting, G. S.; Fuchs, R.; Chambers, E. J. *J. Chem. Soc. Perkin Trans. II* **1990**, 291.
30. Famini, G. R.; Penski, C. A.; Wilson, L. Y. *J. Phys. Org. Chem.* **1992**, *5*, 395.
31. Leo, A. *J. Chem. Rev.* **1993**, *93*, 1281.
32. Cramer, C. J.; Famini, G. R.; Lowrey, A. H. *Acc. Chem. Res.* **1993**, *26*, 599.
33. Bodor, N.; Huang, M.-J.; Harget, A. *J. Mol. Struct. (Theochem)* **1994**, *390*, 259.
34. Tapia, O. *J. Math. Chem.* **1992**, *10*, 139.
35. Olivares del Valle, F. J.; Aguilar, M. A. *J. Mol. Struct. (Theochem)* **1993**, *280*, 25.
36. Tomasi, J.; Persico, M. *Chem. Rev.* **1994**, *94*, 2027.
37. Cramer, C. J.; Truhlar, D. G. In *Quantitative Treatments of Solute/Solvent Interactions*, Politzer, P., Murray, J. S., Eds.; Elsevier: Amsterdam, 1994; p. 9.
38. Cramer, C. J.; Truhlar, D. G. *Rev. Comp. Chem.* **1995**, *6*, 1.
39. Rivail, J.-L.; Rinaldi, D. In *Computational Chemistry, Review of Current Trends, Vol. 1*; Leszczynski, J., Ed.; World Scientific: Singapore, 1996; p. 139.
40. Orozco, M., Alhambra, C.; Barril, X.; Lopez, J. M.; Busquets, M. A.; Luque, F. J. *J. Mol. Modeling* **1996**, *2*, 1.
41. Gao, J. *Rev. Comp. Chem.* **1996**, *7*, 119.
42. Cramer, C. J.; Truhlar, D. G. In *Solvent Effects and Chemical Reactivity*; Tapia, O., Bertrán, J., Eds.; Kluwer: Dordrecht, 1996; p. 1.
43. Takahashi, O.; Sawahata, H.; Ogama, Y.; Kikuchi, O. *J. Mol. Struct. (Theochem)* **1997**, *393*, 141.
44. Baldridge, K.; Klamt, A. *J. Chem. Phys.* **1997**, *106*, 6622.
45. Zhan, C.-G.; Bentley, J.; Chipman, D. M. *J. Chem. Phys.* **1998**, *108*, 177.
46. Cramer, C. J.; Truhlar, D. G. *J. Am. Chem. Soc.* **1991**, *113*, 8305.
47. Cramer, C. J.; Truhlar, D. G. *Science* (Washington, DC) **1992**, *256*, 213.
48. Cramer, C. J.; Truhlar, D. G. *J. Comp. Chem.* **1992**, *13*, 1089.
49. Cramer, C. J.; Truhlar, D. G. *J. Comp.-Aided Mol. Des.* **1992**, *6*, 629.
50. Storer, J. W.; Giesen, D. J.; Hawkins, G. D.; Lynch, G. C.; Cramer, C. J.; Truhlar, D. G. In *Structure and Reactivity in Aqueous Solution*, Cramer C. J., Truhlar, D. G., Ed.; American Chemical Society: Washington, DC, 1994; p. 24.
51. Liotard, D. A.; Hawkins, G. D.; Lynch, G. C.; Cramer, C. J.; Truhlar, D. G. *J Comp. Chem.* **1995**, *16*, 422.
52. Barrows, S. E.; Dulles, F. J.; Cramer, C. J.; Truhlar, D. G.; French, A. D. *Carb. Res.* **1995**, *276*, 219.
53. Chambers, C. C.; Hawkins, G. D.; Cramer, C. J.; Truhlar, D. G. *J. Phys. Chem.* **1996**, *100*, 16385.
54. Hawkins, G. D.; Cramer, C. J.; Truhlar, D. G. *J. Phys. Chem.* **1996**, *100*, 19824.

55. Giesen, D. J.; Chambers, C. C.; Hawkins, G. D.; Cramer, C. J.; Truhlar, D. G. In *Computational Thermochemistry: Prediction and Estimation of Molecular Thermodynamics*; Irikura, K., Frurip, D. J., Eds.; American Chemical Society Symposium Series 677: Washington D.C., in press.
56. Chambers, C. C.; Giesen, D. J.; Hawkins, G. D.; Vaes, W. H. J.; Cramer, C. J.; Truhlar, D. G. In *Computer-Assisted Drug Design*; Truhlar, D. G.; Howe, W. J.; Blaney, J.; Hopfinger, A. J.; Dammkoehler, R. A., Eds.; Springer: New York, in press.
57. Hawkins, G. D.; Cramer, C. J.; Truhlar, D. G. *J. Phys. Chem. B* to be published (SM5.2R model).
58. Zhu, T.; Li, J.; Hawkins, G. D.; Cramer, C. J.; Truhlar, D. G. to be published (SM5.42R/DFT parameterizations).
59. Li, J.; Hawkins, G. D.; Cramer, C. J.; Truhlar, D. G. *Chem. Phys. Lett.* to be published (first SM5.42R/HF paramterization).
60. Giesen, D. J.; Storer, J. W.; Cramer, C. J.; Truhlar, D. G. *J. Am. Chem. Soc.* **1995**, *117*, 1057.
61. Giesen, D. J.; Cramer, C. J.; Truhlar, D. G. *J. Phys. Chem.* **1995**, *99*, 7137.
62. Giesen, D. J.; Gu, M. Z.; Cramer, C. J.; Truhlar, D. G. *J. Org. Chem.* **1996**, *61*, 8720.
63. Giesen, D. J.; Chambers, C. C.; Cramer, C. J.; Truhlar, D. G. *J. Phys Chem.* **1997**, *101*, 2061.
64. Giesen, D. J.; Cramer, C. J.; Truhlar, D. G. *Theor. Chem. Acct.* **1997**, *98*, 85.
65. Hawkins, G. D.; Cramer, C. J.; Truhlar, D. G. to be published (SM5.0R-organic model).
66. Bucher, M.; Porter, T. L. *J. Phys.Chem.* **1986**, *90*, 3406.
67. Still, W. C.; Tempczyk, A.; Hawley, R. C.; Hendrickson, T. *J. Am. Chem. Soc.* **1990**, *112*, 6129.
68. Zerner, M. C. *Rev. Comp. Chem.* **1991**, *2*, 313.
69. Hehre, W. J.; Radom, L.; Schleyer, P. v. R.; Pople, J. A. *Ab Initio Molecular Orbital Theory*; Wiley: New York, 1986.
70. Laird, B. B.; Ross, R. B.; Ziegler, T., Eds. *Chemical Applications of Density Functional Theory*; American Chemical Society: Washington, DC, 1996.
71. Yomosa, S. *J. Phys. Soc. Japan* **1973**, *35*, 1738.
72. Rivail, J.-L.; Rinaldi, D. *Chem. Phys.* **1976**, *18*, 233.
73. Tapia, O. *Theor. Chem. Acta* **1978,** *47*, 157.
74. Storer, J. W.; Giesen, D. J.; Cramer, C. J.; Truhlar, D. G. *J. Comp.-Aided Mol. Des.* **1995**, *9*, 87.
75. Li, J.; Zhu, T.; Cramer, C. J.; Truhlar, D. G. *J. Phys. Chem. A* in press.
76. Bachrach, S. M. *Rev. Comp. Chem.* **1993**, *5*, 171.
77. Lee, B.; Richards, R. M. *J. Mol. Biol.* **1971**, *55*, 379.
78. Hermann, R. B. *J. Phys. Chem.* **1972**, *76*, 2754.
79. Eisenberg, D.; McLachlan, A. D. *Nature* **1986**, *319*, 199.
80. Abraham, M. H. *Chem. Soc. Rev.* **1993**, 73.
81. Abraham, M. H. *et al. J. Chem. Soc. Perkin Trans. II* **1989**, 699.
82. Abraham, M. H. *J. Phys. Org. Chem.* **1993**, *6*, 660.

83. Abraham, M. H. In *Quantitative Treatments of Solute/Solvent Interactions*; Politzer, P., Murray, J. S.; Eds.; Elsevier: Amsterdam, 1994; p. 83.
84. Pauling, L. *The Nature of the Chemical Bond*; 3rd ed.; Cornell University Press: Ithaca, 1960; p. 93.
85. Reichardt, C. In *Solvents and Solvent Effects in Organic Chemistry*; VCH: New York, 1990.
86. Gutmann, V.; Resch, G. *Lecture Notes on Solution Chemistry*; World Scientific: Singapore, 1995.
87. Murray, J. S.; Brinck, T.; Lane, P.; Paulsen, K.; Politzer, P. *J. Mol. Struct. (Theochem)* **1994**, *307*, 55.
88. Mulliken, R. S. *J. Chem. Phys.* **1955**, *23*, 1833.
89. Löwdin, P.-O. *J. Chem. Phys.* **1950**, *18*, 365.
90. Breneman, C. M.; Wiberg, K. B. *J. Comput. Chem.* **1989**, *11*, 361.
91. Dewar, M. J. S.; Zoebisch, E. G.; Healy, E. F.; Stewart, J. J. P. *J. Am. Chem. Soc.* **1985**, *107*, 3902.
92. Dewar, M. J. S.; Zoebisch, E. G. *J. Mol. Struct. (Theochem)* **1988**, *180*, 1.
93. Dewar, M. J. S.; Yuan, Y.-C. *Inorg. Chem.* **1990**, *29*, 3881.
94. Stewart, J. J. P. *J. Comp. Chem.* **1989**, *10*, 221.
95. Easton, R. E.; Giesen, D. J.; Welch, A.; Cramer, C. J.; Truhlar, D. G. *Theo. Chim. Acta* **1996**, *93*, 281.
96. Becke, A. D. *Phys. Rev. A* **1988**, *38*, 3098.
97. Perdew, J. P.; Burke, K.; Wang, Y. *Phys. Rev. B* **1996**, *54*, 6533.
98. Lee, C.; Yang, W.; Parr, R. G. *Phys. Rev. B* **1988**, *37*, 785.
99. Becke, A. D. *J. Chem. Phys.* **1993**, *98*, 5648.
100. Stephens, P. J.; Devlin, F. J.; Chabalowski, C. F.; Frisch, M. J. *J. Phys. Chem.* **1994**, *98*, 11623.
101. Cabani, S.; Gianni, P.; Mollica, V.; Lepori, L. *J. Solution Chem.* **1981**, *10*, 563.
102. Leo, A. J. *Pamona Medchem Database*; BioByte Corp.: Claremont, CA, 1994.

Chapter 9

Rational Approaches to Inhibition of Human Osteoclast Cathepsin K and Treatment of Osteoporosis

Sherin S. Abdel-Meguid [1], Baoguang Zhao [1], Ward W. Smith [1], Cheryl A. Hanson [2], Judith LaLonde [1], Thomas Carr [3], Karla D'Alessio [2], Michael S. McQueney [2], H.-J. Oh [3], Scott K. Thompson [3], Daniel F. Veber [3], and Dennis S. Yamashita [3]

[1] Department of Macromolecular Sciences, [2] Department of Protein Biochemistry, and [3] Department of Medicinal Chemistry, SmithKline Beecham Pharmaceuticals, 709 Swedeland Road, King of Prussia, PA 19406

Novel, potent and selective human osteoclast cathepsin K inhibitors have been designed based on knowledge derived from the crystal structure of papain bound to a tripeptide aldehyde. Cathepsin K is a thiol protease belonging to the papain superfamily. Unlike previously known crystal structures of that family of enzymes in which ligands bind to the nonprime side of the active site, our papain structure shows the ligand in the prime direction. This observation and the identification of key interactions between the protein and the ligand inspired the design of a novel class of inhibitors spanning both sides of the active site. The crystal structure of the first member of this class bound to cathepsin K confirmed our design hypothesis. Inhibitors of cathepsin K are potential drugs for the treatment of osteoporosis.

Recent success in the rational design of novel, potent HIV-1 protease inhibitors and the subsequent verification that they are highly effective drugs, has confirmed the important role of rational design in the drug discovery process. Most of these drugs were designed based on knowledge derived from the crystal structures of HIV protease, renin and other aspartyl proteases (*1,2*). Many other examples of rational drug design are now available (*3*). Here, we will describe the structure-based design of one chemical class of cathepsin K inhibitors that are potential drugs for the treatment of osteoporosis and we will show how the crystal structure of an inhibitor of cathepsin K bound to papain inspired the rational design process.

Bone Remodeling and Osteoporosis

Bone remodeling is a normal and dynamic process involving deposition and resorption of bone matrix. Bone is formed by mature osteoblast cells, while osteoclasts are responsible for bone resorption. Osteoclasts are multinuclear giant cells that solubilize mineralized bone matrix through secretion of proteolytic enzymes into an extracellular, sealed, low pH compartment on the bone surface. It is believed that osteoporosis, a disease characterized by low density, high porosity and brittleness of bone, results from imbalance between bone formation (osteoblasts) and resorption (osteoclasts).

Cathepsin K and its Role in Osteoporosis

Cathepsin K is a recently discovered member of the papain superfamily of cysteine proteinases that is selectively and highly expressed in osteoclasts (*4,5*). It is secreted as a 314 amino acid proenzyme containing a 99 amino acid leader sequence (*6*). The proenzyme self-processes at low pH to generate the mature form (*7*). The crystal structures of cathepsin K in the presence and absence of bound ligands have been determined (*8,9*). The enzyme folds into two subdomains separated by the active site cleft, a characteristic of the papain family of cysteine proteases.

Cathepsin K is believed to play an important role in bone resorption and is a potential therapeutic target for treatment of diseases involving excessive bone loss such as osteoporosis. This is supported by two pieces of evidence. One, it has been known for over a decade that classical thiol protease inhibitors such as E-64 and leupeptin inhibit bone resorption (*10,11*). Two, defects in the gene encoding cathepsin K have been linked recently to pycnodysostosis, a disease characterized by skeletal defects such as dense, brittle bones, short stature and poor bone remodeling (*12*).

Papain as a Surrogate for Cathepsin K

The absence of sufficient cathepsin K for crystallographic structure determination early-on in this study compelled us to search for a suitable model. Papain, having 46% identical amino acid sequence to cathepsin K, was chosen because of the availability of its structure in the presence and absence of ligands. A number of crystal structures of papain with bound inhibitors had been reported (*13, 14, 15*). The inhibitors in all of these structures were found to bind on the nonprime side of the active site (Figure 1a). Using these structures, we modeled a number of our di- and tri-peptide aldehyde inhibitors into the nonprime side of the active site of papain and into a homology model of cathepsin K derived from papain. These modeling studies did not explain our SAR data which showed strong preference for the presence of a Cbz or other aromatic moiety at the amino terminus of these peptides. Thus, to rationally design inhibitors of cathepsin K it was necessary to obtain crystal structures using our own inhibitors. Again, papain was selected because it is commercially available in large quantities (ICN Biomedicals #1009-24) and because its crystallization and crystallographic studies are well documented (*16*).

1a.

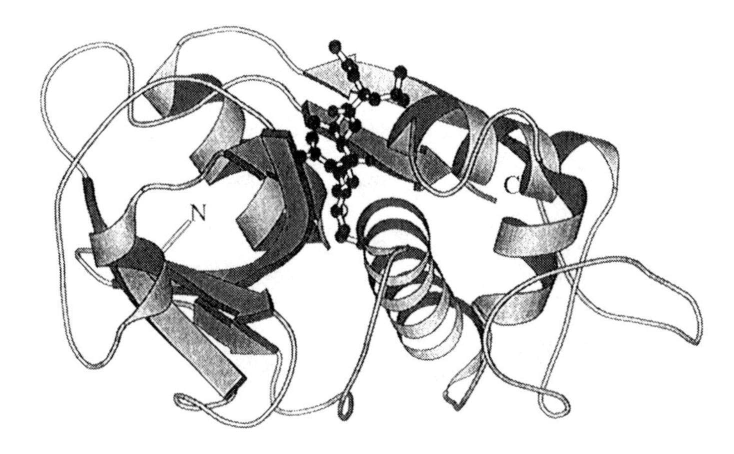

Papain - Leupeptin

1b.

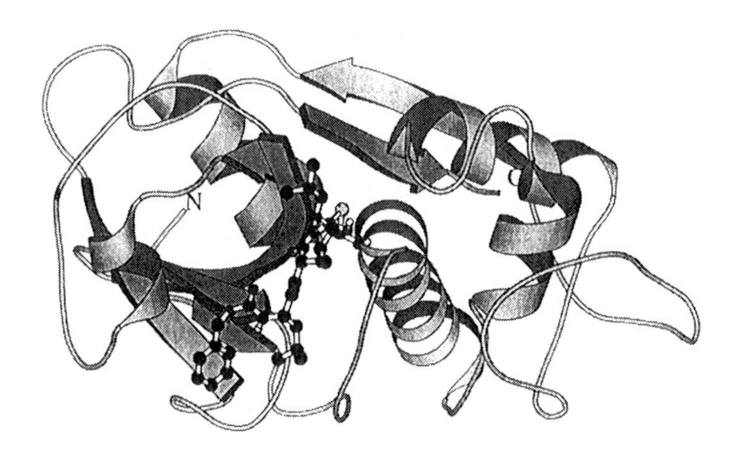

Papain - Cbz-Leu-Leu-Leu-Aldehyde

Figure 1. a) Ribbon drawing of the crystal structure of the complex of papain with leupeptin. b) Ribbon drawing of the papain complex with the peptide aldehyde inhibitor Cbz-Leu-Leu-Leu-OH. The figure was prepared with MOLSCRIPT (*22*).

Crystal Structure of Papain/Cbz-Leu-Leu-Leu Aldehyde

We have determined the crystal structure of papain bound to the Cbz-Leu-Leu-Leu aldehyde whose chemical structure is shown in Figure 2. The structure of the complex was obtained using crystals grown by vapor diffusion from a solution of 0.1 M Tris-HCl at pH 8.5 containing 0.5 M sodium citrate and 20% PEG 600. The crystals belong to the monoclinic space group C2, with a=100.5 Å, b=50.7 Å, c=62.3 Å, β=99.9° and one molecule/asymmetric unit. They grow in a space group different from those previously reported for papain/ligand complexes. The crystal structure of the papain/Cbz-Leu-Leu-Leu aldehyde was solved by the molecular replacement method (*17*), using the structure of papain (PDB code 1PIP; *18*) as a starting model.

Surprisingly, the inhibitor in our structure was found to bind on the prime side of the active site (Figure 1b). A major point of interaction between the inhibitor and the protein was an edge-to-face interaction between the phenyl ring of the inhibitor and the indole ring of Trp181 (Figure 3). This tryptophan is conserved between papain and cathepsin K.

In order to ensure that the novel binding mode observed in the crystal structure of the papain/Cbz-Leu-Leu-Leu aldehyde was not an artifact of crystallization, we produced crystals of papain bound to leupeptin under exactly the same conditions as those used for the papain/Cbz-Leu-Leu-Leu aldehyde. The crystals were isomorphous, and our structure of the papain/leupeptin complex was nearly identical to that previously reported (Figure 1a), with the inhibitor bound to the nonprime subsites (S1 to S3; *19*) of the enzyme.

Design of a Novel Cathepsin K Inhibitor Based on the Crystal Structure of the Papain/Cbz-Leu-Leu-Leu aldehyde

The observations that inhibitors containing Cbz or other aromatic groups at the amino terminus bind to the prime side of the active site and that such binding may be facilitated by the interaction with Trp181 led to the design of novel inhibitors spanning both sides of the active site (Figure 4). The prototype of this class of inhibitors was a symmetric inhibitor that resulted from an overlay of the Cbz-Leu-Leu-Leu aldehyde and leupeptin papain crystal structures. The two inhibitors were merged computationally by replacing their aldehyde functions with a single ketone (Figure 4). The resulting model of a ketone-containing inhibitor was further simplified by removal of the side chains on both sides of the ketone moiety. This was necessary since the arginyl and leucyl sidechains occupied the same region of space. Furthermore, a homology model of cathepsin K derived from the structure of papain suggested that Trp184 of cathepsin K (Trp177 in papain), a highly conserved residue within the papain superfamily, would form a better aromatic-aromatic interaction with the Cbz moiety. Thus, the hypothetical inhibitor was shortened by one Leu residue from the right side (Figure 4), resulting in a yet smaller molecule. A second Cbz moiety was introduced on the left side (Figure 4), as a final step to make the inhibitor truly symmetric. This was done not to mimic any symmetry in the active site (there is none), but rather to simplify the chemical synthesis of this initial member of a new class of inhibitors. This Cbz group was also hypothesized to reach to Tyr67 on the

2a.

Leupeptin

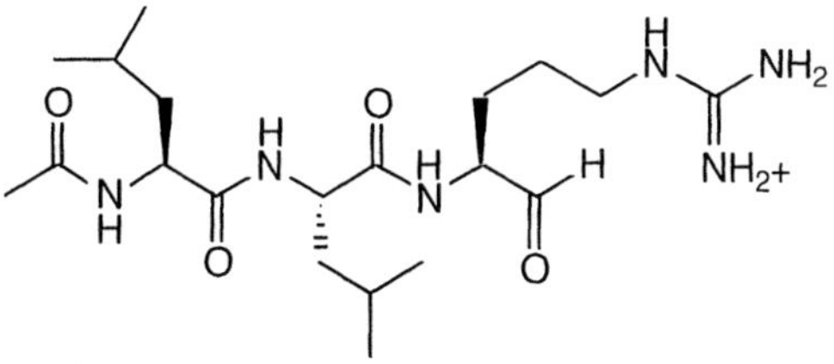

Leu-Leu-Arg-aldehyde

2b.

Cbz-tripeptide aldehyde

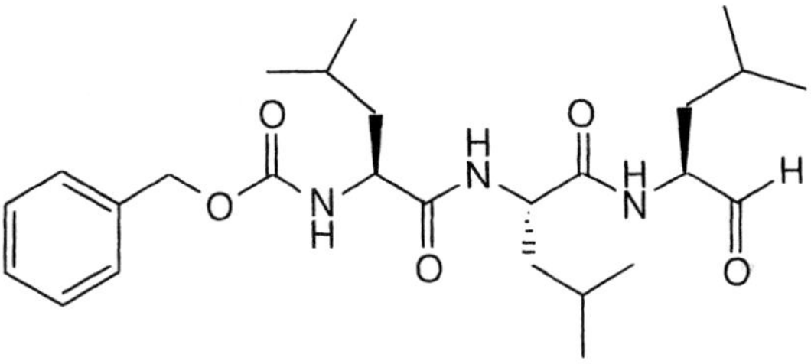

Cbz-Leu-Leu-Leu-aldehyde

Figure 2. Chemical structure of a) leupeptin and b) Cbz-Leu-Leu-Leu-aldehyde.

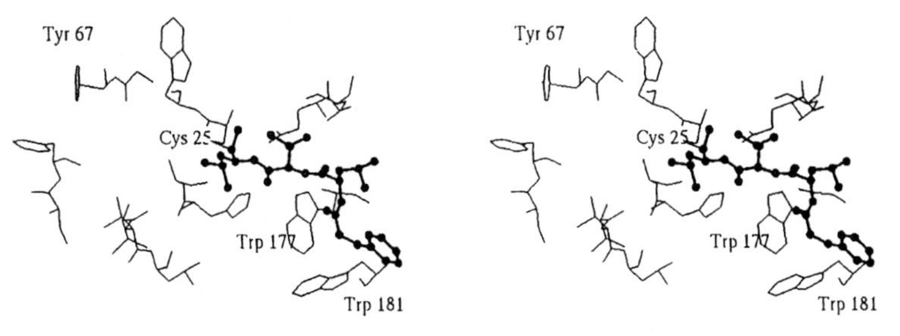

Figure 3. Stereo view of the active site of papain bound to Cbz-Leu-Leu-Leu-OH. Inhibitor atoms are drawn as ball-and-stick.

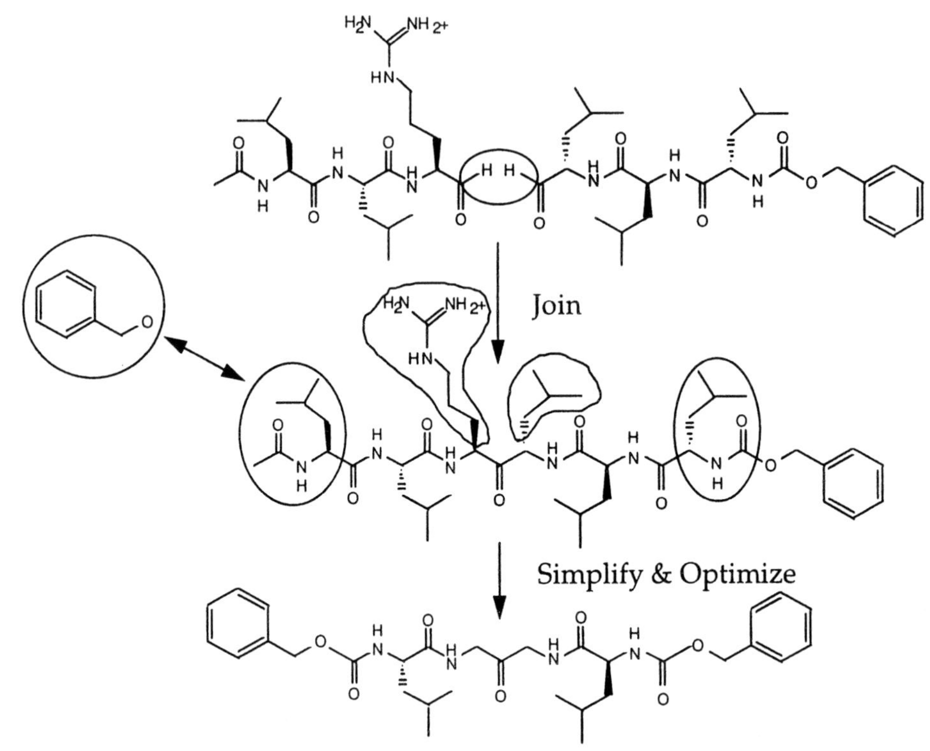

Figure 4. Schematic drawing of the design of the symmetric diacylaminomethyl ketone inhibitor based on the crystal structures of papain bound to leupeptin and to Cbz-Leu-Leu-Leu-aldehyde.

nonprime side of the cathepsin K active site for additional aromatic-aromatic interaction. The resulting diacylaminomethyl ketone (1,3-bis[[N-[(phenylmethoxy)carbonyl]-L-leucyl]amino]-2-propanone) is shown in Figure 4.

Binding of the Novel, Symmetric Diacylaminomethyl Ketone to Cathepsin K

The novel diacylaminomethyl ketone is a selective, competitive, reversible inhibitor of cathepsin K with a K_i of 23 nM (*20*). Spanning both sides of the active site has allowed for enhanced potency and selectivity by taking simultaneous advantage of interactions on the nonprime and prime sides of the active site, and by allowing the use of a less reactive electrophilic carbon for attack at the cysteine. Yamashita *et al.* (*20*) have shown that this diacylaminomethyl ketone is a relatively poor inhibitor of papain, cathepsin L, cathepsin B and cathepsin S, with $K_{i,app}$ of 10,000 nM, 340 nM, 1,300 nM and 890 nM, respectively.

To confirm our design hypothesis, we have determined the crystal structure of diacylaminomethyl ketone bound to cathepsin K. Data for the structure determination were obtained from crystals grown using vapor diffusion from a solution of 10% isopropanol, 0.1 M $NaPO_4$-citrate at pH 4.2. These crystals belong to the tetragonal space group $P4_32_12$, with a=57.7 Å, c=131.1 Å, and the unit cell contains one molecule/asymmetric unit. The crystal structure was determined using the molecular replacement method (*17*) and a model consisting of all atoms from the crystal structure reported by Zhao *et al.* (*8*).

The diacylaminomethyl ketone inhibitor binds in the cathepsin K active site as predicted. It spans both sides of the active site (Figure 5) and makes a number of key interactions with the enzyme (Figure 6). The phenyl groups on both ends of the inhibitor engage Trp184 and Tyr67 in a face-face and edge-face interaction, respectively (Figure 7). The crystal structure clearly shows the inhibitor covalently attached to the enzyme at the sulfur atom of Cys25 (the active site cysteine) as expected. The P2 leucyl sidechain of the inhibitor fits snugly in the hydrophobic S2 pocket defined by residues Met68, Leu209, Ala134, Ala163 and Tyr67 (Figure 6). Hydrogen bonding interactions are seen between ND1 of His162, NE2 of Gln19 and the backbone amide nitrogens of Cys25 and Gly66, all of which donate a hydrogen to oxygen atoms of the inhibitor. The remainder of the inhibitor interacts poorly or not at all with the enzyme indicating potential for further optimization of this class of inhibitors.

Conclusions

The design hypothesis generated from the crystal structures of papain bound to Cbz-Leu-Leu-Leu aldehyde and to leupeptin resulted in the design and synthesis of a novel, potent, reversible and selective symmetric inhibitor. Confirmation was achieved through crystallographic structure determination of the resulting diacylaminomethyl ketone inhibitor bound to cathepsin K. This in turn led to the generation of numerous novel inhibitors of cathepsin K that span both sides of the active site as described by Yamashita *et al.* (*20*) and Thompson *et al.* (*21*). These new inhibitors were then optimized through iterative cycles of structure-based design. Although the papain

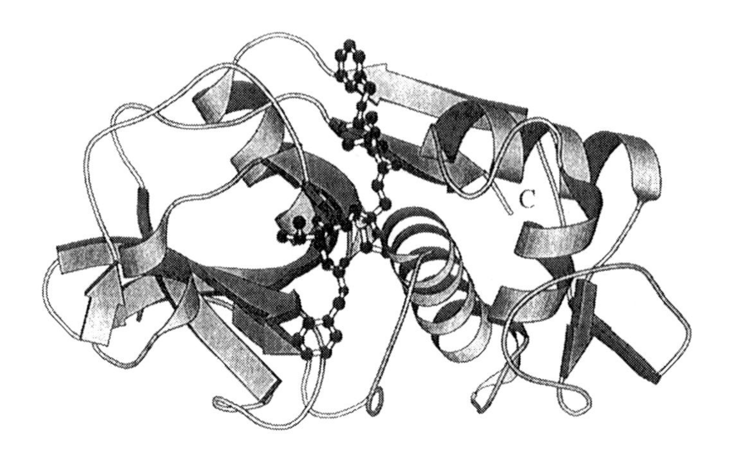

Human Cathepsin K - Diacylaminomethyl Ketone

Figure 5. Ribbon drawing of the crystal structure of the complex of human cathepsin K with the symmetric diacylaminomethyl ketone.

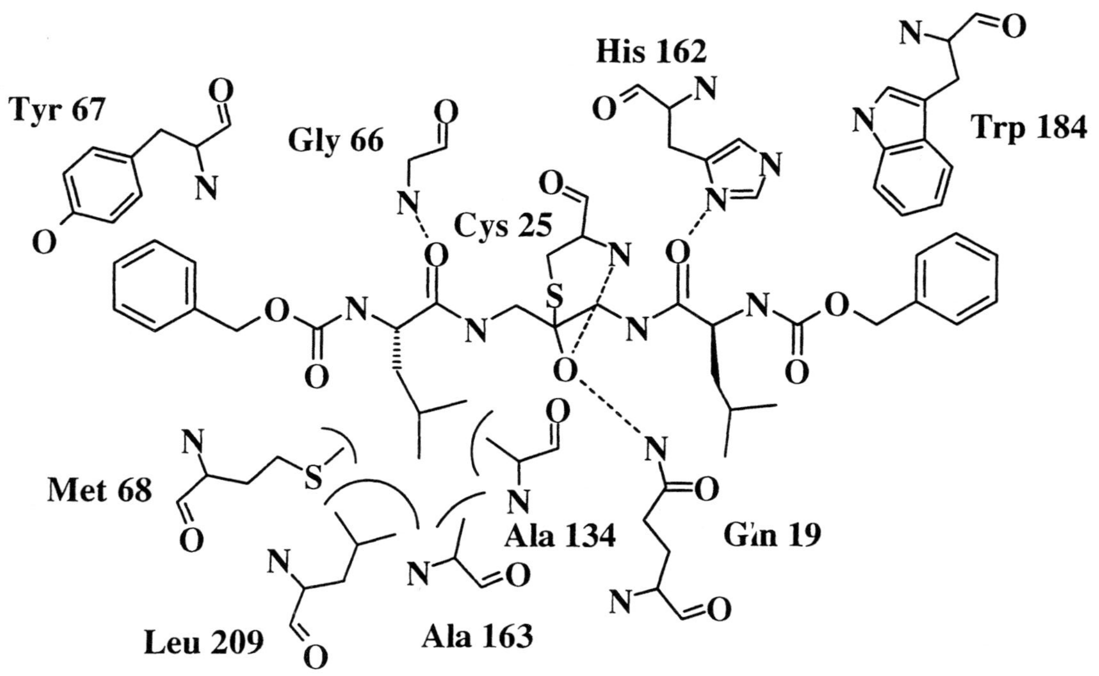

Figure 6. Schematic view of the interactions in the active site of cathepsin K with the symmetric diacylaminomethyl ketone inhibitor.

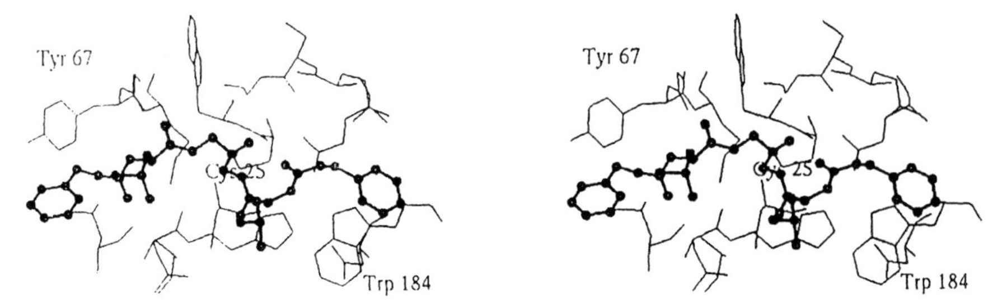

Figure 7. Stereo view of the active site of the complex of human cathepsin K with the symmetric diacylaminomethyl ketone. Inhibitor atoms are drawn as ball-and-stick.

structures provided pivotal insights for the design of the first inhibitors, further generations of inhibitors designed based on this preliminary insight required optimization through numerous crystal structures of cathepsin K, the actual target, with bound ligands.

Finally, we conclude that use of surrogate enzymes can lead to important insights for the rational design of novel inhibitors, but optimization requires knowledge of the structure of the target molecule preferably bound to inhibitors. This is reminiscent of studies to identify renin inhibitors where surrogates such as endothiapepsin, rhizopuspepsin and penicillopepsin structures were used to design inhibitors of renin (*1*).

References

1. Abdel-Meguid, S. S. *Medicinal Research Reviews* 1993, 13, 731-778.
2. Wlodawer, A., Erickson, J. W. *Annu. Rev. Biochem.* **1993**, *62*, 5434-5485.
3. Babine, R. E., Bender, S.L. *Chem. Rev.* **1997**, *97*, 1359-1472.
4. Drake, F. H., Dodds, R. A., James, I. E., Connor, J. R., Debouck, C., Richardson, S., Lee-Rykaczewski, E., Coleman, L., Rieman, D., Barthlow, R., Hastings, G., Gowen, M. *J. Biol. Chem.* **1996**, *271*, 12511-12516.
5. Bromme, D., Okamoto, K. *Biol. Chem. Hoppe-Seyler* **1995**, *376*, 379-384.
6. Bossard, M. J., Tomaszek, T. A., Thompson, S. K., Amegadzie, B. Y., Hanning, C. R., Jones, C., Kurdyla, J. T., McNulty, D. E., Drake, F. H., Gowen, M., Levy, M. A. *J. Biol. Chem.* **1996**, *271*, 12517-12524.
7. McQueney, M. S., Amegadzie, B. Y., D'Alessio, K., Hanning, C. R., McLaughlin, M. M., McNulty, D., Carr, S. A., Ijames, C., Kurdyla, J., Jones, C. S. *J. Biol. Chem.* **1997**, *272*, 13955-13960.
8. Zhao, B., Janson, C. A., Amegadzie, B. Y., D'Alessio, K., Griffin, C., Hanning, C.R., Jones, C., Kurdyla, J., McQueney, M., Qiu, X., Smith, W. W., Abdel-Meguid, S. S. *Nat. Struct. Biol.* **1997**, *4*, 109-111.
9. McGrath, M. E., Klaus, J. L., Barnes, M. G., Bromme, D. *Nat. Struct. Biol.* **1997**, *4*, 105-109.
10. Delaisse, J. M., Boyde, A., Maconnachie, E., Ali, N. N., Sear, C. H. J., Eeckhout, Y., Vaes, G. *Bone* **1980**, *8*, 305-313.
11. Delaisse, J. M., Eeckhout, Y., Vaes, G. *Biochem. Biophys. Res. Commun.* **1984**, *125*, 441-447.
12. Gelb, B.D., Shi, G.P., Heller, M., Weremowicz, S., Morton, C., Desnick, R.J., Chapman, H.A. *Genomics* **1997**, *41*, 258-262.
13. Drenth, J., Kalk, K.H. & Swen, H.M. *Biochemistry* **1976** *15*, 3731-3738.
14. Varughese, K.I., Ahmed, F.R., Carey, P.R., Hasnain, S., Huber, C.P. & Storer, A.C. *Biochemistry* **1989**, *28*, 1330-1332.
15. Yamamoto, D., Matsumoto, K., Ohishi, H., Ishida, T., Inoue, M., Kitamura, K. and Mizuno, H. *J. Biol. Chem.* **1991**, *266*, 14771-14777.
16. Yamamoto, A., Tomoo, K., Doi, M, Ohishi, H., Inoue, M., Ishida, T., Yamamoto, D., Tsuboi, S., Okamoto, H., Okada, Y. *Biochemistry*. **1992**, *31*, 11305-11309.
17. Brunger, A. T., Kuriyan, J., Karplus, M. *Science* **1987**, *235*, 458-460.
18. Bernstein, F. C., Koetzle, T. F., Williams, G. J., Meyer, E. E. J., Brice, M. D., Rogers, J. R., Kennard, O., Shimanouchi, T., Tasumi, M. *J. Mol. Biol.* **1977**, *112*, 535-542.
19. Schechter, I., Berger, A. *Biochem. Biophys. Res. Commun.* **1967**, *27*, 157-162.
20. Yamashita, D. S., Smith, W. W., Zhao, B., Janson, C. A., Tomaszek, T. A. Bossard, M. J., Levy, M. A., Marquis, R. W., Oh, H-J., Ru, Y., Carr, T. J.,

Thompson, S. K., Ijames, C. F., Carr, S. A., McQueney, M., D'Alessio, K. J., Amegadzie, B. Y., Hanning, C. R., Abdel-Meguid, S. S., DesJarlais, R. L., Gleason, J. G., Veber, D. F. *J. Amer. Chem. Soc., in press.*

21. Thompson, S. K., Halbert, S. M., Bossard, M. J., Tomaszek, T. A., Levy, M. A., Meek, T. D., Zhao, B., Smith W. W., Janson, C. A., D'Alessio, K. J., McQueney, M., Abdel-Meguid, S. S., DesJarlais, R. L., Briand, J., Sarkar, S. K., Huddleston, M., Ijames, C., Carr, S., Garnes, K. T., Shu, A., Heys, J. R., Bradbeer, J., Zembryki, D., Lee-Rykaczewski, L., James, I. E., Lark, M., Drake, F. H., Gowen, M., Gleason J. G., Veber, D.F. *Proc. Natl. Acad. Sci. USA, in press.*
22. Kraulis, P. *J. Appl. Crystallogr.* **1991**, *24,* 946-950.

Chapter 10

Building a Hypothesis for Nucleotide Transport Inhibitors

K. Raghavan [1], Scott D. Kahn [2], and John K. Buolamwini [3]

[1] **Molecular Simulations, Inc., 1804 N. Naper Boulevard, Suite 424, Naperville, IL 60563**
[2] **Molecular Simulations, Inc., 9685 Scranton Road, San Diego, CA 92121**
[3] **Department of Medicinal Chemistry, School of Pharmacy, University of Mississippi, University, MS 38677**

Structure-activity relationship studies of nucleoside transport inhibitors have revealed a diverse group of compounds with potent inhibitory activity against the major mammalian equilibrative nucleoside transporter, the *es* transporter. Inhibitors of the *es* transporter have potential therapeutic applications for adenosine potentiation in heart disease and stroke, as well as for anticancer and antiviral chemotherapy. Computational techniques have been applied to derive a pharmacophore hypothesis of generalized chemical interaction features that can be used to search 3D molecular databases to identify novel inhibitors. The methodology of feature-based hypothesis generation and the results are presented .

Nucleoside transporters (NT) are integral membrane glycoproteins required for the cellular uptake of physiological nucleosides and their analogs (1). Nucleoside transport inhibitors have potential therapeutic uses in many areas including heart disease, anti-cancer and anti-viral chemotherapy. Since most of the compounds that are known to inhibit NT do not meet the requisite pharmacological profiles , there is need for the discovery of better inhibitors. Among the known, potent inhibitors, NBMPR [N^6-(4-nitrobenzyl)thioinosine] and dipyridamole are most commonly used. The *es* (equilibrative inhibitor-sensitive) transporter is sensitive to inhibition by NBMPR whereas the *ei* (equilibrative inhibitor-insensitive) transporter is relatively resistant. The *es* transporter, which is the focus of this study, is by far the major nucleoside transporter of most mammalian cells examined to date. It has been known that the inhibition by NBMPR arises from specific, high affinity binding (K_d = 0.1-1.0 nM) at the transporter protein(2). Recently, the cDNA of the *es* transporter protein has been cloned(3).

As is a common case in drug discovery, the 3-dimensional structure of the *es* transporter protein is not known, requiring an indirect approach to be taken to analyze known inhibitors followed by the generation of pharmacophore hypotheses that capture the binding requirements of potentially improved inhibitors. The binding studies on an array of structures including a number of analogs of 6-N-(4-Nitrobenzyl) Adenosine are available (1) and have been used in this study.

In an effort to gain better understanding of nucleoside transport inhibitors and find novel inhibitors, a study of pharmacophore analysis on known inhibitors has been carried out. The resulting hypothesis was used to search structural databases of molecular structures, and has yielded a set of new and structurally diverse compounds satisfying the pharmacophore features derived to be responsible for activity against the nucleoside transporter. The process of hypothesis generation and the database results are presented in this article.

Method

Conformational Analysis. The molecules considered in the study are given in Figure 1. For each molecule, conformations were generated using Catalyst software (4) with the poling method (5-7). Briefly, poling is a direct method for promoting conformational variation. During the stage of geometry optimization, this method introduces a penalty function (poling function) that penalizes any conformer that is too close to another conformer in the set. Mathematically, the potential energy function is modified so that there is a repulsive wall at the location of previously generated conformations. This effectively forces the search away from the location of this pole (ie, an existing conformation), and in so doing ensures broad coverage of low energy conformations within a specified energy threshold. For the purpose of identifying pharmacophores, it is important to consider energetically accessible conformations in the neighborhood of the lowest energy conformation as the precise binding conformation is not known. In fact, it is possible that the binding conformation can be quite different from the lowest energy conformation (8).

Pharmacophore Development. Catalyst hypothesis can consider molecular features such as, hydrogen bond (HB) donors and acceptors, positive and negative ionizable centers, positive and negative charge centers, aliphatic and aromatic hydrophobic centers and aromatic rings (9). In addition, molecular substructures and user defined features can also be included in the generation of catalyst hypothesis. The features used in this study, based on visually examining the molecules are: HB donor, HB acceptor, Hydrophobic center, and Ring Aromatic. The common feature hypothesis generation algorithm (10) begins by identifying 3D configurations of features (ie, a pharmacophore) common to the molecules. A molecule matches a configuration if it possesses a set of features and a conformation such that the set of features can be superimposed with the corresponding locations of each feature. A set of features is considered superimposed if each feature lies within a specified distance (tolerance) from the corresponding ideal location. Each molecule in the set is considered as reference molecule, and contribute to the initial set of pharmacophore patterns to be considered across the entire set of compounds. The requirement that all molecules match all features in a configuration can be relaxed. Certain molecules can be specified to miss one of the features in a configuration being evaluated. It is also possible to permit a specified number of appropriately indicated molecules to completely miss a common pharmacophore configuration. In this regard, a complete miss is defined when two or more features must be omitted from a configuration for a molecule to map all remaining features. This ability to allow complete misses helps account for "active" molecules that may bind at different sites, or that simply do not require all of the specified interactions for their activity. The configurations are identified by an exhaustive search, starting with small sets of features and extending them until no larger common configuration exists.

Molecular conformations within 10 kcal/mol from the minimum energy conformation were used for the subsequent hypothesis generation step. Figure 1 lists all the molecules used in this study and the number of conformations used; the 10 kcal/mol energy cutoff resulted in 19 to 103 conformations. The common features hypothesis method (also known as HipHop (10)) in Catalyst was used to generate potential pharmacophore hypothesis for the nucleoside transport inhibitors.

Results and Discussion

The hypothesis, shown in figure 2, contains one hydrogen bond donor, three hydrogen bond acceptors, and an aromatic ring with the relationship in 3D space as shown. The features, hydrogen bond donor and acceptor are depicted as two spheres, with the smaller sphere representing the location of the heavy atom in the ligand and the larger sphere representing the location of the receptor atom involved in the hydrogen bonding, thus considering the directionality of the hydrogen bond as well. The aromatic ring is also shown with two spheres and an arrow in order to define the orientation of the plane of the ring, shown by the normal to the plane.

Figure 3 shows two molecules from the study set, NBMPR and 2'-deoxy-6-N-(4-nitrobenzyl)adenosine mapped on to the generated pharmacophore hypothesis. The SAR data show that a nitro substituent in the 4 position of the 6-position benzyl group and the 3' hydroxyl group are important for the high affinity binding to the *es* transporter (1). The hypothesis derived recognizes these chemical features. The conformations selected for these molecules by Catalyst are 2.39 and 5.11 kcal/mol from their corresponding low energy conformations. The fit values represent how well the chemical features map within the centers of location/tolerance in the hypothesis. This hypothesis was used to search different databases to find molecules that contains these chemical features. The databases used are NCI (National Cancer Institute's open database), Derwent, Maybridge and ACD(Available Chemicals Directory) (11). The NCI database that was searched contains 123219 structures. The search with the hypothesis resulted in 186 structures ranging in molecular weight from 316.27 to 1664.88. Consideration of the low molecular weight of known inhibitors, a subset of these hits with a molecular weight less than 605 narrows the set to 125 structures requiring further analysis. The search results contain a number of interesting structures that can be tested for activity against nucleoside transporters. For example, figure 4 shows a molecule with a totally new ring system connecting the nitro benzyl group with the sugar ring. This is a new family that can be considered. There were also molecules with pyridyl ring in place of nitrobenzyl

1a.

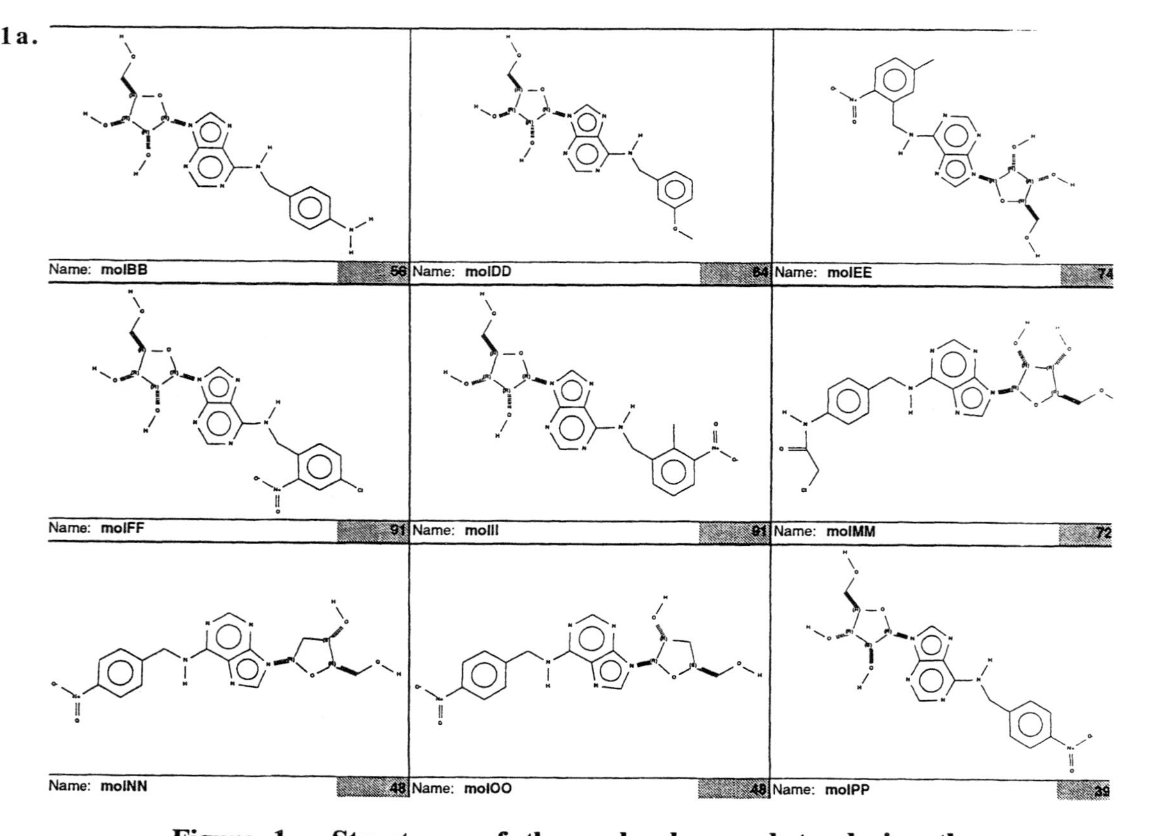

Figure 1. Structures of the molecules used to derive the common feature hypothesis.

1b.

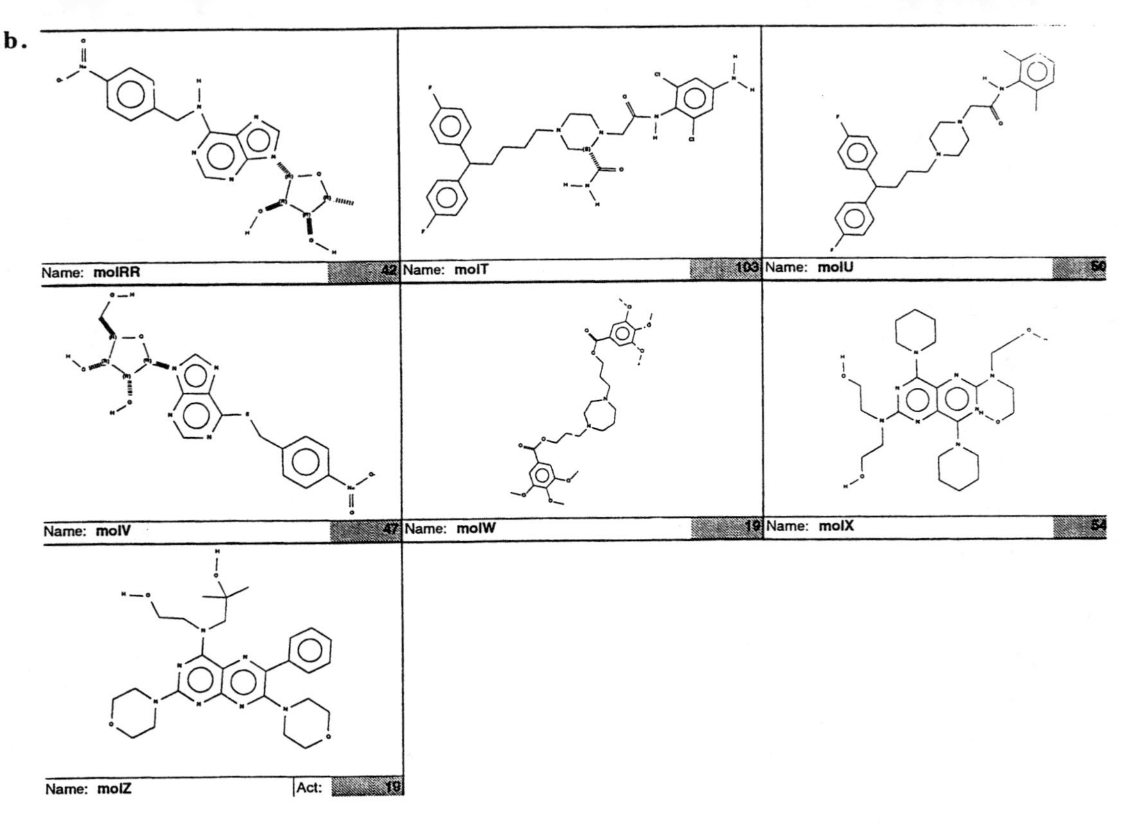

Figure 1. Structures of the molecules used to derive the common feature hypothesis.

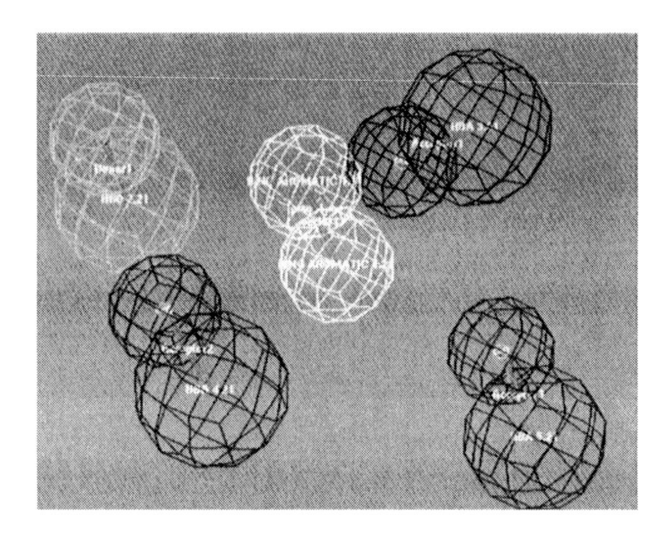

Figure 2. A Common feature hypothesis for NT inhibitors derived by Catalyst. Hydrogen bond donor (magenta), hydrogen bond acceptors (green) and ring aromatic (yellow) are the features in the hypothesis.
(Figure is printed in color in color insert.)

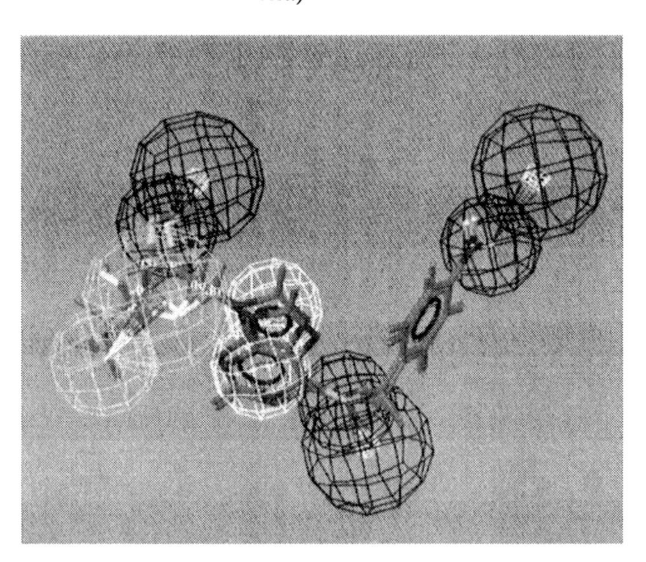

NBMPR N^6–(4–nitrobenzyl)thioinosine (fit=5.0 ; E_{conf}=2.39 kcal/mol)
2'–deoxy–6–N–(4–nitrobenzyl)adenosine (fit=3.67;E_{conf}=5.11 kcal/mol)

Figure 3. NBMPR and 2'-deoxy-6-N-(4-nitrobenzyl) adenosine mapped onto the hypothesis.
(Figure is printed in color in color insert.)

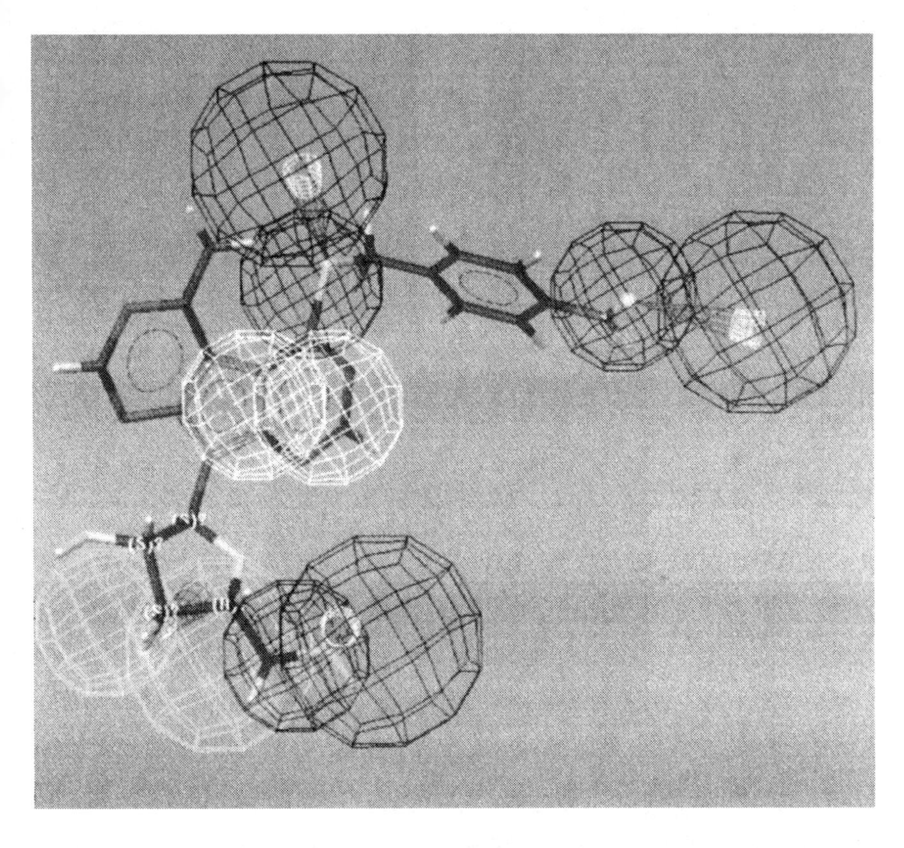

Figure 4. A hit from a search on the NCI database (267224)

(Figure is printed in color in color insert.)

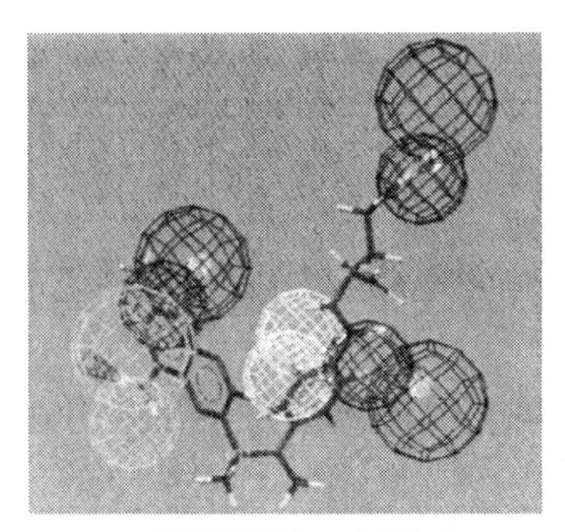

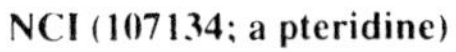

NCI (107134; a pteridine)

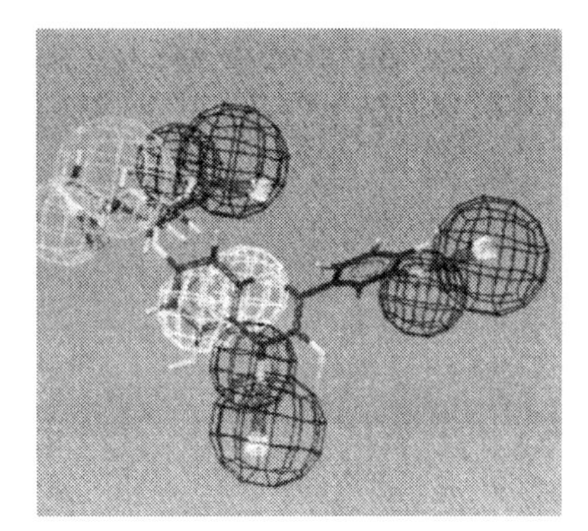

NCI (115917; a flavone)

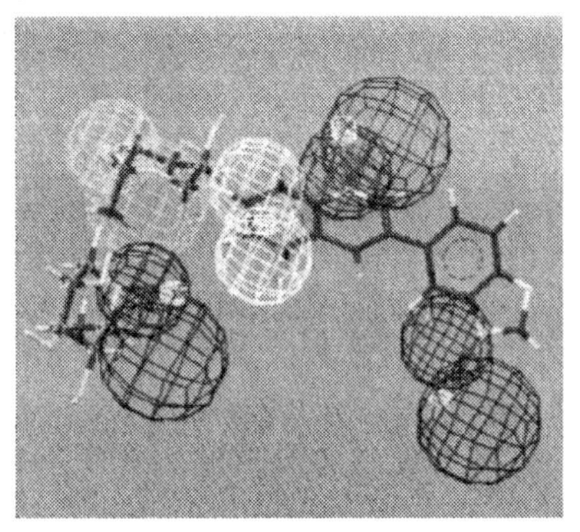

NCI (127487; isoflavone analog)

Figure 5. Example of pterine, flavone and isoflavone analogs from the NCI database mapped onto the pharmacophore hypothesis.

(Figure is printed in color in color insert.)

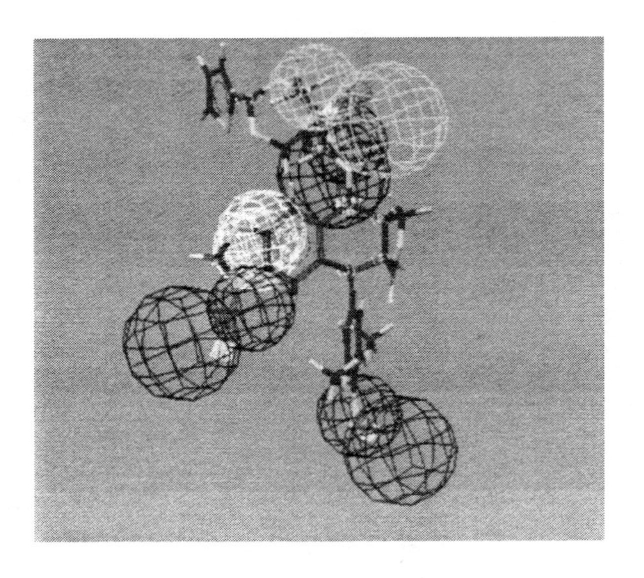

NCI (141540)
Etoposide (VP–16)

NCI (122819)
Teniposide (VM–26)

Figure 6. Mapping of the chemical features of etoposide and teniposide on the common feature hypothesis.

(Figure is printed in color in color insert.)

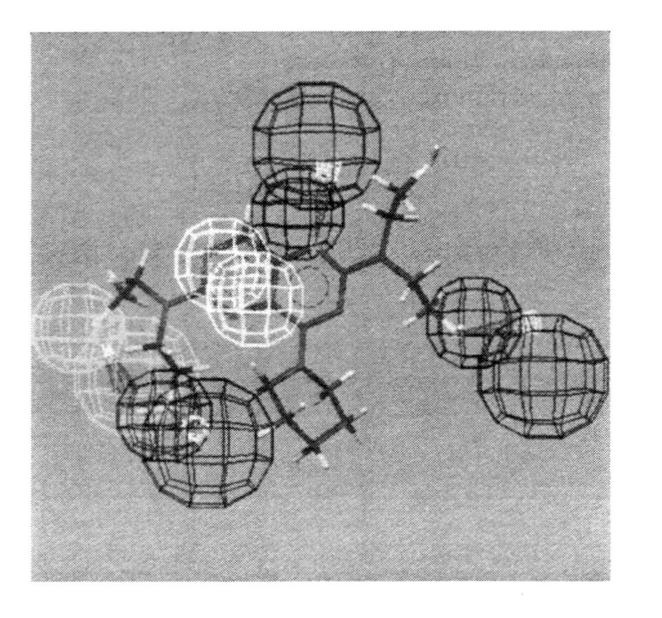

Mopidamole

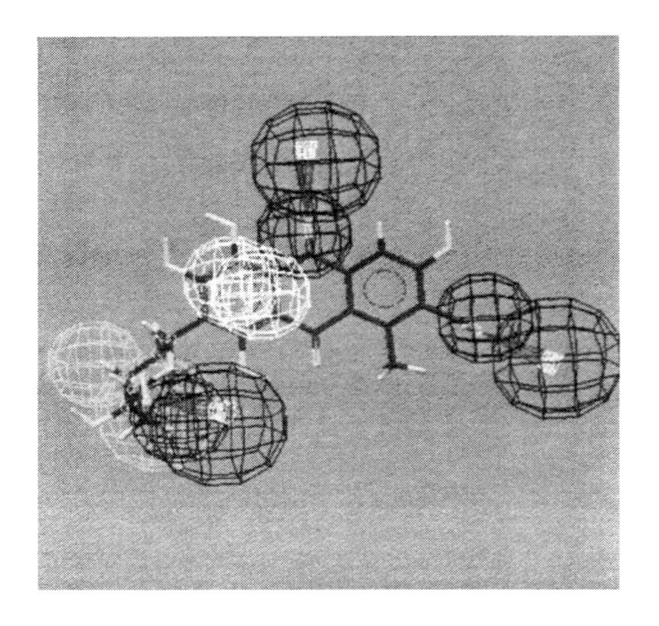

Carminate

Figure 7. Mopidamole and Carminate from the database of Derwent World Drug Index having the proposed pharmacophoric features.
(Figure is printed in color in color insert.)

group, and it has been known that such compounds posses activity in the micromolar range(12). The search on the NCI database included a number of pteridine (figure 5) analogs. Pteridine class of molecules are already known to inhibit nucleoside transport(13). Those analogs that have not been tested can be tested for improved activity. In addition, a number of flavones and isoflavones (figure 5) came out in the search from NCI database. To our knowledge this class of molecules have not yet been tested for activity against nucleoside transport. It would be interesting to test this class of molecules as these are natural products and known to be less toxic. Topoisomerase II inhibitors, etoposide and teniposide (figure 6) also appeared in the list from NCI. It is very encouraging to see that this class of structures are identified by the hypothesis. Because both etoposide and teniposide have been shown to be selective for the es-transporter (14). Some structures with selenium instead of sulfur or NH- linking the nitrobenzyl group also appeared as NCI database hits.

A search on the Maybridge database (53404 compounds) resulted in only seven hits. Mostly these are flavone and isoflavone family. The Derwent World Drug Index (49661 compounds) was also searched resulting in a number of interesting structures, including Daidzin, Genistin (isoflavones), Liquirtin (a flavanone), members of pteridine family of compounds and also anthraquinones. For example, figure 7 shows mopidamole and carminate. Mopidamole, a pyrimidopyrimidine, is already known to be an inhibitor of the nucleoside transporter and was not included in the study set used to generate the hypothesis(15). Also, the molecule carminate is of interest. Recently, a Chinese group has isolated two antibiotics from a fungus found in Antartican soil, and shown them to inhibit thymidine and uridine transport with IC_{50}'s at low micromolar range and also potentiate the anticancer activity of antimetabolites in cultured tumor cells(16). One of these antibiotics is a 3,9-dihydroxy-1-methoxy-7-methylanthraquinone. Carminate is a derivative of this molecule with a glycosylation. The hypothesis suggests that glycosylation of the anthraquinone analog would still meet the functional feature requirements and can be tested.

Finally the database of Available Chemicals Directory containing 230000 compounds was searched producing 276 hits. Of these, there were only 100 molecules with molecular weight less than 500. The results once again include a number of flavones in addition to other families including nucleoside analogs.

Summary and Conclusion

The structures of a set of nucleoside transport inhibitors have been analyzed and a pharmacophore hypothesis developed using the common feature hypothesis generation method (HipHop) in Catalyst. The hypothesis generated contains the following chemical features: one hydrogen bond donor, three hydrogen bond acceptors and an aromatic ring. The hypothesis was used to search four different databases of structures (NCI, MayBridge, Derwent World Drug Index and Available Chemicals Directory). The database search results produced a number of interesting hits suggesting new classes of molecules that can be tested for activity against nucleoside transport. The database hits also included molecules that are already known to be NT inhibitors, but were not included in the original study set. This increases the confidence level that the hypothesis captured the necessary chemical features that may be important for NT inhibition. Among the new classes of molecules suggested by the hypothesis are the natural products, flavones and isoflavones and some anthraquinone analogs. We are currently in the process of acquiring and testing these compounds for NT binding/inhibition. The results will also be used in the refinement of the hypothesis to increase the specificity/selectivity.

Literature Cited

1. (a) Buolamwini, J.K., *Current Medicinal Chemistry*, **1997**,4, 35.
   (b) Robins, M.J., Asakura, J., Kaneko, M., Shibuya, S., Jakobs, E.S., Agbanyo, F.R., Cass, C.E., and Paterson, A.R.P., *Nucleosides and Nucleotides*, **1994**, 13, 1627.
2. Paterson, A.R.P., Jacobs, E.S., Harley, E.R., Fu, N.-W., Robins, M.J., Cass, C.E., *In Regulatory Functions of Adenosine*; Berne, R.M., Rall, T.W., Rubio, R., Eds.; Martinus Nijhoff: The Hague, **1983**; pp. 203-220.
3. Griffiths. M., Beaumont, N., Yao, S.Y.M., Sundaram, M., Boumah, C.E., Davies, A., Kwong, F.Y.P., Coe, I., Cass, C.E., Young, J.D., and Baldwin, S.A., *Nature Medicine*, **1997**, 3, 89.
4. Catalyst software, version 3.1(**1997**), Molecular Simulations Inc., San Diego, CA.
5. Smellie, A., Teig, S. and Towbin, P., *J. Comp. Chem.*, **1995**, 16, 171.
6. Smellie, A., Kahn, S.D., and Teig, S.L., *J. Chem. Inf. Comput. Sci.*, **1995**, 35, 285.
7. Smellie, A., Kahn, S.D., and Teig, S.L., *J. Chem. Inf. Comput. Sci.*, **1995**, 35, 295.
8. Nicklaus, M.C., Wang S., Driscoll. J.S., and Milne, G.W.A., Bioorganic & medicinal chemistry, **1995**, 3, 411.
9. Greene, J.; Kahn, S.; Savoj, H.; Sprague, P.; Teig, S., *J. Chem. Inf. Comput. Sci.*, **1994**, *34*, 1297-1308.
10. Barnum, D., Greene, J., Smellie, A. and Sprague, P., *J. Chem. Inf. Comput. Sci.*, **1996**, 36, 563.
11. These are multiconformer databases associated with the Catalyst software derived from the corresponding original databases for use within Catalyst.
12. Paul, B., Chen, M.F., Paterson, A.R.P., *J. Med. Chem.*, **1975**, 18, 968.
13. Chen, H.-X., Bamberger, U., Heckel, A., Guo, X., Cheng, Y.-C., Cancer Res., **1993**, 53, 1974.
14. Hammond, J.R., *Nucleosides and Nucleotides*, **1991**, 10, 103.
15. Ramu, A., *In Resistance to Antineoplastic Drugs.*; Kessel, D., Ed., CRC Press: Boca Raton, **1986**; pp. 63-80.
16. Su., J. Zheng, Y.S., Qi, C.Q., Hu, J.L. *Cancer Chemother. Pharmacol.*, **1995**, 36, 149.

Chapter 11

Unified Pharmacophoric Model for Cannabinoids and Aminoalkylindoles Derived from Molecular Superimposition of CB_1 Cannabinoid Receptor Agonists CP55244 and WIN55212-2

Joong-Youn Shim,[1] Elizabeth R. Collantes,[1,3] and William J. Welsh,[1] and Allyn C. Howlett[2]

[1] Department of Chemistry and Center for Molecular Electronics, University of Missouri-St. Louis, St. Louis, MO 63121
[2] Department of Pharmacological and Physiological Science, Saint Louis University School of Medicine, St. Louis, MO 63104

Novel superimposition models were developed based on 3-D pharmacophore mapping of the two highly potent CB_1 cannabinoid receptor agonists: the cannabinoid CP55244 and the aminoalkylindole (AAI) WIN55212-2. The superimposition models so derived confirm earlier speculation about certain key pharmacophoric elements common to both the cannabinoids and AAIs. The present models also provide insight into the curious observation that the C1 hydroxyl group of certain cannabinoids may be unnecessary for the cannabinoid activity. To test the validity of our superimposition models, Comparative Molecular Field Analysis (CoMFA) was employed to construct a 3D-QSAR using a mixed training set composed of twelve cannabinoids and twelve AAIs.

Despite their obvious structural dissimilarities, the cannabinoids and aminoalkylindoles (AAIs) have been shown by numerous studies (*1-5*) to exhibit similar *in vitro* and *in vivo* cannabimimetic activities. This body of evidence suggests that the cannabinoids and AAIs interact with the same cannabinoid receptor and share at least some regions in common when bound to the receptor to elicit the cannabinoid activity. Thus, it is highly possible that they compete for the same binding regions of the CB_1 cannabinoid receptor. In support of this hypothesis, binding studies (*1,6*) have revealed that the potent cannabinoid [^{3}H]CP55940 was displaced by WIN55212-2 and other AAIs and that, conversely, [^{3}H]WIN55212-2 was displaced by Δ^9-tetrahydrocannabinol (Δ^9-THC) and other cannabinoids (Figure 1). Furthermore, irreversible covalent binding of an isothiocyanato AAI to the CB_1 cannabinoid receptor obstructed any subsequent interactions with the cannabinoid agonist [^{3}H]CP55940 (*7*).

[3] Current address: Monsanto Life Science Company, St. Louis, MO 63167

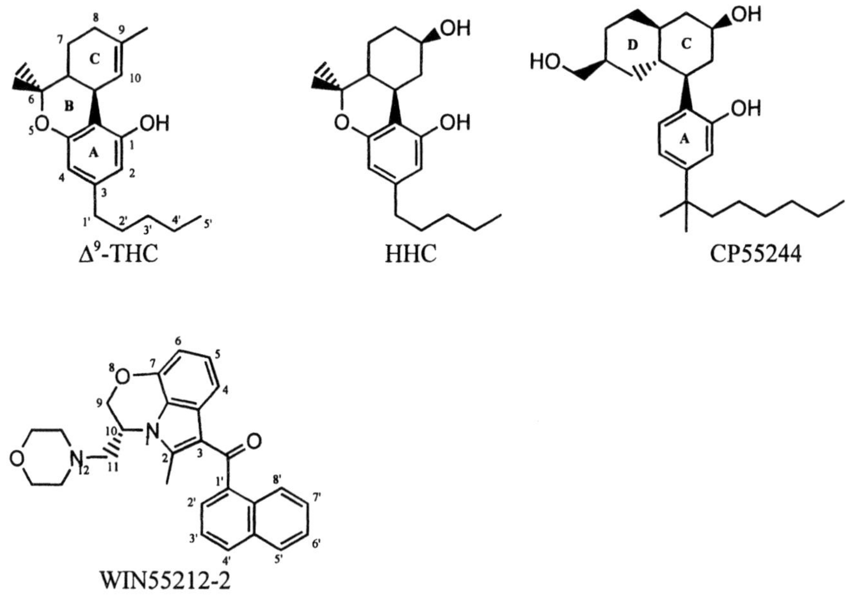

Δ^9-THC

HHC

CP55244

WIN55212-2

Figure 1. The cannabinoids and AAIs used for developing superimposition models.

Structure-activity relationship (SAR) studies of the cannabinoids (*8-11*) and the AAIs (*12-14*) revealed pharmacophoric elements common to both classes of compounds (see Figure 1): (1) a lipophilic and/or sterically bulky group (i.e., the C3 side chain in the cannabinoids and the C3 aroyl group in the AAIs) that appears to be a structural prerequisite for cannabinoid activity (*10*); (2) a polar oxygen atom (i.e., the C1 hydroxyl group of the phenolic A-ring in the cannabinoids and the C3 carbonyl oxygen in the AAIs) that may form a hydrogen bond with the receptor (*15*) and (3) the cyclic ring system (i.e., the cyclohexyl C-ring in the cannabinoids and the indole ring in the AAIs). Another polar group, specifically the C9 hydroxyl group of the cyclohexyl C-ring of 9-nor-9β-hydroxyhexahydrocannabinol (HHC), a 9-hydroxylated analog of Δ^9-THC, in the cannabinoids and the nitrogen atom of the aminoalkyl side chain in the AAIs, is also considered an important pharmacophoric element.

In order to understand the similarity in cannabimimetic activity of these structurally quite distinct cannabinoids and AAIs and to identify common pharmacophoric features, some efforts have been made to superimpose cannabinoids and AAIs (*16-18*). Two distinct superimposition models, hereafter known as the Huffman model and the Makriyannis model, have evolved from these superimposition studies. The Huffman model (*17*), developed by superimposing a structurally modified analog of WIN55212-2 with Δ^9-THC (Figure 1), assumed a common functionality between the C3 side chain of the cannabinoids and the N1 aminoalkyl side chain of the AAIs. Huffman et al. noted a similarity between the N1 aminoalkyl side chain in the AAIs and the C3 side chain in the cannabinoids in terms of the sensitivity of activity with side-chain length. Their superimposition model ignored the benzene ring in the indole moiety of WIN55212-2 with respect to overlaying it with any part of Δ^9-THC, thus implying that this benzene ring is unimportant. This decision is questionable, as evidenced by the sharp decrease both in vivo and in vitro activities of a series of pyrrole analogues (*19*) versus the corresponding AAI analogues. The Huffman model was derived without precise structural information, and suffers from a lack of consideration of the polar nature of the heterocyclic N1 aminoalkyl side chain of the AAIs (*3*).

The Makriyannis model (*18*) was derived by superimposition of HHC with WIN55212-2 whose structure was ascertained from interpretation of 2D-NMR spectra and MD simulations. In the Makriyannis model, the C3 aroyl group of the AAI is superimposed on the C3 side chain of the cannabinoid, and the N atom of the N1 side chain of the AAI was positioned to nearly coincide with the hydroxyl group of the cyclohexyl C-ring of the cannabinoid. Nevertheless, the rationale for choosing specific atoms to fit for superimposition remains unclear.

Although largely incompatible, the Huffman and Makriyannis models both align the C1 hydroxyl group of the cannabinoids with the C3 aroyl oxygen of the AAIs. This point of agreement is consistent with the notion of a common hydrogen-bonding interaction with the corresponding region of the receptor's binding site (*15*). Eissenstat et al. (*12*) recently proposed a model in which the C1 hydroxyl group of the cannabinoids overlays the N1 side chain of the AAIs. They based their choice on the observation that the C9 hydroxyl group of the cannabinoids functioned differently from the morpholino N in the pravadoline series (*14*). As yet, however, no unified superimposition model for both cannabinoids and AAIs has been generally accepted. Consequently, it was felt that the cannabinoid/AAI superimposition models warranted further investigation and validation.

In the present work, novel superimposition models were developed based on 3-D pharmacophore mapping of two highly potent CB_1 cannabinoid receptor agonists CP55244 and WIN55212-2 using starting conformations that corresponded to those ascertained by Tong et al. (*20*) and Shim et al. (*21*), respectively. CP55244 is uniquely different from Δ^9-THC and HHC in that it possesses an additional pharmacophoric element which is not found in the classical ABC tricyclic cannabinoids (*11*). The D-ring methanol extension (comparable to the hydroxypropyl in CP55940) forms a potential hydrogen bonding site which may confer the extremely high potency exhibited by CP55244. The superimposition models so derived confirm earlier speculation about certain key pharmacophoric elements common to both the cannabinoids and AAIs. The present models also provide insight into the curious observation that the C1 hydroxyl group of certain cannabinoids may be unnecessary for the cannabinoid activity (*22*).

Computational Methods

The highly potent CP55244 (K_i = 0.11 nM) and WIN55212-2 (K_i = 1.1 nM) were selected to represent the cannabinoids and AAIs, respectively (*11, 21*). CP55940 (**32** in Figure 2), which is structurally similar to CP55244 but without the D-ring, and WIN55212-2 have been used as the radioligands to measure binding potency of other cannabinoids and AAIs [for WIN55212-2, see references *6* and *12*; for CP-55940, see references *10, 11,* and *21*]. The conformations of the cannabinoids and AAIs were taken from our previously derived CoMFA models (*20, 21*). For WIN55212-2, additional conformations were explored by conducting a systematic search of the torsion angles ω1(C2=C3-C=O) and ω2(O=C-C1'-C2').

The DISCO module [DIStance COmparison (DISCO) technique (*23*)], accessed through the molecular modeling program Sybyl (version 6.2) (*24*), was employed to extract the common pharmacophoric elements from the cannabinoids and AAIs. DISCO first identifies certain predefined pharmacophoric features, i.e., hydrophobic center, donor site, acceptor site, donor atom, acceptor atom, for each compound and then generates superimposition models by matching common features. Based on these superimposition models, the corresponding pharmacophoric elements were identified. Superimposition models of CP55244 and WIN55212-2 were compared and evaluated using the following criteria: (1) root-mean-square (RMS) fit of corresponding pharmacophoric elements, (2) proper orientation and overlap of the C3 dimethylheptyl side chain of CP55244 with the C3 aroyl moiety of WIN55212-2 (which was deemed critical for tight binding), (3) the number of pharmacophoric elements, and (4) the degree of overlap of molecular volumes. Probably because of the difficulty in assigning a hydrophobic center for the C3 dimethylheptyl side chain of CP55244, DISCO failed to designate this moiety as "hydrophobic". Instead, superimposition models were chosen that exhibited proper orientation and overlap of these hydrophobic moieties. WIN55212-2 was used as the reference compound for fitting as it contains a greater number of pharmacophoric features than CP55244. The superimposition models selected by DISCO were further refined by fitting WIN55212-2 to CP55244 using the "field fit" option in Sybyl. After field fitting for 500 iterations using the simplex algorithm in which the partial atomic charges for the electrostatic interactions were calculated using the Gasteiger-Marsili formalism (*25*) were included, WIN55212-2 was then subjected to full geometry optimization. To compare with the present superimposition models, the Huffman and Makriyannis models were reconstructed by superimposing CP55244 and WIN55212-2 in the conformations considered in the present report using the same alignment atoms as described in the respective original papers.

To test the validity of our superimposition models, Comparative Molecular Field Analysis (CoMFA) (*27*) was employed to construct a 3D-QSAR using a mixed training set composed of twelve cannabinoids and twelve AAIs (Table I and Figure 2). In view of the structural dissimilarity of CP55244 and WIN55212-2, different alignment rules were chosen for the cannabinoids and AAIs, specifically: the phenolic oxygen, the C9 hydroxyl oxygen, and the C1' of the C3 side chain of CP55244 for the cannabinoids; and the C5, C_α of the N1 side chain, and carbonyl oxygen at C3 of WIN55212-2 for the AAIs. The CoMFA models were then validated by predicting the pK_i for a test set composed of eighteen cannabinoids and two AAIs.

Results and Discussion

Superimposition of CP55244 and WIN55212-2. DISCO (*23*) was employed to help identify the corresponding pharmacophoric elements in the cannabinoids (represented by CP55244) and the AAIs (represented by WIN55212-2). DISCO found two separate AAI conformers designated Z and C that differ with respect to the torsion angle ω1(C2=C3-C=O). The value of ω1 is –152.8° in the Z form and 29.2° in the C form (Figure 3). Superimposition models based on each form are illustrated in Figure 4 and summarized in Table II. With WIN55212-2 in the Z form, DISCO identified five pharmacophoric features: (i) two around the C1 phenolic oxygen of CP55244 and the C3 carbonyl oxygen of WIN55212-2 (oxygen as the acceptor atom and a donor site), (ii) one hydrophobic center for the C-ring of CP55244 and the benzene ring of the indole of WIN55212-2, and (iii) two around the D-ring hydroxyl group of CP55244 and the morpholino oxygen of WIN55212-2 (oxygen as an acceptor atom and a donor site). With WIN55212-2 in the alternative C form, DISCO identified three pharmacophoric features: (i) two around the C1 phenolic oxygen of CP55244 and the C3 carbonyl oxygen of WIN55212-2 (oxygen as an acceptor atom and two donor sites), and (ii) one around the C9 hydroxyl oxygen of CP55244 and the morpholino nitrogen of WIN55212-2 as a donor atom. Both models displayed a high degree of overlap between the C3 side chain of CP55244 and the C3 aroyl moiety of WIN55212-2, consistent with the notion that a hydrophobic moiety is important for cannabimimetic activity (*10, 12-14, 26*). In addition, both models insinuate that addition of hydrophobic substituents to the second ring of the naphthyl group (i.e., 6' or 7' position) in WIN55212-2 enhance binding potency.

Analysis of the present two superimposition models (hereafter called the Z and C models) provides some interesting comparisons. The molecular volume overlap is only slightly larger for the Z model (156 $Å^3$) than for the C model (142 $Å^3$). In the Z model, the morpholino oxygen of WIN55212-2 is aligned with the D-

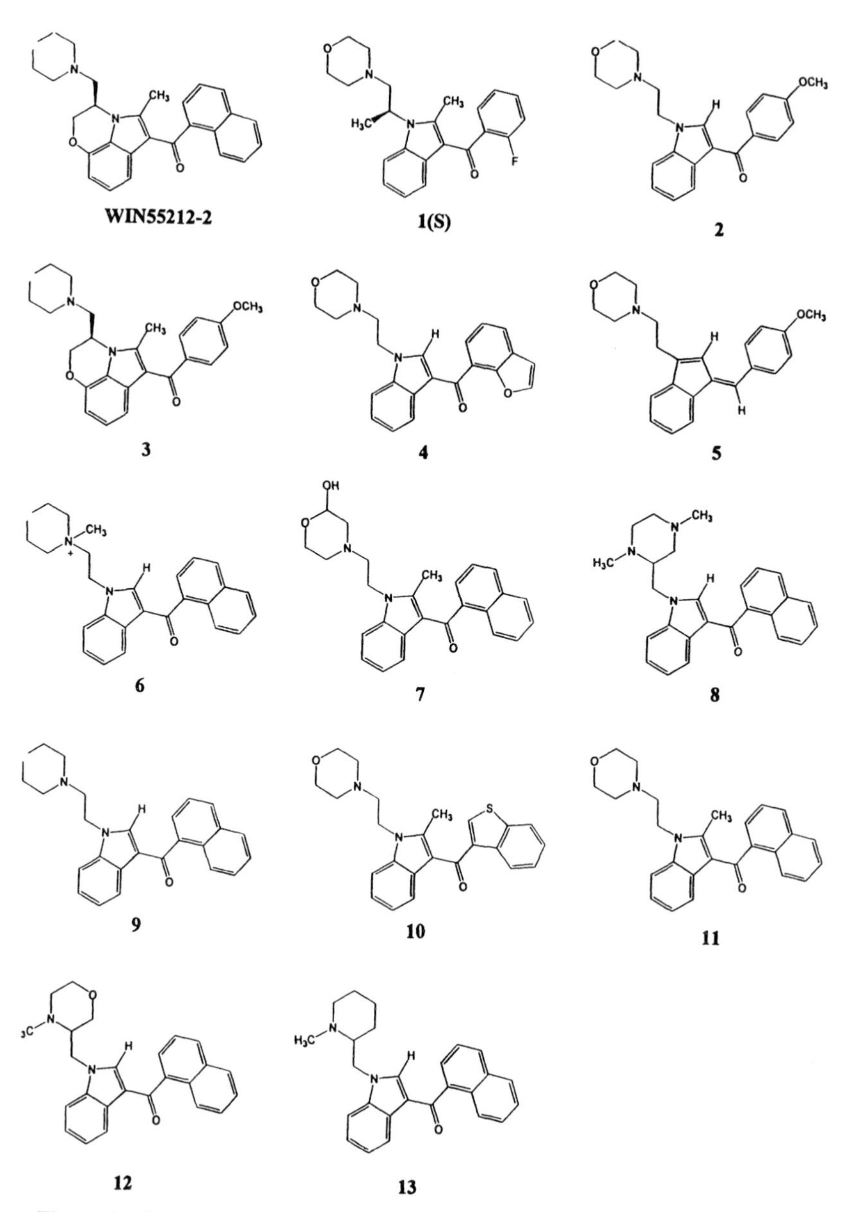

Figure 2. Structures of cannabinoids and AAIs used as the training set to construct CoMFAs. Also shown are compounds **8**, **12**, and **16**, **17**, **20**, **21**, **24**, **25**, **27-35**, **37**, **39**, and **40** which belonged to the test set.

R	Compound
tert-butyl	**14**
1,1-dimethylpropyl	**15**
1,1-dimethylbutyl	**16**
1,1-dimethylpentyl	**17**
1,1-dimethylhexyl	**18**
1,1-dimethylheptyl	**19**
1,1-dimethyloctyl	**20**
1,1-dimethylnonyl	**21**
1,1-dimethyldecyl	**22**
1,1-dimethylundecyl	**23**

R_1	R_2	Compound
methyl	H	**24**
ethyl	H	**25**
n-propyl	H	**26**
n-propyl	methyl	**27**
n-butyl	H	**28**
allyl	H	**29**
hydroxymethyl	H	**30**
hydroxyethyl	H	**31**
hydroxypropyl	H	**32** (CP55940)
hydroxybutyl	H	**33**
methoxypropyl	H	**34**
aminomethyl	H	**35**

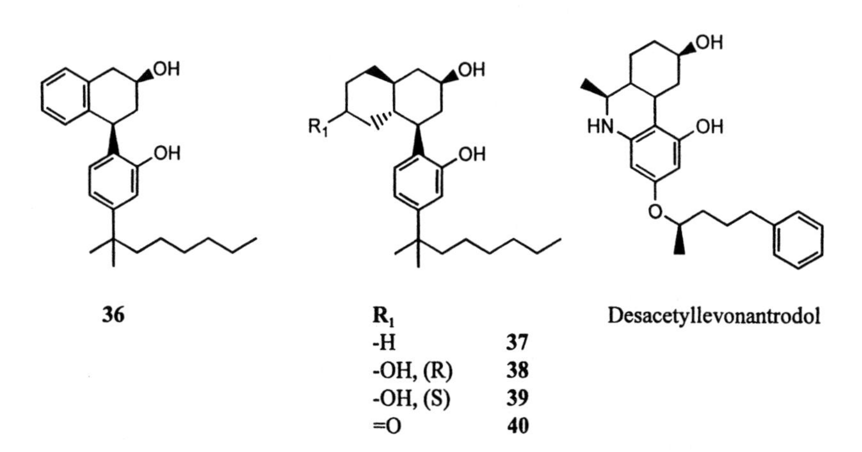

R_1	
-H	**37**
-OH, (R)	**38**
-OH, (S)	**39**
=O	**40**

Figure 2. *Continued.*

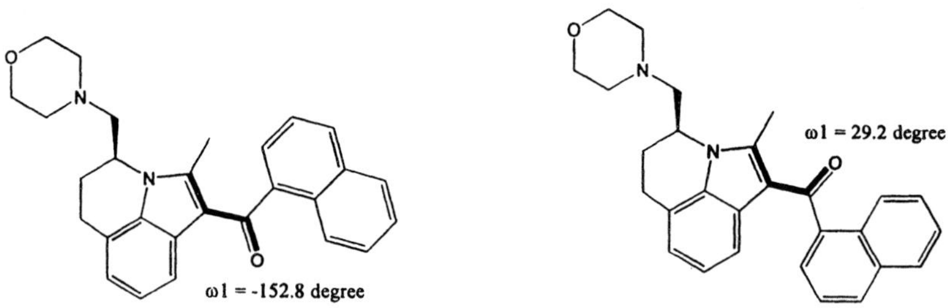

Figure 3. Illustration of the AAI WIN55212-2 in both the Z form (left) and C form (right).

Table I. Observed versus calculated pK_i values for the training set compounds.

Compounds		Z model		C model	
	pK_i, obs'd	pK_i, calc'd	residual	pK_i, calc'd	residual
AAIs					
WIN55212-2	-0.04	0.01	-0.05	-0.03	-0.01
1(S)	-2.95	-3.00	0.05	-3.11	0.16
2	-2.55	-2.51	-0.04	-2.43	-0.12
3	-1.74	-1.71	-0.03	-1.79	0.05
4	-1.74	-1.80	0.06	-1.73	-0.01
5	-1.32	-1.26	-0.06	-1.22	-0.10
6	-3.41	-3.47	0.06	-3.46	0.05
7	-2.65	-2.53	-0.12	-2.53	-0.12
9	-1.68	-1.69	0.01	-1.76	0.08
10	-1.48	-1.45	-0.03	-1.51	0.03
11	-1.45	-1.73	0.28	-1.55	0.10
13	0.12	0.25	-0.13	0.20	-0.08
Cannabinoids					
CP55244	0.96	0.90	0.06	0.95	0.01
Desacetyllevonantrodol	0.21	0.16	0.05	0.23	-0.02
Δ^9-THC	-1.01	-1.00	-0.01	-1.05	0.04
14	-2.48	-2.62	0.14	-2.52	0.04
15	-2.61	-2.37	-0.24	-2.39	-0.22
18	-1.28	-1.08	-0.20	-0.99	-0.29
19	-0.34	-0.63	0.29	-0.65	0.31
22	-1.67	-1.74	0.07	-1.79	0.12
23	-2.23	-2.11	-0.12	-2.18	-0.05
26	-0.11	-0.11	0.00	-0.17	0.06
36	0.00	0.06	-0.06	0.03	-0.03
38	-0.63	-0.66	0.03	-0.64	0.01

Table II. Comparison of the Z and C superimposition models.

	Z model	C model
WIN55212-2		
ω1(C2=C3-C=O)	-152.8 °	29.2 °
ω2(O=C-C1'-C2')	-133.5 °	-134.9 °
Sybyl Energy (kcal/mol)	41.223	40.621
DISCO results		
Pharmacophoric features in terms of CP55244	• C1 hydroxyl oxygen i. 1 acceptor atom ii. 1 donor site • C ring centroid iii. 1 hydrophobic center • D ring hydroxyl oxygen iv. 1 acceptor atom v. 1 donor site	• C1 hydroxyl oxygen i. 1 acceptor atom ii. 2 donor sites • C9 hydroxyl oxygen iii. 1 donor atom
RMS fit	0.35Å	0.35Å
Overlap volume (CP55244 / WIN55212-2)	156 $Å^3$ / 330 $Å^3$ (48 %)	142 $Å^3$ / 330 $Å^3$ (43 %)
CoMFA statistics		
r^2_{cv}	0.464 (6 pc)	0.404 (6 pc)
r^2	0.988	0.989
SE	0.146	0.142
F	239	250

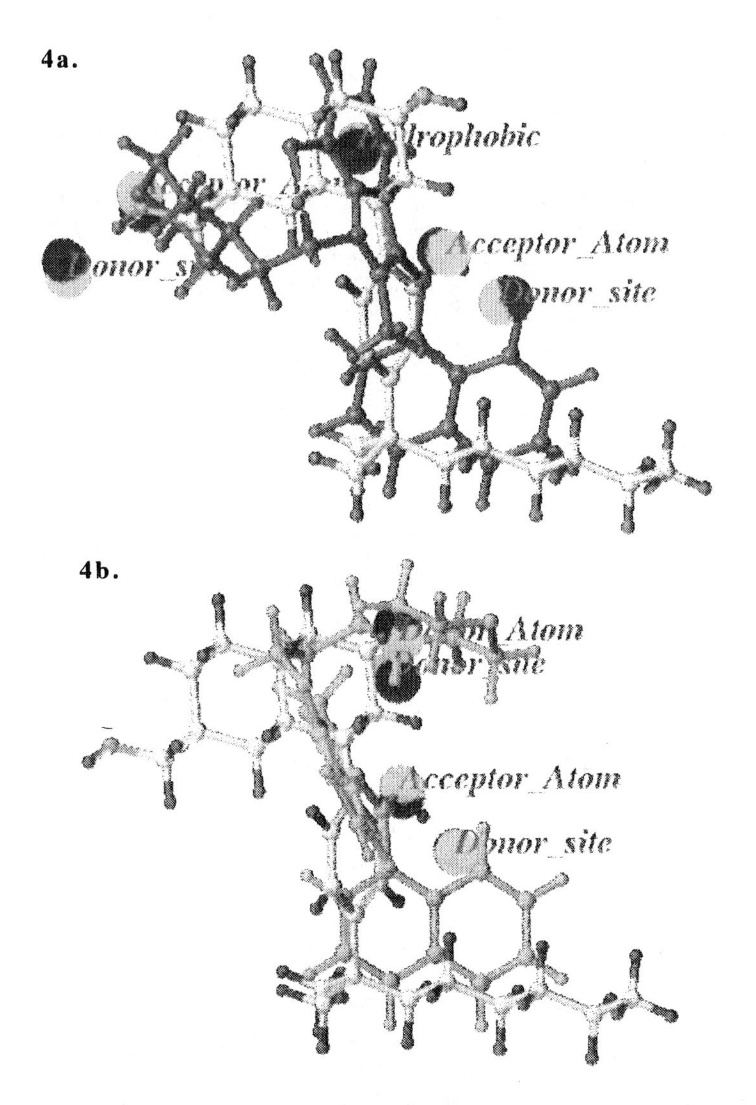

Figure 4. DISCO-derived Z and C superimposition models for CP55244 and WIN55212-2. The pharmacophoric features of the reference molecule (WIN55212-2) are represented as red spheres and those of CP55244 are represented as blue spheres. CP55244 is colored by atom type while WIN55212-2 is colored in purple (4a: Z model) and in orange (4b: C model). (Figure is printed in color in color insert.)

ring hydroxyl group of CP55244. In the C model, the morpholino nitrogen of WIN55212-2 is aligned with the C-ring hydroxyl group of CP55244. The C model bears a likeness to the Makriyannis model (*18*); however, the corresponding Z model shows no similarity to the Huffman model (*17*). Comparison of the lipophilic potentials of CP55244 and WIN55212-2 (Figure 5) reveals that the C3 side chain of CP55244 and the C3 aroyl moiety of WIN55212-2 are the most lipophilic parts of the respective molecules. Hence, it was considered reasonable to superimpose these two lipophilic moieties.

Restricting the motion of the hydroxypropyl moiety attached to the C-ring in the AC-bicyclic cannabinoids (e.g., CP55940 (**32**)) or of the N1 side chain in the AAIs (e.g., WIN55212-2) has been shown to increase cannabinoid activity. Representative of the unconstrained forms, the K_i values for CP55940 (**32**) and **11** are 1.14 nM (*10,11*) and 28 nM (*21*), respectively. In contrast, the K_i values for the corresponding constrained analogues CP55244 and WIN55212-2 are 0.11 nM and 1.1 nM (*11, 21*), respectively. Specifying the Z model, these observations suggest that constraining corresponding (i.e., overlapping) moieties of the cannabinoids and AAIs will promote favorable interaction with the CB_1 cannabinoid receptor and enhance binding potency (*11*). Based on the extensive SAR studies of AC-bicyclic and ACD-tricyclic nonclassical cannabinoids (i.e., CP55940, CP55244 and their derivatives), some workers (*10, 28*) have proposed that substitution of a unique hydrophilic moiety on the lipophilic D-ring is important for enhanced cannabinoid activity. In the Z model, this hydrophilic moiety in the cannabinoids is aligned with the hydrophilic morpholino oxygen of the AAIs which, in turn, is separated from the indole ring by a largely lipophilic string of atoms (i.e., $-CH_2-CH_2-N-CH_2-CH_2-$).

The C model, which bears a resemblance to the Makriyannis model (*18*), seems to more properly address the importance of the C-ring C9 hydroxyl group in the cannabinoids and its similarity to the morpholino nitrogen in the AAIs. However, it has been suggested that the C9 hydroxyl group of the cannabinoids may be not essential for potent binding (*10, 22*) and, furthermore, that the cannabinoid C9 hydroxyl and the AAI morpholino nitrogen interact with different receptor binding sites (*12*). Noting that the sensitivity of K_i to the length of the lipophilic alkyl N1 substituent for a series of AAIs was similar to that observed for the cannabinoids with respect to variation in length of the lipophilic C3 alkyl side chain, Huffman et al. (*17*) decided to overlay these two lipophilic groups in their superimposition model. This choice would imply that the hydrophilic side chain of the AAIs (e.g., the O atom in the morpholino ring of WIN55212-2) is not critical. However, the Z model could explain why AAIs with an N1 side chain of four to seven carbons exhibit high potency. Earlier workers have proposed that a specific hydrophobic region of the receptor borders the B and D rings of cannabinoids (*10, 11, 28*). Consistent with this notion, the Z model superimposes the N1 side chain of the AAIs on the hydrophobic substituents attached to the B and D rings of the cannabinoids. By virtue of its ability to resolve this apparent inconsistency, the Z model may be superior to alternative superimposition models in terms of accommodating the structurally dissimilar cannabinoids and AAIs inside the same critical binding site of the CB_1 cannabinoid receptor.

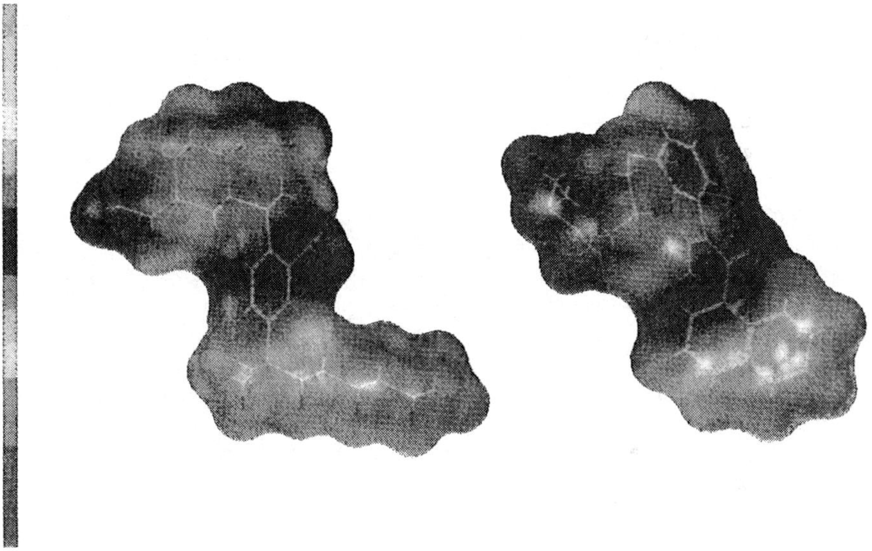

Figure 5. Lipophilicity potentials for CP55244 and WIN55212-2 (Z form) on the Connolly molecular surfaces. The most lipophilic part is the C3 side chain in CP55244 and the C3 aroyl moiety in WIN55212-2.
(Figure is printed in color in color insert.)

Construction of 3D-QSAR models. CoMFA was employed to build 3D-QSAR models (Figure 6) using a mixed training set of 12 cannabinoids and 12 AAIs. Separate CoMFA models were constructed for both the C and Z models. Our intention was to compare the statistical quality of the 3D-QSARs for the Z and C models for purposes of selecting one over the other. Unfortunately, equally strong linear correlations were obtained for the C model (r^2 = 0.989) and the Z model (r^2 = 0.988). On the other hand, the cross-validated r^2 (r_{CV}^2) was somewhat higher for the Z model (0.464) than for the corresponding C model (0.404). Both CoMFA models showed slightly greater contributions from the steric fields than the electrostatic fields (55 % vs. 45 % for the C model, and 53 % vs. 47 % for the Z model). Except for cannabinoids **27** and **35**, the CoMFA-predicted pK_i values were within one log unit of the corresponding observed values for a test set composed of 16 cannabinoids and 2 AAIs (Table III).

An attractive feature of CoMFA is its ability to provide visual representations of the principal steric and electrostatic fields that contribute to a 3D-QSAR model in terms of color-coded coefficient contour maps. The steric contour map (Figure 6) indicates that introducing a lipophilic side chain at C3 of the cannabinoids is consistent with high activity, in agreement with the known SAR. Substitution with a lipophilic group on the B or D ring in the cannabinoids, corresponding to the lipophilic N1 side chain in the AAIs, is also consistent with enhanced activity. A sterically unfavorable region appears between the indole ring and morpholino side chain of the AAIs, suggesting that the corresponding region of the receptor binding site is conformationally restricted for the AAIs. This same region of the receptor is believed to interact with the A ring of the cannabinoids which, however, is less conformationally restricted. This distinction between the cannabinoids and the AAIs with regard to their interaction with this region of the receptor may in part explain any observed differences in their activity profiles.

The CoMFA electrostatic contour map (Figure 6) reveals a significant contribution to activity from the electrostatic fields in the vicinity of the heterocyclic ring of the N1 side chain in the AAIs. On the contrary, the absence of electrostatic fields around the corresponding C-ring substituents in the cannabinoids indicates little or no correlation with ligand binding affinity.

Proposed Cannabinoid Receptor Map. Based on our superimposition models and the known SAR for the cannabinoid and AAIs, we have constructed a pharmacophoric map for the cannabinoid CB_1 receptor appropriate to both the cannabinoids and AAIs (illustrated in Figure 7 for the Z model). Similar to the one proposed by Howlett et al. (*28*) from the SAR for bi- and tricyclic nonclassical cannabinoids, this receptor map depicts the pharmacophoric elements required for cannabimimetic activity including those common to both the cannabinoid and AAI compounds. The map also shows those pharmacophoric elements that are specific for each compound, such as a lipophilic receptor site near the benzene ring of the AAIs and a hydrophilic receptor site next to the C9 hydroxyl of the cannabinoids.

Inspection of our proposed CB_1 cannabinoid receptor map reveals that WIN55212-2 could be accommodated inside the binding site in either the C and Z models. Both conformations of WIN55212-2 seem capable satisfying those interactions with the receptor deemed necessary for tight binding. In fact, the

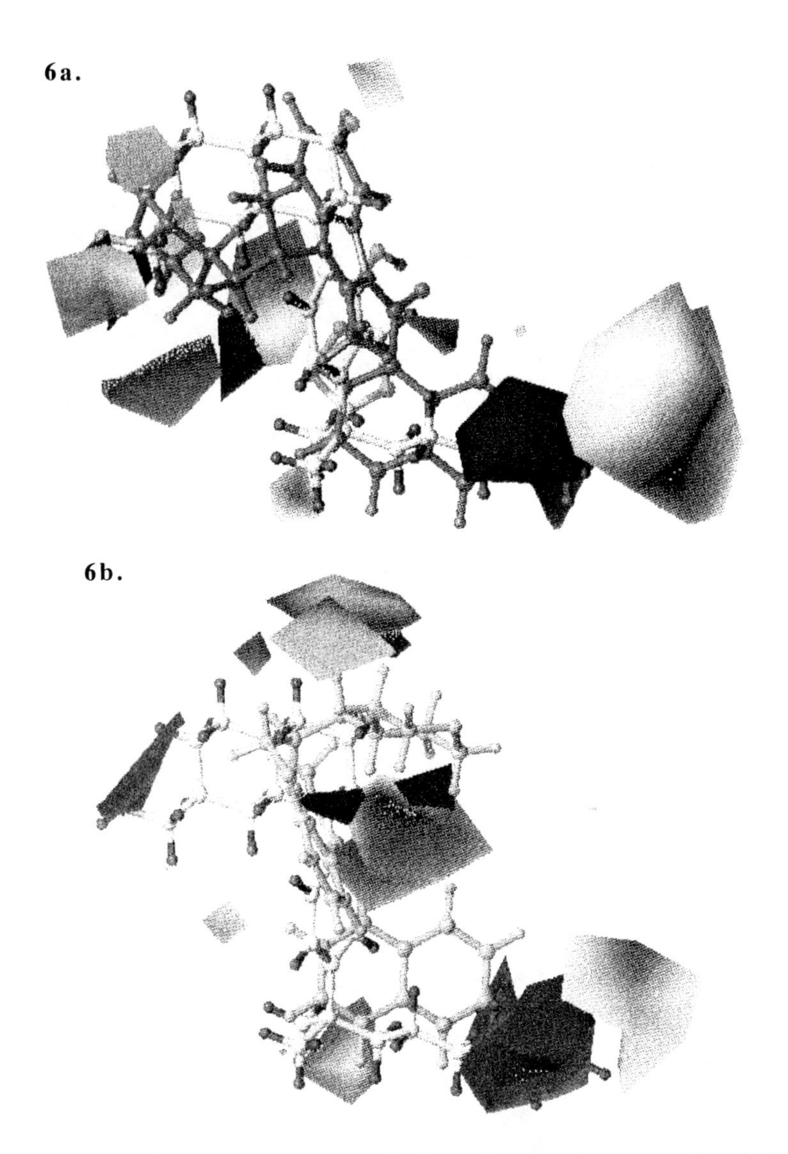

Figure 6. CoMFA steric maps (sterically favored regions in green and sterically disfavored regions in yellow) and electrostatic maps (positive charge favored in blue and negative charge favored in red) of the Z and C superimposition models. CP55244 and WIN55212-2 are colored in orange and by atom type. CP55244 is colored by atom type while WIN55212-2 is colored in purple (6a: Z model) and in orange (6b: C model).

(Figure is printed in color in color insert.)

Table III. Observed versus CoMFA-predicted pK_i values for the test set compounds.

Compounds	pK_i, observed	pK_i, predicted	
		Z model	C model
AAIs			
8	-1.96	-2.30	-1.83
12	-1.03	-0.23	-0.71
Cannabinoids			
16	-2.51	-1.83	-1.86
17	-1.88	-1.38	-1.51
20	0.08	-0.84	-1.08
21	-0.88	-1.26	-1.35
24	-0.32	-0.20	-0.37
25	-0.31	-0.19	-0.34
27	-1.95	-0.51	-0.18
28	-0.41	0.12	0.18
29	-1.15	-0.34	-0.17
30	-1.16	-0.37	-0.30
31	-0.75	-0.40	-0.27
32 (CP55940)	0.86	-0.05	-0.17
33	-0.09	-0.17	-0.20
34	-0.52	0.01	-0.21
35	-2.08	-0.67	-0.27
37	-1.01	-0.33	-0.34
39	-0.50	-0.14	-0.33
40	-0.49	-0.23	-0.48

region of the receptor binding site in contact with the C and D rings of the cannabinoids and with the indole and N1 side chain of the AAIs appears to possess the proper distribution of hydrophilic and lipophilic sites with respect to the C6-C7 axis of the cannabinoids and the C3-C(carbonyl) of the AAIs (i.e., complementary hydrophilic-lipophilic sites on top left and top right of Figure 7). So either conformation of WIN55212-2 could function accordingly.

Why do cannabinoids without the phenolic hydroxyl still retain potency? In a recent study, Huffman et al. (*22*) examined certain cannabinoids that exhibit strong binding affinity for the CB_1 cannabinoid receptor even though the C1 hydroxyl group is missing. This finding conflicts with the extensive body of SAR data on the cannabinoids (*15, 26, 29*). Nevertheless, Huffman and his coworkers explained the exceptionally high potency of these cannabinoids by proposing that the CB_1 cannabinoid receptor forms a hydrogen bond with the C9 hydroxyl group in place of the C1 hydroxyl group. Melvin et al. (*10*) also observed the retention of in vivo biological activity for a series of CP-55940 analogues in which the phenolic C1 hydroxyl group is missing. These latter workers hypothesized that the hydroxypropyl group on the C-ring side chain of CP55940 (**32**) compensates for the reduced binding affinity that would be expected from the loss of the phenolic hydroxyl group.

Based on recent molecular modeling studies of the same cannabinoids examined by Huffman et. al. (*22*), we now propose that the pyran O atom (e.g., O5 in Δ^9-THC, Figure 1) or the O atom in the C-ring side chain of the AC-bicyclic cannabinoids (e.g., CP55940) could mimic the C1 hydroxyl group by occupying the same receptor binding site. In fact, rotation of the cannabinoid molecule about the C3 side chain would align these oxygen atoms with the position believed to be occupied by the phenolic C1 hydroxyl inside the receptor binding site. This argument would imply that the lipophilic C3 side chain and the C1 hydroxyl group (or its counterpart), but not the C9 substituent, are indeed essential for activity. As suggested by our Z and C models, no significant difference in the interaction with the CB_1 cannabinoid receptor was found by rotating about the torsion angle $\omega1$(C2=C3-C=O) of WIN55212-2. This operation is tantamount to rotation about the C3 side chain in the cannabinoids. A similar concept was proposed by Thomas et al. (*30*) in a recent study of the pharmacophore of anandamides in the relation to the cannabinoids, in which the pyran oxygen of the B ring was selected as a pharmacophoric element. An interesting test of our hypothesis would be to measure the binding potency of cannabinoids that lacked both the C1 phenolic hydroxyl group and any oxygen atom capable of mimicking its functionality. We would predict poor binding affinity for such compounds.

Acknowledgments

This work was supported in part by National Institute on Drug Abuse (NIDA) grants R01-DA06312 and K05-DA00185 to ACH.

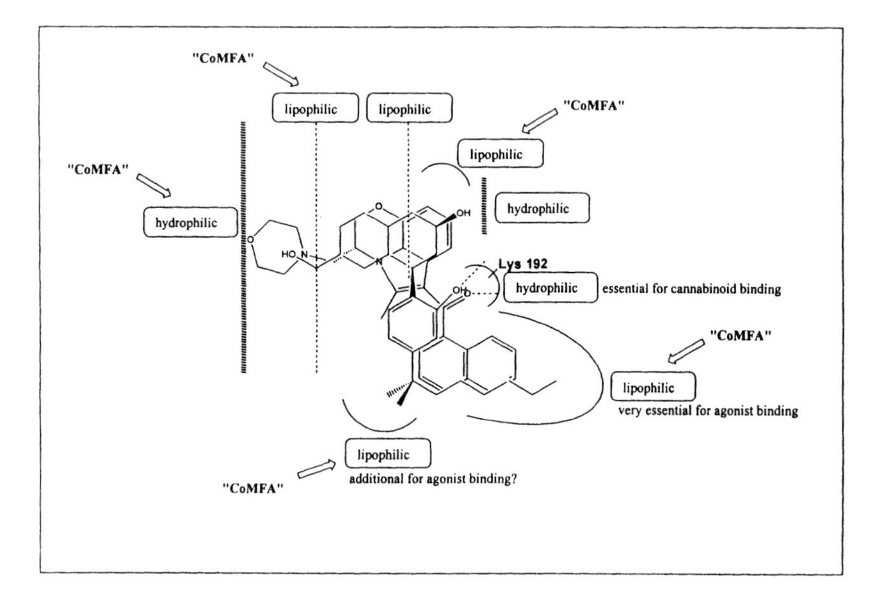

Figure 7. Putative pharmacophoric model of the CB_1 cannabinoid receptor showing possible interactions with both cannabinoid and AAI agonists (Z conformation).

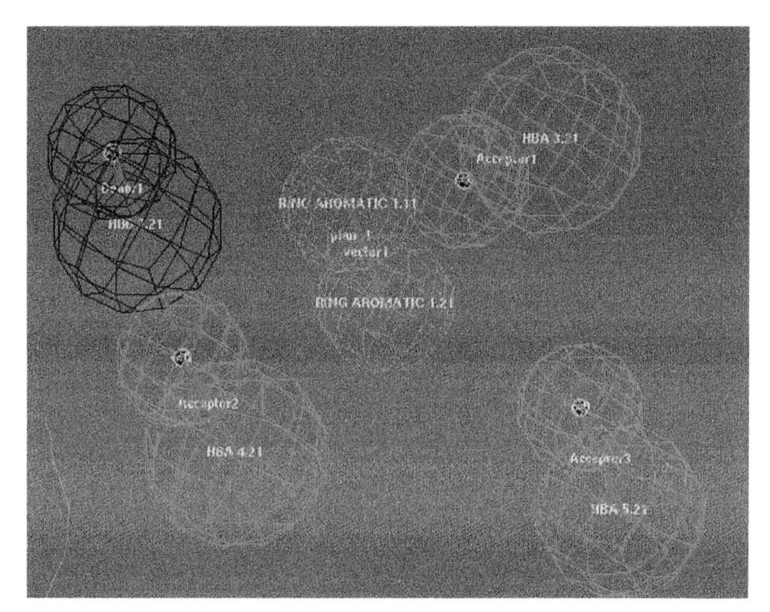

Figure 2. A Common feature hypothesis for NT inhibitors derived by Catalyst. Hydrogen bond donor (magenta), hydrogen bond acceptors (green) and ring aromatic (yellow) are the features in the hypothesis.

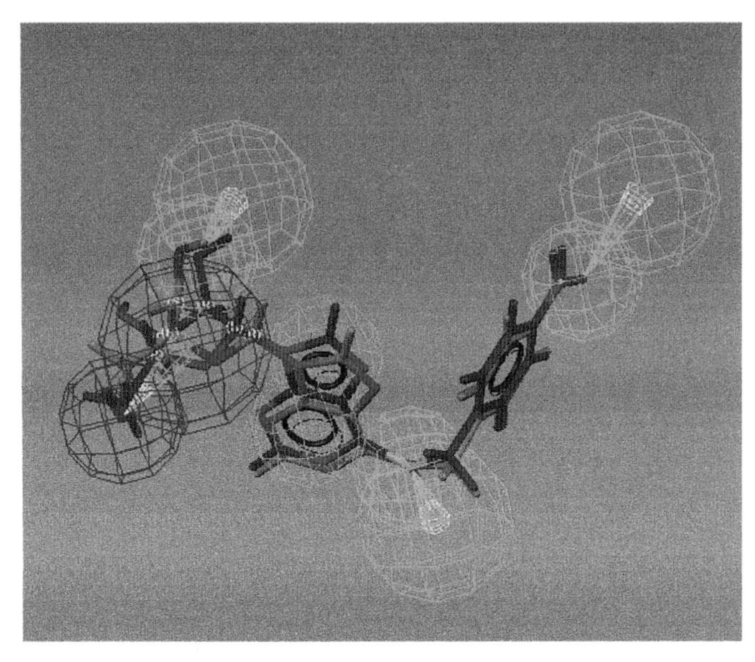

NBMPR N^6–(4–nitrobenzyl)thioinosine (fit=5.0 ; E_{conf}=2.39 kcal/mol)
2'–deoxy–6–N–(4–nitrobenzyl)adenosine (fit=3.67;E_{conf}=5.11 kcal/mol)

Figure 3. NBMPR and 2'-deoxy-6-N-(4-nitrobenzyl) adenosine mapped onto the hypothesis.

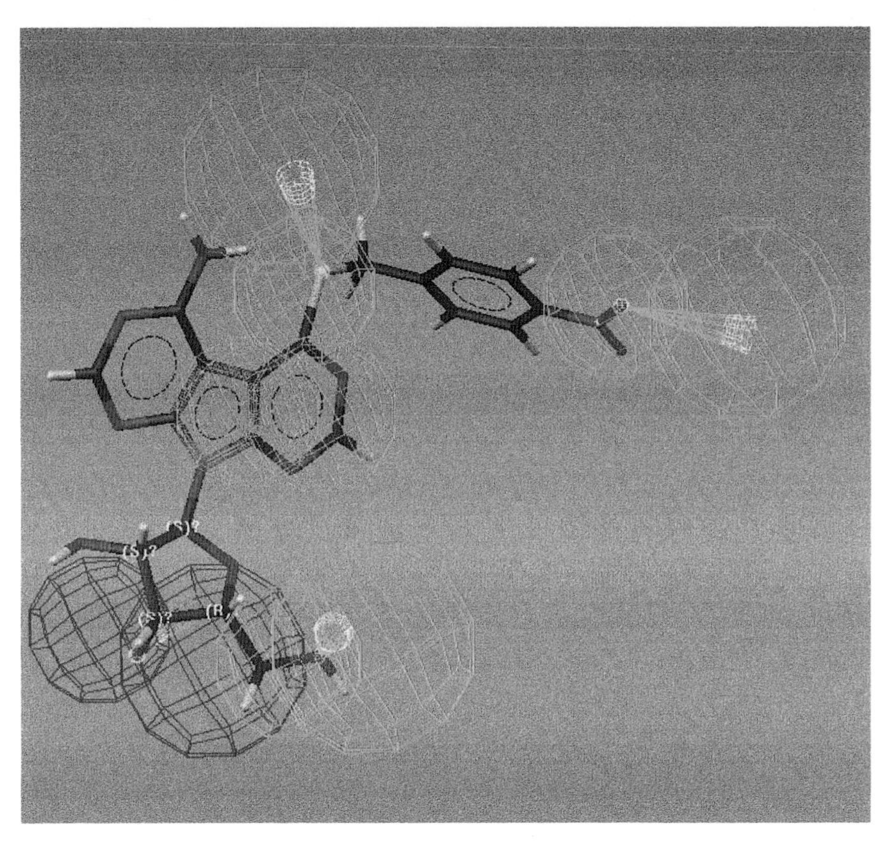

Figure 4. A hit from a search on the NCI database (267224)

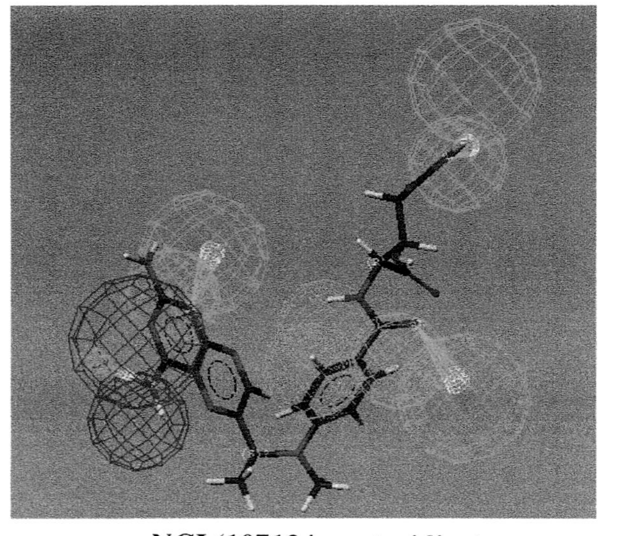

NCI (107134; a pteridine)

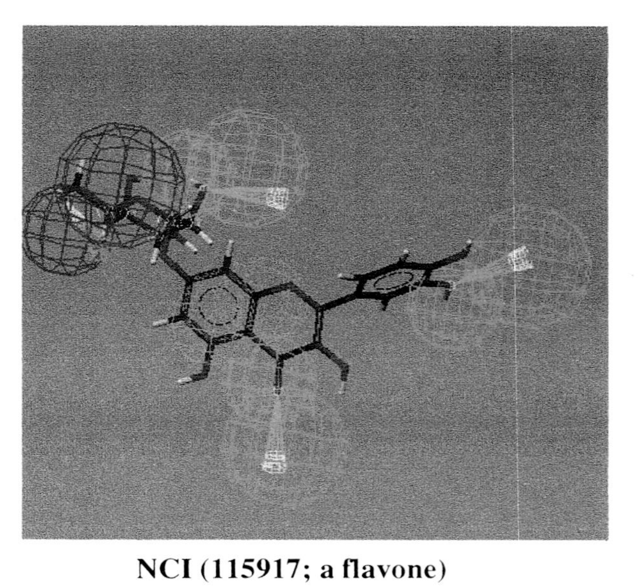

NCI (115917; a flavone)

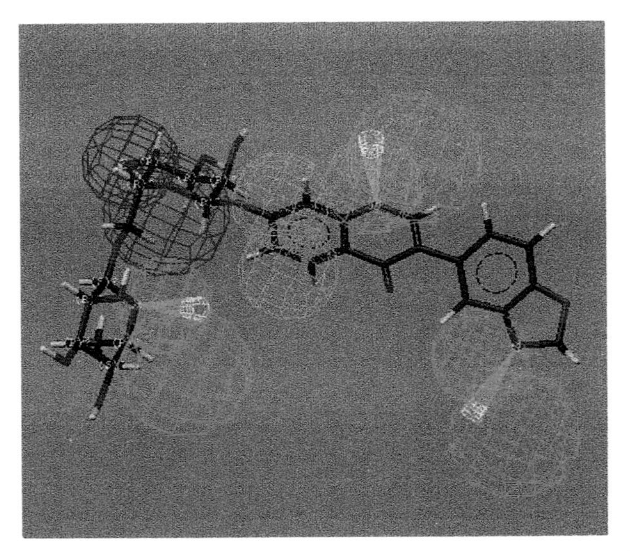

NCI (127487; isoflavone analog)

Figure 5. Example of pterine, flavone and isoflavone analogs from the NCI database mapped onto the pharmacophore hypothesis.

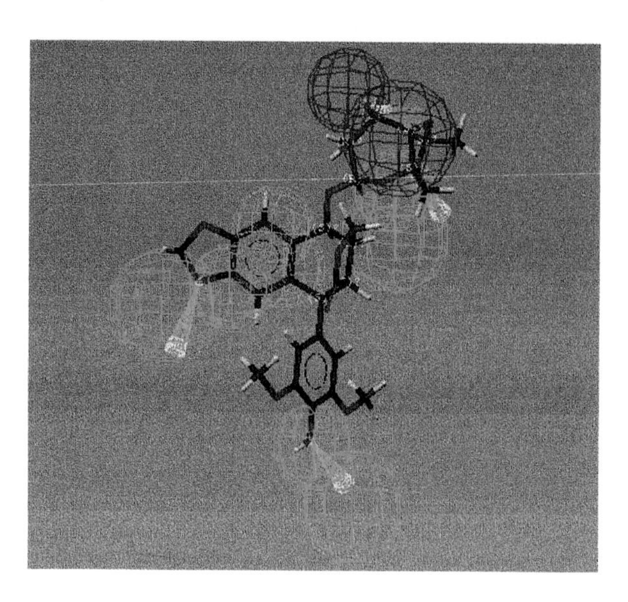

NCI (141540)
Etoposide (VP–16)

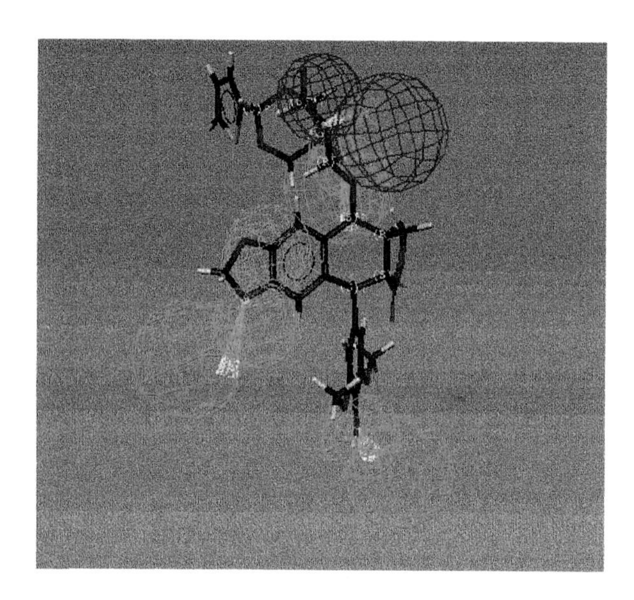

NCI (122819)
Teniposide (VM–26)

Figure 6. Mapping of the chemical features of etoposide and teniposide on the common feature hypothesis.

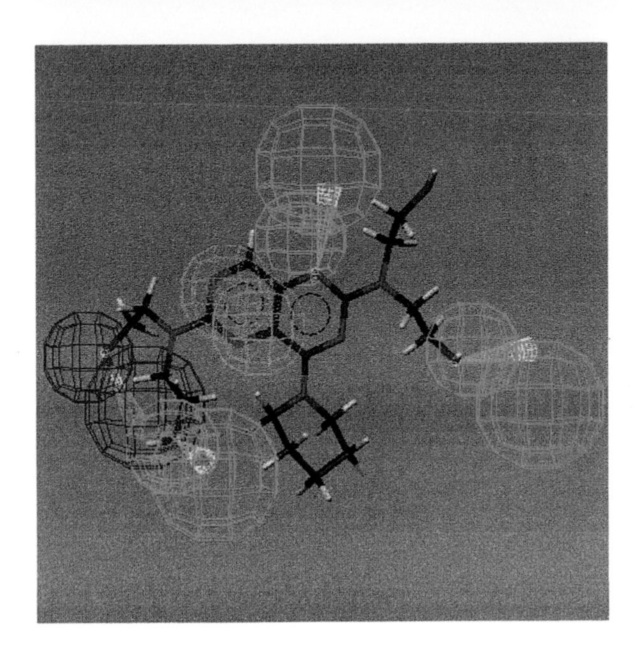

Mopidamole

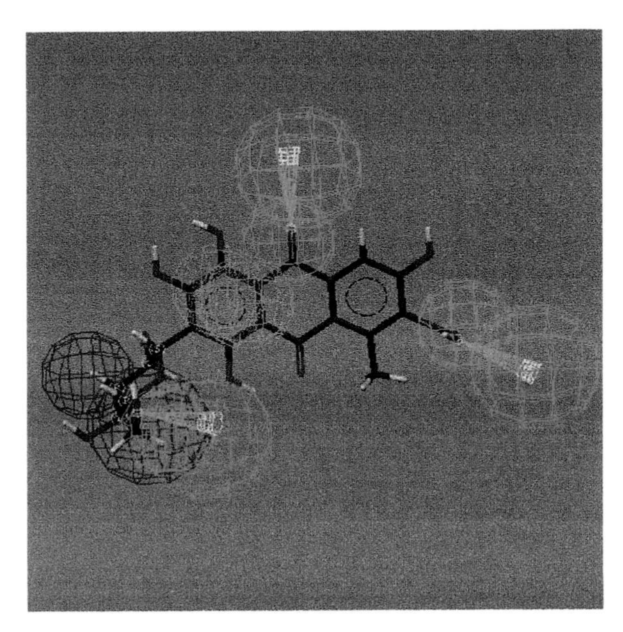

Carminate

Figure 7. Mopidamole and Carminate from the database of Derwent World Drug Index having the proposed pharmacophoric features.

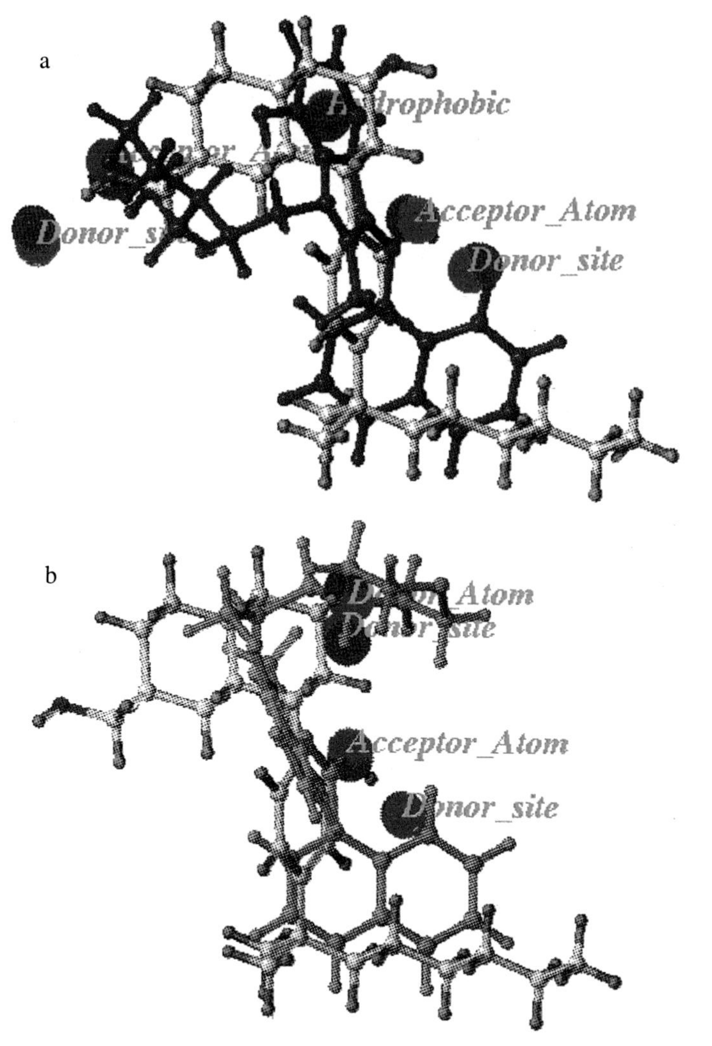

Figure 4. DISCO-derived Z and C superimposition models for CP55244 and WIN55212-2. The pharmacophoric features of the reference molecule (WIN55212-2) are represented as red spheres and those of CP55244 are represented as blue spheres. CP55244 is colored by atom type while WIN55212-2 is colored in purple (4a: Z model) and in orange (4b: C model).

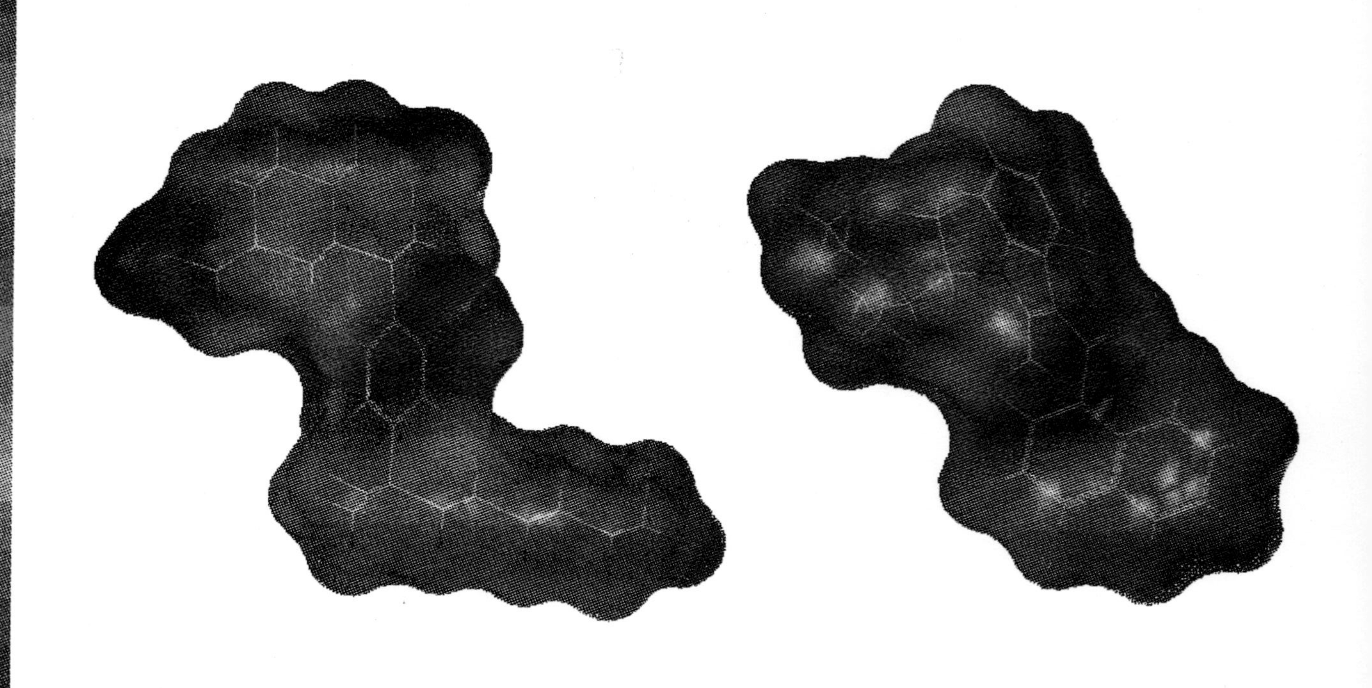

Figure 5. Lipophilicity potentials for CP55244 and WIN55212-2 (Z form) on the Connolly molecular surfaces. The most lipophilic part is the C3 side chain in CP55244 and the C3 aroyl moiety in WIN55212-2.

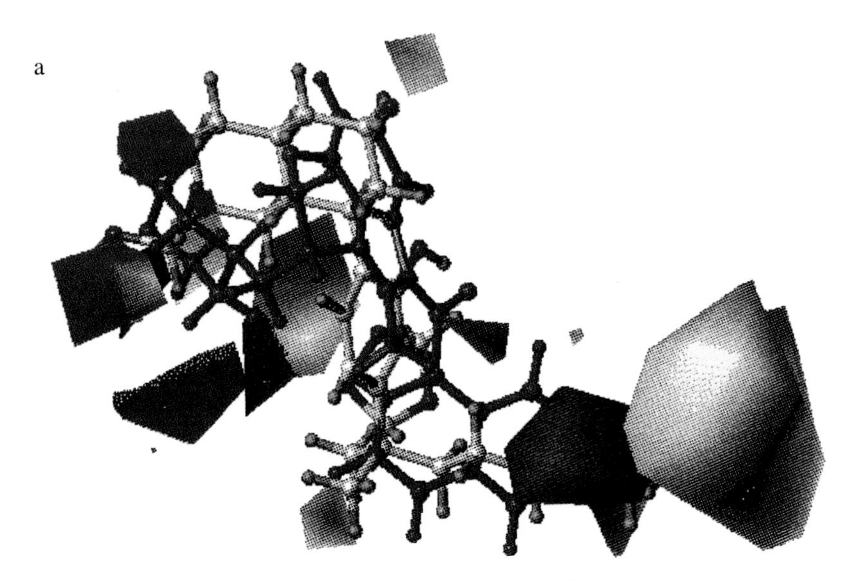

Figure 6. CoMFA steric maps (sterically favored regions in green and sterically disfavored regions in yellow) and electrostatic maps (positive charge favored in blue and negative charge favored in red) of the Z and C superimposition models. CP55244 and WIN55212-2 are colored in orange and by atom type. CP55244 is colored by atom type while WIN55212-2 is colored in purple (6a: Z model) and in orange (6b: C model).

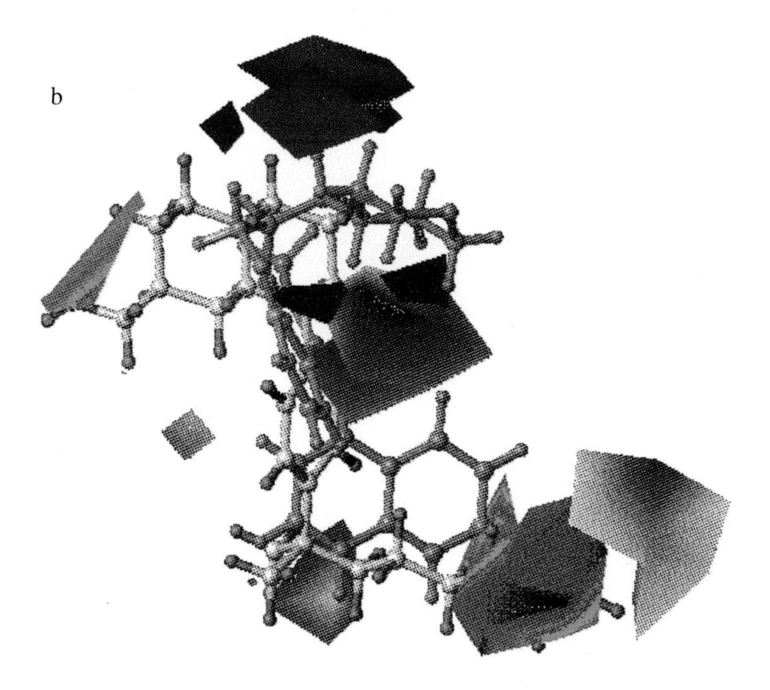

Figure 6. *Continued*.

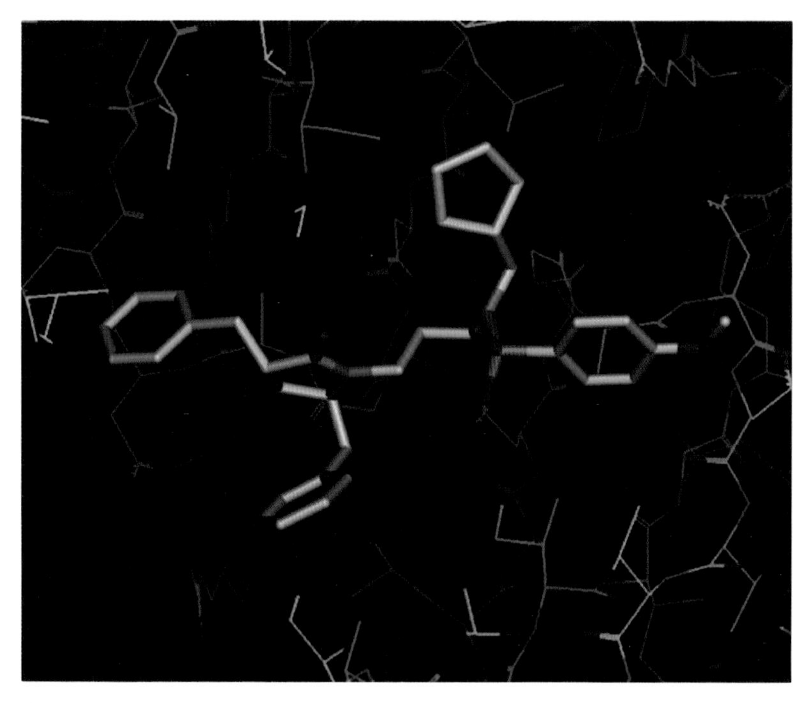

Figure 11. Crystal structure of compound D6 in the active site of HIV-1 protease. The inhibitor is shown in thick sticks. All atoms are colored according to the atom-types (C: green, O: red, N: blue, S: yellow).

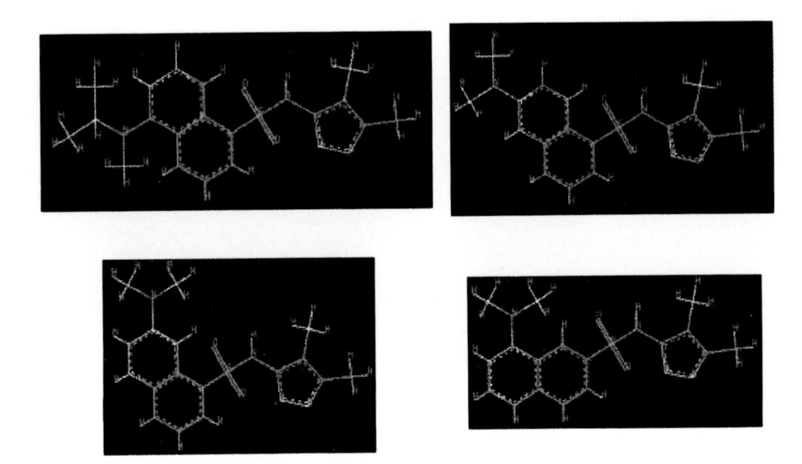

Figure 4. HQSAR model interpretation for four members of the Endothelin data set.

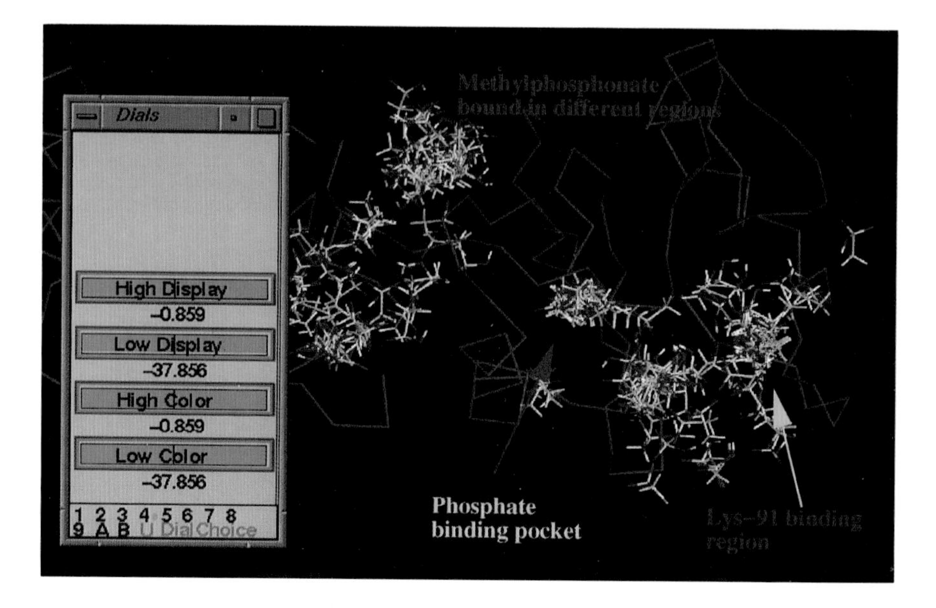

Figure 6. Analysis of the MCSS results of the N-terminus SH_2 domain of SHPTP2 with methanephosphonic acid as the ligand.

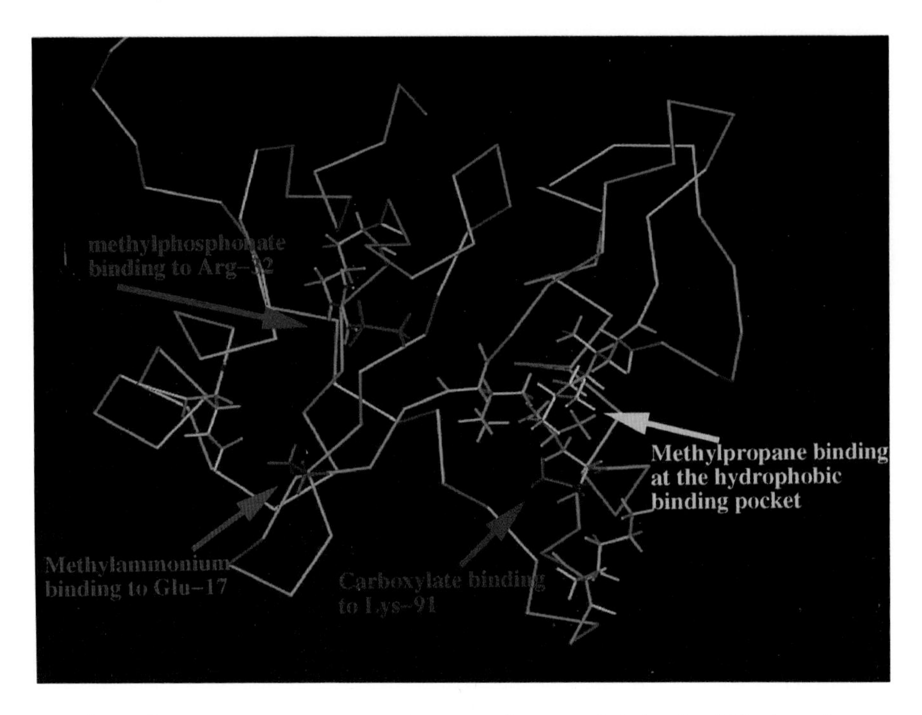

Figure 7. An MCSS deduced pharmacophore model for the SH_2 domain phosphotyrosine binding pocket of SHPTP2 protein.

Literature Cited

1. Ward, S. J.; Baizman, E.; Bell, M.; Childers, S.; D'Ambra, T.; Eissenstat, M.; Estep, K.; Haycock, D.; Howlett, A.; Luttinger, D.; Miller, M.; Pacheco, M. In *Problems of Drug Dependence*; Harris, L. S., Ed.; NIDA Research Monograph, National Institute on Drug Abuse: Rockville, MD **1990**; 425-426.
2. Felder, C. C.; Veluz, J. S.; Williams, H. L.; Briley, E. M.; Matsuda, L. *Mol. Pharmacol.* **1993**, *42*, 838-845.
3. Howlett, A. C.; Berglund, B.; Melvin, L. S. *Curr. Pharmaceut. Des.* **1995**, *1*, 343-354.
4. Mechoulam, R.; Breuer, A.; Jarbe, T. U. C.; Hiltunen, A. J.; Glaser, R. *J. Med. Chem.* **1990**, *33*, 1037-1043.
5. Martin, B. R.; Compton, D. R.; Thomas, B. F.; Prescott, W. R.; Little, P. J.; Razdan, R. K.; Johnson, M. R.; Melvin, L. S.; Mechoulam, R.; Ward, S. J. *Pharmacol. Biochem. Behav.* **1991**, *40*, 471-478.
6. Kuster, J. E.; Stevenson, J. I.; Ward, S. J.; D'Ambra, T. E.; Haycock, D. A. *J. Pharmacol. Exp. Ther.* **1993**, *264*, 1352-1363.
7. Yamada, K.; Rice, K. C.; Flippen-Anderson, J. L.; Eissenstat, M. A.; Ward, S. J.; Johnson, M. R.; Howlett, A. C. *J. Med. Chem.* **1996**, *39*, 1967-1974.
8. Razdan, R. K. *Pharmacol. Rev.* **1986**, *38*, 75-149.
9. Melvin, L. S.; Johnson, M. R. In *Structure-Activity Relationships of the Cannabinoids*; NIDA Research Monograph 79; National Institute on Drug Abuse: Rockville, MD, 1987; 31-47.
10. Melvin, L. S.; Milne, G. M.; Johnson, M. R.; Subramaniam, B.; Wilken, G. H.; Howlett, A. C. *Mol. Pharmacol.* **1993**, *44*, 1008-1015.
11. Melvin, L. S.; Milne, G. M.; Johnson, M. R.; Wilken, G. H.; Howlett, A. C. *Drug Design and Discovery* **1995**, *13*, 155-166.
12. Eissenstat, M. A.; Bell, M. R.; D'Ambra, T. E.; Alexander, E. J.; Daum, S. J.; Ackerman, J. H.; Gruett, M. D.; Kumar, V.; Estep, K. G.; Olefirowicz, E. M.; Wetzel, J. R.; Alexander, M. D.; Weaver, III, J. D.; Haycock, D. A.; Luttinger, D. A.; Casiano, F. M.; Chippari, S. M.; Kuster, J. E.; Stevenson, J. I.; Ward, S. J. *J. Med. Chem.* **1995**, *38*, 3094-3105.
13. D'Ambra, T. E.; Estep, K. G.; Bell, M. R.; Eissenstat, M. A.; Josef, K. A.; Ward, S. J.; Haycock, D. A.; Baizman, E. R.; Casiano, F. M.; Beglin, N. C.; Chippari, S. M.; Grego, J. D.; Kullnig, R. K.; Daley, G. T. *J. Med. Chem.*, **1992**, *35*, 124-135.
14. Bell, M. R.; D'Ambra, T. E.; Kumar, V.; Eissenstat, M. A.; Herrmann, Jr. J. L.; Wetzel, J. R.; Rosi, D.; Philion, R. E.; Daum, S. J.; Hlasta, D. J.; Kullnig, R. K.; Ackerman, J. H.; Haubrich, D. R.; Luttinger, D. A.; Baizman, E. R.; Miller, M. S.; Ward, S. J. *J. Med. Chem.* **1991**, *34*, 1099-1110.
15. Reggio, P. H.; Greer, K. V.; Cox, S. M. *J. Med. Chem.* **1989**, *32*, 1630-1635.
16. Reggio, P. H.; Panu, A. M.; Miles, S. *J. Med. Chem.* **1993**, *36*, 1761-1771.
17. Huffman, J. W.; Dai, D.; Martin, B. R.; Compton, D. R. *Bioorganic & Medicinal Chem. Lett.* **1994**, *4*, 563-566.
18. Xie, X.-Q.; Eissenstat, M.; Makriyannis, A. *Life Sci.* **1995**, *56*, 1963-1970.

19. Lainton, J. A. H.; Huffman, J. W.; Martin, B. R.; Compton, D. R. *Tetrahedron Lett.* **1995**, *36*, 1401-1404.
20. Tong W.; Collantes, E. R.; Howlett, A. C.; Welsh, W. J. *J. Med. Chem.* (in press).
21. Shim, J. Y.; Collantes, E. R.; Welsh, W. J.; Subramaniam, B.; Howlett, A. C.; Eissenstat, M. A.; Ward, S. J. *J. Med. Chem.* (submitted).
22. Huffman, J. W.; Yu, S.; Showalter, V.; Abood, M. E.; Wiley, J. L.; Compton, D. R.; Martin, B. R.; Bramblett, R. D.; Reggio, P. H. *J. Med. Chem.* **1996**, *39*, 3875-3877.
23. DISCO: Martin, Y.; Bures, M. G.; Danaher, E. A.; DeLazzer, J.; Lico, I.; Pavlik, P. A. *J. Comp.-Aid. Mol. Des.* **1993**, *7*, 83-102.
24. The molecular modeling program Sybyl is a product of Tripos, Inc., St. Louis, MO 63144, USA.
25. Gasteiger, J.; Marsili M. *Tetrahedron* **1980**, *36*, 3219-3228.
26. Rapaka, R. S., Makriyannis, A. *Structure-Activity Relationships of the Cannabinoids*; NIDA Research Monograph 79; National Institute on Drug Abuse: Rockville, MD, 1987.
27. Cramer, R. D.; Patterson, D. E.; Bunce, J. D. *J. Am. Chem. Soc.* **1988**, *110*, 5959-5967.
28. Howlett, A. C.; Johnson, M. R.; Melvin, L. S.; Milne, G. M. *Mol. Pharmacol.* **1988**, *33*, 297-302.
29. Johnson, M. R; Melvin, L. S.; Althuis, T. H.; Bindra, J. S.; Harbert, C. A.; Milne, G. M.; Weissman, A. *J. Clin. Pharmacol.* **1981**, *21*, 271S-282S.
30. Thomas, B. F.; Adams, I. B.; Mascarella, S. W.; Martin, B. R.; Razdan, R. K. *J. Med. Chem.* **1996**, *39*, 471-479.

Chapter 12

Structure-Based Design of Novel Conformationally Restricted HIV Protease Inhibitors

B. G. Rao, C. T. Baker, J. T. Court, D. D. Deininger, J. P. Griffith, E. E. Kim, J. L. Kim, B. Li, S. Pazhanisamy, F. G. Salituro, W. C. Schairer, and R. D. Tung

Vertex Pharmaceuticals Incorporated, 130 Waverly Street, Cambridge, MA 02139

In our efforts to discover a new generation of HIV protease inhibitors, which are structurally distinct and have minimal cross resistance to our current clinical candidate, *amprenavir* (VX-478), we designed a set of potent compounds bearing a novel backbone structure. Structural and modeling analysis of initial leads showed that these inhibitors bind in a strained conformation. We obtained dramatic improvement in binding by relieving the backbone conformational strain of the inhibitor. In this report, we will present our structure-based design approaches as well as the enzymatic results on these novel and highly potent inhibitors.

HIV-1 protease inhibitors (PIs) have revolutionized treatment of individuals with HIV and AIDS (*1*). Four PIs are already on the market and a fifth PI, *amprenavir* (with the chemical name of (3S)-tetrahydro-3-furyl N-((1S,2R)-3-(4-amino-N-isobutylbenzenesulfonamido)-1-benzyl-2-hydroxypropyl) carbamate, also formerly known as VX-478 and 141W94) is undergoing advanced phase III clinical trials (*2*). *amprenavir* is a small molecular weight (506 Da), potent (K_i = 0.6 nM; IC_{90} = 40 nM), and synthetically accessible HIV protease inhibitor that emerged from a focused application of structure-based design approaches along in coordination with the disciplines of medicinal chemistry and pharmacology *(3,4)*. The chemical structure of *amprenavir* is shown in Figure 1. In spite of the availability of potent regimens of PIs in combination with reverse transcriptase inhibitors (RTIs), a cure is not in sight and patients may have to be on these drug regimens for long time. Additionally, the currently available drug regimens are not well-tolerated by many patients and are very difficult to comply with. Related to these issues is the problem of drug resistance, which is a very compelling and immediate problem, since it can directly compromise the therapeutic efficacy of a given treatment regimen. Therefore, there is a need to design a new generation of PIs, which are more potent, easy to take and are not cross-resistant to currently used PIs. Partaledis *et al.* (*5*) had shown using cell culture passage experiments that *amprenavir* is resistant to HIV-1 protease with a unique set of mutations at L10F, M46I, I47V and I50V. It has also been shown that I50V is absolutely necessary for resistance against amprenavir by in cellular and enzymatic assays (6). We have further characterized the biophysical basis of the resistance of these mutants by X-ray crystal structure and computational methods (7). The results of this analysis is that the mutation of Ile to Val at residue 50 causes loss in hydrophobic interactions with *amprenavir*; mainly the loss of interaction between the terminnal (CD1) methyl group of I50 and the P_2' aryl sulfonamide group of the inhibitor contributes most to the loss of binding. Based on these results, we have utilized the following four design concepts for the design of next generation inhibitors: (a) Maintain strong interactions with the catalytic aspartates. (b) Maintain strong flap water interactions. (c) Space out P_1 and P_1' branching, so that inhibitors will have less hydrophobic contact with the centrally located I50 and I84 side-chains. (d) Minimize interactions of the P_2/P_2' groups with I50/I50' side-chains.

Linear Carbamate based Inhibitors

The compound A in Figure 2 is a close analog of *amprenavir* with a Ki of < 0.1 nM. The shifting of P_1 benzyl from the C_α to the P_1 amide nitrogen spaces out the P_1 and P_1' branching. Also, such a compound is expected to maintain strong interactions with the catalytic dyad (D25 and D25') and with the flap water. Therefore, the new compound B satisfies the first three design principles stated earlier. This compound showed good activity (Ki = 600 nM), but it is much less potent than the parent compound. A preliminary modeling study (see modeling details in the next section) of this compound suggested that the benzyl group is too short to fill the P_1 pocket. The substitution of P_1 benzyl with phenethyl lead to increased potency (Compound B1, Ki = 40 nM). Also, this compound is more potent than its diastereomer B2 (Ki = 170 nM), suggesting that the preferred configuration of the central hydroxyl is the same as the parent compound. The chemical structures of these two compounds are shown in Figure 3. However, the best of these compounds is still > 400-fold weaker than the parent compound A. The new inhibitors are expected to be weaker due to two obvious reasons: (a) loss of the hydrogen bond interaction with G27 carbonyl of the enzyme, since the P_1 amide has no hydrogen bond donor, and (b) added flexibility of the main-chain and longer P_1 side-chain. But these factors alone are not expected to account for the >400 fold loss in binding.

Modeling and Structural Analysis of Linear Carbamates

Modeling. In order to understand the weaker binding of the linear carbamate inhibitors, we modeled compound B1 in the active site of HIV-1 protease. The crystal structure of compound A complexed with HIV-1 protease (E. E. Kim, unpublished results) was used for modeling the bound conformation of compound B1. The program QUANTA (Version 4.0b, Molecular Simulations Inc., Burlington, MA, 1992) was used for model building. Energy minimization was carried out with CHARMM force field within QUANTA program. As the prime-side of compounds A and B1 are common, the bound conformation of A in the crystal structure was

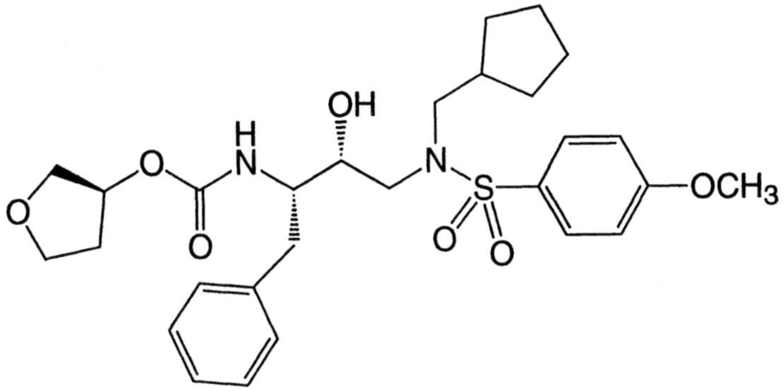

Figure 1. Chemical structure of *amprenavir* (VX-478).

Compound A

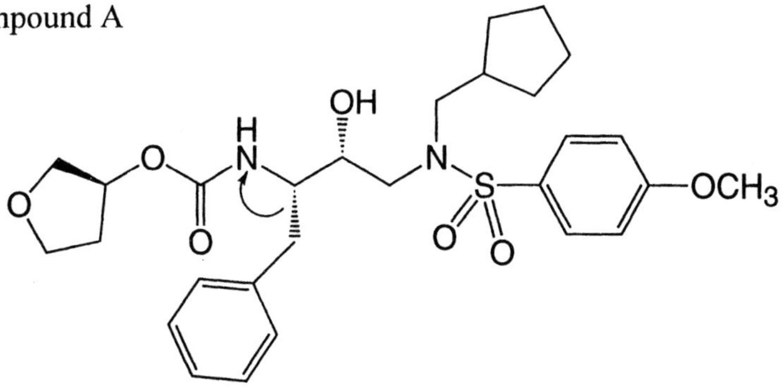

Compound B

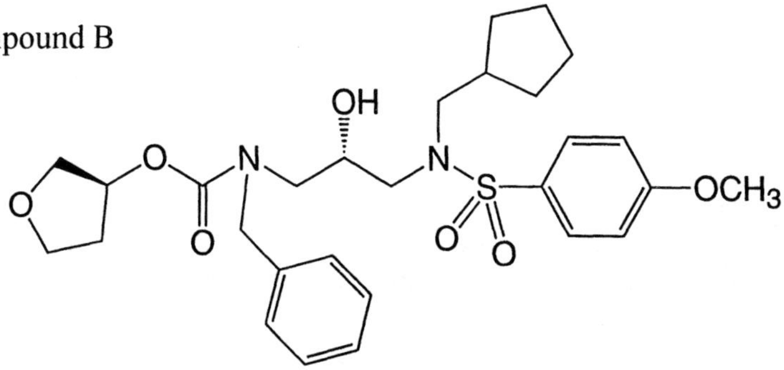

Figure 2. Chemical structures of compound A (a close analog of *amprenavir*, previously described as VB-11,328 in reference 6) and compound B (the new carbamate inhibitor, derived from compound A by shifting the P_1 benzyl from C_α position to the amide N of the same residue).

Compound B1

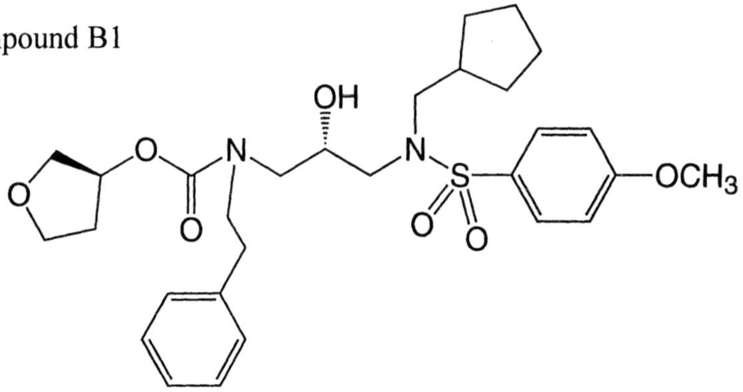

Compound B2

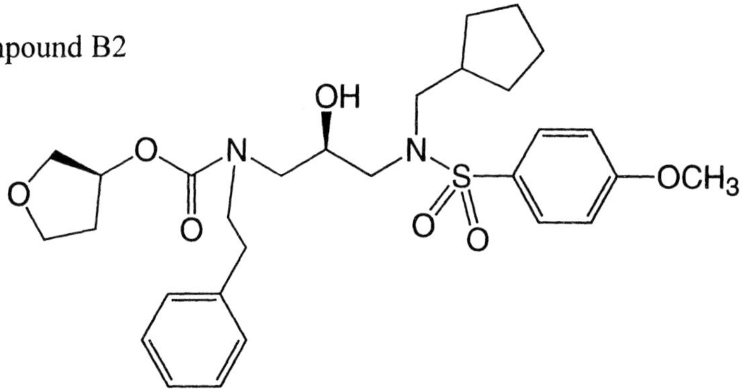

Figure 3. The two stereoisomers (Compound B1 and Compound B2) of a linear carbamate inhibitor with Phenethyl side-chain at P_1.

adopted for compound B1. The phenethyl and the THF groups on the non-prime side of compound B1 was modeled into the S_1 and S_2 pockets, respectively, by manipulating the main-chain and side-chain torsions. This model was energy minimized using two different approaches. In the first approach, the enzyme coordinates and the flap water were held fixed, and all atoms of the inhibitor were minimized first by steepest descents method for 200 steps, followed by 1000 steps of Adopted Basis Newton Raphson method in QUANTA. In the second approach, the enzyme coordinates were fixed, but flap water and inhibitor were allowed to move during minimization with a distance constraint between flap water and the P_2 carbonyl to mimic a hydrogen bond. The two step minimization process was applied in this case also. The two approaches resulted in two energetically similar models.

The two models, superimposed with the parent compound A are shown Figures 4 and 5, respectively. Firstly, it may be noted that the compound B1 does not make hydrogen bond with G27 carbonyl of the enzyme in either models and it has two additional rotatable bonds. In model 1, the THF group overlays quite well with the same group of compound A, but the aromatic ring of the P_1 group is pushed out of the pocket. In the second model, the P_1 ring is closer to the same group of compound A, but the THF groups do not overlap as well. Further, the P_2 carbonyl in model 1 is not oriented to make the flap water interaction, whereas it was forced to make flap water interaction in the second model. In either case, the inhibitor is likely to loose significant binding due to (a) non-optimal interactions of P_1 and P_2 groups, (b) weak flap water interactions, (c) lack of interaction with G27 carbonyl group of the enzyme and (d) addition of two rotatable bonds. This analysis of the models is consistent with the higher Ki for compound B1. However, the modifications of these inhibitors suggested from these two models yielded only moderate improvements in binding (results not shown).

Structural Analysis. In view of these difficulties in understanding this series of compounds by modeling, we attempted crystallization of several potent compounds in this linear carbamate series. The crystal structures described in this paper were obtained by the following procedure: Purified wild-type protease was refolded at 5° C by rapid dilution from 7 M urea, into a buffer containing 25 mM sodium formate, 50 mM DTT and a 5-fold molar excess of inhibitor. The complex was concentrated and washed extensively in 15 mM sodium acetate, 5 mM DTT buffer (pH 5.4). Hexagonal rod shaped crystals grew at room temperature in about a week by vapor diffusion against an ammonium sulfate reservoir as described previously (3), in space group $P6_1$. All data were collected at room temperature from one crystal each of the various complexes, using a Rigaku R-axis IIc image plate area detector (Molecular Structure Co., Woodlands, TX). In all cases, all measured reflections beyond 8.0 Å resolution were included in the structure refinement, which was carried out using X-PLOR (*8*). The complex structures were refined using the slow-cool algorithm with starting coordinates from an isomorphous structure of HIV-1 protease in complex with VX-478 (3). The program QUANTA was used for model building.

The first crystal structure obtained in this series of compounds complexed with HIV-1 protease is of VB-13,674 (Ki = 17 nM), which is a close analog of compound

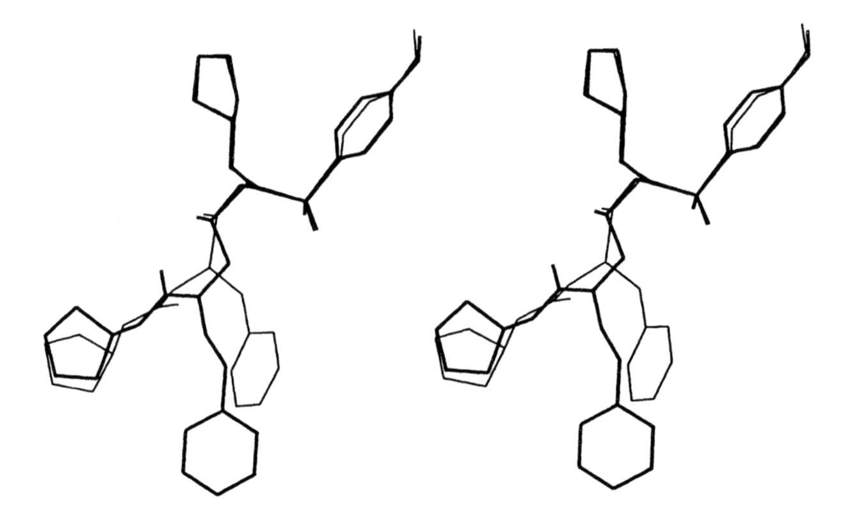

Figure 4. A stereo diagram of the first model of compound B1 (thick line) overlapped with the crystal structure of compound A (thin line) in the active site of HIV-1 protease.

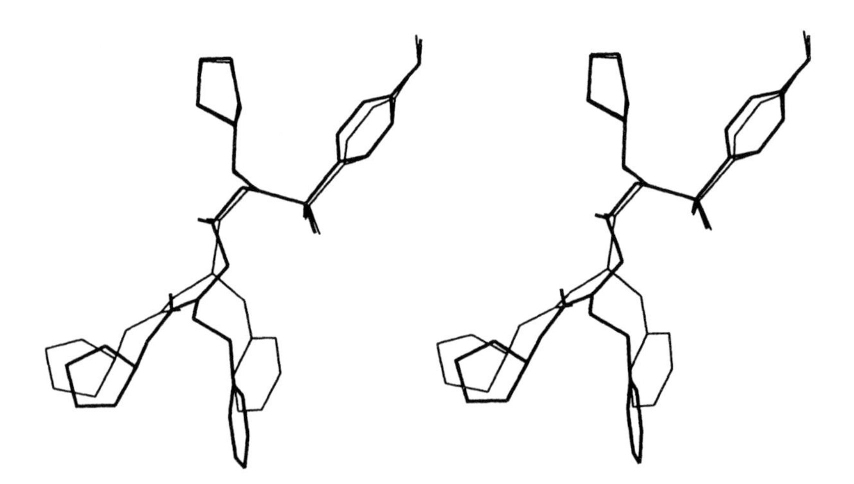

Figure 5. A stereo diagram of the second model of compound B1 (thick line) overlapped with the crystal structure of compound A (thin line) in the active site of HIV-1 protease.

B1. VB-13,674 has pyridyl methyl carbamate at P_2 instead of tetrahydrofuranyl carbamate in compound B1. The electron density of the inhibitor at 2.2 Å resolution is shown in Figure 6. A comparison of this structure with the structure of the parent compound shows that these two structures overlap quite well (Figure 7). Surprisingly, the P_1 group occupies almost the same position as the P_1 group of compound A and also maintains the flap water interactions. On both these counts, the two models are in disagreement with the crystal structure. Hence, it was not at once clear from the structure the reasons for weaker binding of the linear carbamate based inhibitors with HIV-1 protease. Further analysis of the structure showed that the P_2-P_1 main-chain of the inhibitor is quite distorted in the crystal structure: The P_1-P_2 amide bond is twisted from planarity by about 35 degrees and the N-C-C-OH torsion is in an eclipsed conformation (Figure 7). These two factors in addition to the loss of hydrogen bond of the inhibitor with G27 carbonyl appear to be responsible for >400 fold loss in binding.

As models maintained the low energy conformation of the two bonds, they did not predict the crystal structure conformation. The crystal structure revealed that the penalty for distortion of the main-chain is less than the gain in binding due to interactions with the flap water and the hydrophobic residues in the S_1 pocket. This reality is obviously not captured by the modeled structures, reflecting the deficiency of current force fields. This result, therefore, suggests that we have to be mindful of such exceptions while using the docking and Ki prediction algorithms using current force fields.

Design of Cyclic Lactam-based Inhibitors

More importantly for this study, these structural and modeling results lead us to new approaches in the modifications of these inhibitors that would minimize the main chain distortions to gain binding. One of these approaches is the design of cyclic lactams:

A close examination of the crystal structure of VB-13,674 showed that the carbamate oxygen of the P_2 group and the C_α methelene group of the P_1 side-chain are close in space. This suggested that these two positions can be cyclized into a ring, thereby correct the amide distortion and decrease the flexibility of the main-chain at the same time. We modeled both 5- and 6-membered ring systems, as lactams, cyclic carbamates and cyclic ureas. All the cyclic models were energy minimized in the active site of HIV-1 protease using the procedures described earlier. Of all the models, 5-membered lactam based inhibitors looked best in terms of both conformations and energetics. The model of a 5-membered lactam based inhibitor, minimized in the active site of the enzyme, is overlapped with the crystal structure of VB-13,674 in Figure 8. It is clear that the model overlays very well with the crystal structure: the backbone is not distorted, its carbonyl maintains flap water interaction and it offers possibilities for filling the S_2 side different substituents on the ring.

The first compound synthesized with a cyclic scaffold is, however, a cyclic carbamate C1 (Figure9). This compound has a Ki of 1.2 μM, and is as potent as the corresponding linear inhibitor C2 (Ki = 1.6 μM) (Figure 9). These results show that

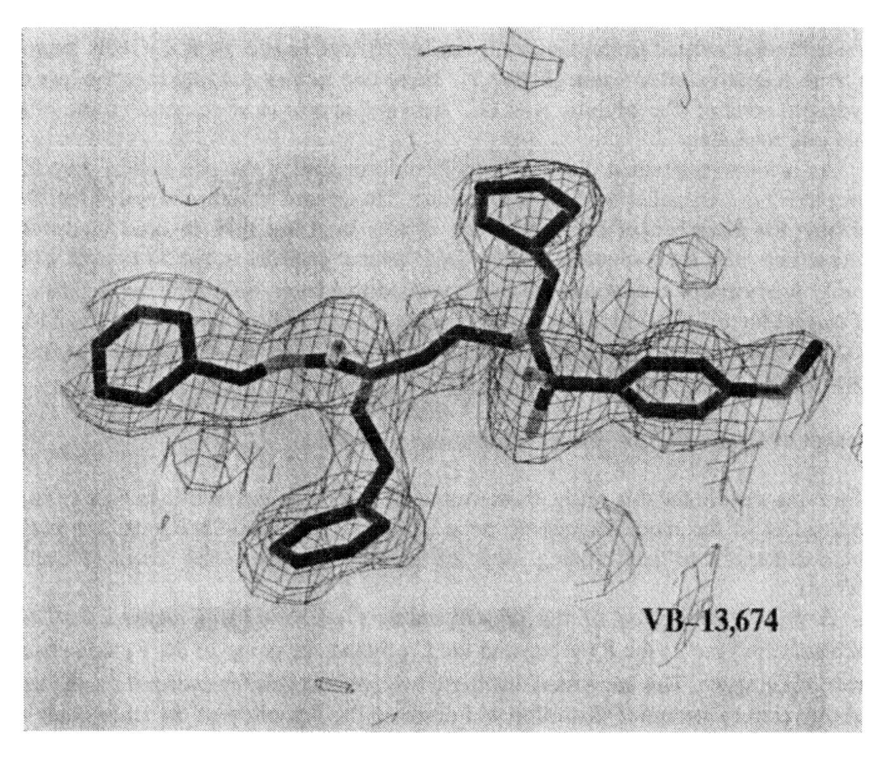

Figure 6. Diagram of the 2|Fo| - |Fc| electron density around VB-13,674, a close analog of compound B1.

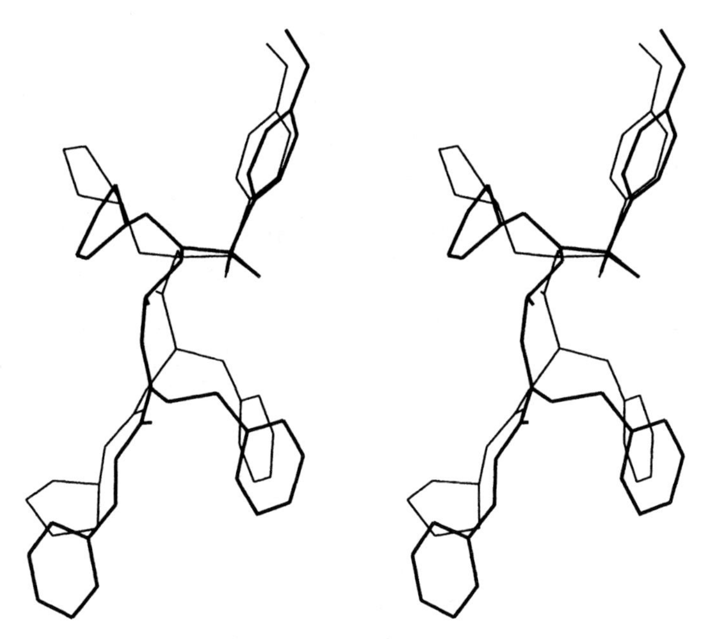

Figure 7. A stero diagram of the overlap of the crystal structures of VB-13,674 (thick line) and compound A (thin line).

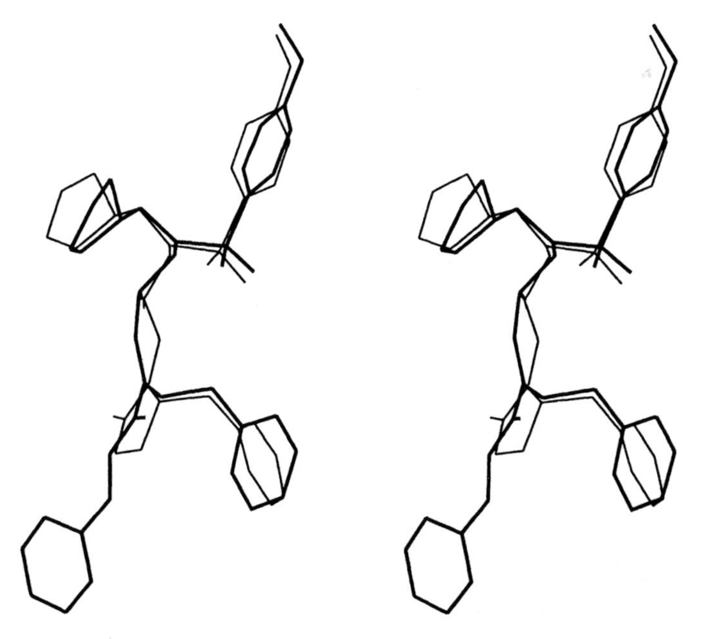

Figure 8. A stereo diagram of the overlap of the model of a cyclic carbamate (thin line), with the crystal structures of VB-13,674 (thick line).

Compound C1

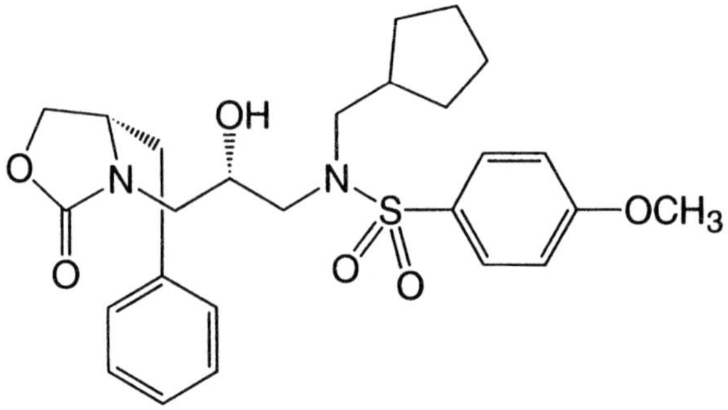

Compound C2

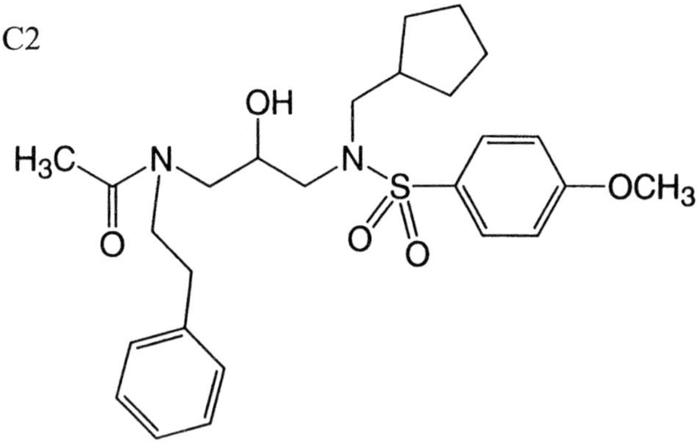

Figure 9. Chemical structures of a cyclic carbamate inhibitor, C1 and the corresponding linear analog, C2.

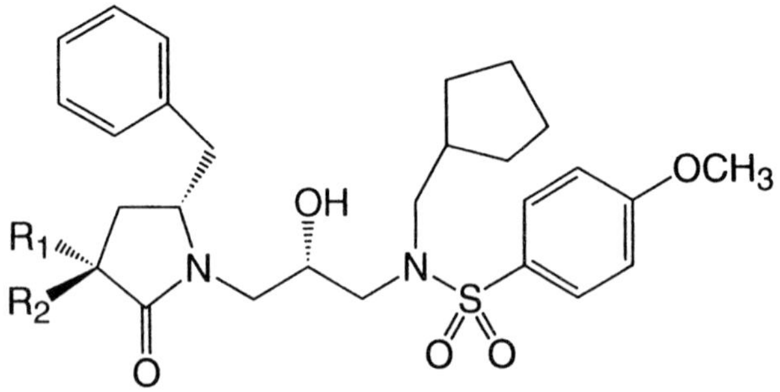

Figure 10. The 5-membered cyclic carbamates with different P_2 side-chains.

novel scaffolding at the P_1 and P_2 lead to compounds with comparable or better potency than the linear carbamate inhibitors. The low potency of the two molecules is due to lack of any P_2 interactions. As this cyclic carbamate scaffold does not provide any possibility to fill the S_2 pocket, this series was not perused further.

The 5-membered lactams were elaborated with different P2 side-chains, as shown in Figure 10. The Ki values of these compounds are given in Table I. The compound

Table I. Ki values of cyclic lactam based inhibitors illustrated in Figure 10.

Name	R_1	R_2	Ki (nM)
D1	H	H	550
D2	Me	H	130
D3	H	Me	60
D4	Me	Me	115
D5	H	Allyl	9
D6	H	Benzyl	0.6

D1 with only hydrogens at R1 and R2 has a Ki of 550 nM. This compound is about 2-fold better in binding than compound C1 (Ki = 1.2 μM). When R1 or R2 is methyl, binding improved further by 4-9 fold. It is also clear that the methyl substituent (R2) trans to P_1 benzyl group offers bigger improvement in binding than the methyl at R1, cis to the P_1 benzyl group. Further elaboration of R2 methyl to allyl jumped the binding by more than 10 fold in compound D5 with Ki = 9 nM. When allyl is substituted with benzyl at R2, the binding improved even more dramatically by about 15 fold in compound D6 with Ki = 0.6 nM. This compound is as potent as *amprenavir* in terms of enzymatic inhibition. The crystal structure of D6 in the active site of HIV-1 protease is shown in Figure 11. It may be seen from the structure that it fills all the four subsites and maintains flap water interaction without any distortion of the main-chain conformation.

However, compound D6, which is novel and potent, does not satisfy one of our gaols of maintaining high level of potency against I50V mutant of HIV-1 protease. In fact, the Ki's of compound D6 against I50V and M46I/I47V/I50V mutants are 43 nM and 135 nM, respectively. This reduction in binding of this compound is as high as that of VX-478 against these two mutants (*6*). As our structural analysis of the mutants suggested (*7*) the reduction of binding results from the interactions of the P_2' group with I50 side-chain. Hence, , the Ki results against mutants are not surprising as the P_2' part of the new inhibitor D6 is similar to that of VX-478.

Conclusions

The structural data and modeling has been utilized successfully to discover HIV-1 protease inhibitors with novel scaffolds with P_1 and P_2 substituents. These compounds are very potent; the best one described has a Ki of 0.6 nM. There are a lot

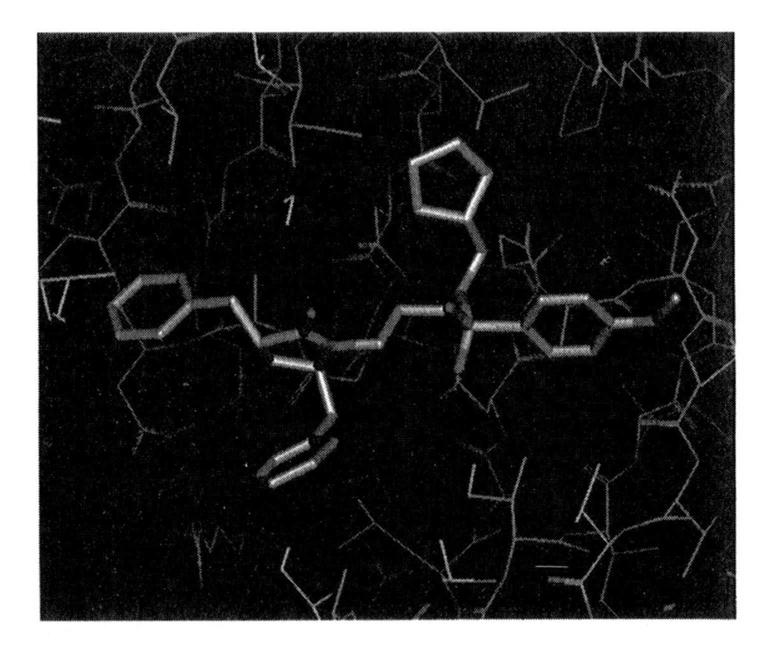

Figure 11. Crystal structure of compound D6 in the active site of HIV-1 protease. The inhibitor is shown in thick sticks. All atoms are colored according to the atom-types (C: green, O: red, N: blue, S: yellow).
(Figure is printed in color in color insert.)

of examples of inhibitor optimization published in the literature which show improvement in binding due to addition/optimization of hydrophobic, hydrogen bonding or electrostatic interactions. The present work is one of the very few studies and perhaps the first one to show substantial increase in binding accompanying the improvement in inhibitor conformation.

We have discovered several more potent inhibitors in this series and related analogs with different P_1-P_2 scaffolds during the first phase of our second generation HIV-1 protease inhibitor discovery program carried out solely at Vertex Pharmaceuticals. However, all these compounds have the same prime side substituents at P_1' and P_2' sites, much similar to *amprenavir.* As one of the goals of the program is to discover inhibitor which are chemically distinct from *amprenavir* series, the second phase of this program focused on changing the prime-side of these new inhibitors. The second phase of the program is being carried out in collaboration with GlaxoWellcome and the newer compounds will be described elsewhere.

Acknowledgments

We would like to thank of all the members of HIV-1 protease discovery team at Vertex for their contributions, and Dr. Vicki Sato for her support and encouragement.

Literature Cited

(*1*) Chrusciel, R. A.; Romines, K. R. *Exp. Opin. Ther. Patents*, **1997**, *7*, 111.
(*2*) Painter, G. R.; Ching, S.; Reynolds D.; St. Clair M.; Sadler B. M.; Elkins M.; Blum, R.; Dornsife, R; Livingston, D. J; Parteledis, J. A.; Pazhanisamy, S; Tung, R. D., Tisdale, M. *Drugs of the Future*, **1996**, ***21***, 347.
(*3*) Kim, E. .E.; Baker; C. T.; Dwyer, M. D.; Murcko, M. A.; Rao, B. G.; Tung, R. D.; Navia, M.A. *J. Am. Chem. Soc.*, **1995**, ***117***, 1181.
(*4*) Navia, M. A.; Sato, V. L.; Tung, R.D. *International Antiviral News*, **1995**, *3*, 143.
(*5*) Partaledis, J. A.; Yamaguchi, K.; Tisdale, M; Blair, E. E.; Falcione, C.; Maschera, B.; Myers, R. E.; Pazhanisamy, S.; Futer, O.; Cullinan, A. B.; Stuver, C.; Byrn, R. A.; Livingston, D. J. *J. Virol.* **1995**, ***69***, 5228.
(*6*) Pazhanisamy, S.; Stuver, C. M.; Cullinan, A. B.; Margolin, N.; Rao, B. G.; Livingston, D. J. *J. Biol. Chem.*, **1996**, *271*, 17979.
(*7*) Rao, B. G.; Dwyer, M. D. ; Deininger, D. D.; Tung, R. D.; Navia, M. A.; Kim, E. E. *Antiviral Therapy*, **1996**, *1(Suppl. 1)*, 13.
(*8*) Brunger, A. T.; Krukowski, A.; Erickson, J. W. *Acta Cryst.*, **1990**, *A46*, 585.

Chapter 13

"New Tricks for an Old Dog": Development and Application of Novel QSAR Methods for Rational Design of Combinatorial Chemical Libraries and Database Mining

A. Tropsha, S. J. Cho, and W. Zheng

The Laboratory for Molecular Modeling, Division of Medicinal Chemistry and Natural Products, School of Pharmacy, University of North Carolina, Chapel Hill, North Carolina 27599

We discuss the development of novel automated variable selection QSAR methods in the context of their application to rational design of targeted combinatorial chemical libraries. The two QSAR methods developed in this laboratory include Genetic Algorithm – Partial Least Squares (GA-PLS) and K-Nearest Neighbors (KNN). Both methods employ multiple topological descriptors of chemical structures and use stochastic optimization algorithms to achieve robust QSAR models, which are characterized by the highest value of cross-validated R^2 (q^2). The GA-PLS method uses a combination of genetic algorithms (GA) and PLS to evolve an initial population of the QSAR equations to the final population with the highest average q^2. The KNN-QSAR method formally employs the active analog principle and predicts the activity of a compound as the average activity of K most chemically similar compounds using the optimized subset of descriptors to characterize the similarity. Both QSAR methods can be used to search for bioactive compounds in the virtual chemical libraries or chemical databases on the basis of either (i) their (high) activity predicted from the QSAR model, or (ii) their similarity to a probe (lead molecule) evaluated using only variables selected by the QSAR model.

Rapid development of combinatorial chemistry and high throughput screening in recent years has provided a powerful alternative to more traditional approaches to lead generation and optimization. In traditional medicinal chemistry, these processes frequently involve the purification and identification of bioactive ingredients of natural, marine, or fermentation products or random screening of synthetic compounds. This is often followed by a series of painstaking chemical modification or total synthesis of promising lead compounds, which are tested in adequate bioassays. On the contrary, combinatorial chemistry involves systematic assembly of a set of "building blocks" to generate a large library of chemically different molecules which are screened simultaneously in various bioassays (1,2). In the case of targeted library design, the lead identification and optimization then becomes generating libraries with structurally diverse compounds, which are similar to a lead compound; the underlying assumption is that structurally similar compounds should exhibit similar biological activities. Conversely, structurally dissimilar compounds should exhibit very diverse biological activity profiles; thus the goal of the diverse library design is to generate libraries with maximum chemical diversity of the constituent compounds (3).

In many practical cases, the exhaustive synthesis and evaluation of combinatorial libraries becomes prohibitively expensive, time consuming, or redundant (4). Recently, we have initiated the development of computational approaches aimed at rational design of combinatorial libraries for both targeted and diverse screening (5,6). Obviously, using all available experimental information about the biological target or pharmacological compounds capable to interact with the target can significantly enhance the rational design of targeted chemical libraries. In many cases, the number of compounds with known biological activity is sufficiently large to develop a viable QSAR model for such dataset. Thus, a pre-constructed QSAR model can be used as a means of selecting virtual library compounds (or actual compounds from existing databases) with high predicted biological activity. Alternatively, if a variable selection method has been employed in developing a QSAR model, the use of only selected variables can improve the performance of the rational library design or databases mining methods based on the similarity to a probe.

In this paper we describe the development and application of two fast, automated algorithms for QSAR based on the principles of variable selection and stochastic optimization. We show that the resulting QSAR models can be effectively used in rational targeted library design and database mining. We conclude that combinatorial chemistry creates previously unforeseen challenges for the field of QSAR analysis.

Methods

General Computational Details. SYBYL molecular modeling package (7) was used to generate chemical structures and convert them to SMILES (8) notation. MolconnX program (9) was used to generate topological indices for all datasets used in this study. All calculations were performed on SGI Indigo^2 workstations.

GA-PLS Method. The algorithm of the GA-PLS method (10) was implemented as follows. *Step 1.* The MolconnX program (9) was applied to generate descriptor variables (460 topological indices) automatically for each data set represented in the SMILES notation. All atom-id dependent descriptors (150 descriptors) and descriptors with zero variance were removed. *Step 2.* An initial population of 100 different random combinations of subsets of these descriptors (parents) was generated as follows. Each parent was described by a string of random binary numbers (i.e. one or zero), with the length (total number of digits) equal to the total number of descriptors selected for each dataset. The value of one in the string implied that the corresponding descriptor was included for the parent, and the value of zero meant that the descriptor was excluded. *Step 3.* For every combination of descriptors (i.e., every parent), a QSAR equation was generated for the training data set using the PLS algorithm (11). Thus, for each parent a q^2 value was obtained. This value was further used to calculate the following fitness function: $[1-(n-1)(1-q^2)/(n-c)]$, where q^2 is the cross-validated R^2, n is the number of compounds, and c is the optimal number of components from PLS analysis. This fitness function was used to guide GA. [See our earlier paper (10) for more discussion on the selection of the fitting function.] *Step 4.* Two parents were selected randomly and subjected to a crossover (i.e., the exchange of the equal length substrings), which produced two offsprings. Each offspring was subjected to a random single-point mutation, i.e. randomly selected one (or zero) was changed to zero (or one) and the fitness of each offspring was evaluated as described above (cf. Step 3). *Step 5.* If the resulting offsprings were characterized by a higher value of the fitness function, then they replaced parents, otherwise, the parents were kept. *Step 6.* Steps *3-5* were repeated until a predefined convergence criterion was achieved. As the convergence criterion we used the difference between the maximum and minimum values of the fitness function. Calculations were terminated when this difference was less than 0.02.

In summary, each parent in this method represents a QSAR equation with randomly chosen variables, and the purpose of the calculation is to evolve from the initial population of the QSAR equations to the population with the highest average value of the fitness function. In the course of the GA-PLS process, the initial number of members of the population (100) is maintained while the average value of the fitness function for the whole population converges to a high number.

QSAR based on the KNN principle. The general KNN technique is a conceptually simple, nonlinear approach to pattern recognition problems. In this method, an unknown pattern is classified according to the majority of the class labels of its K nearest neighbors of the training set in the descriptor space. Two sets of descriptors have been utilized: molecular connectivity indices (MCI) as discussed above and atom pairs descriptors (AP) derived on the basis of the approach initiated by Carhart *et al* (12).

The assumptions underlying KNN-QSAR method are as follows. First, structurally similar compounds should have similar biological activities, and the activity of a compound can be predicted (or estimated) simply as the average of the activities of similar compounds. Secondly, the perception of structural similarity is relative and should always be considered in the context of a particular biological target. Since the physicochemical characteristics of receptor binding site vary from one target to another, the structural features that can best explain the observed biological similarities between compounds are different for different biological end-points. These critical structural features are defined in this work as the topological pharmacophore (TP) for the underlying biological activity. Thus, one of the tasks of building a KNN-QSAR model is to identify the best TP. This is achieved by the "bioactivity driven" variable selection, i.e. by selecting a subset of molecular descriptors that afford a highly predictive KNN-QSAR model. Since the number of all possible combinations of descriptors is huge, an exhaustive search of these combinations is not possible. Thus, a stochastic optimization algorithm, i.e., simulated annealing (SA) has been adopted for the efficient sampling of combinatorial space. Figure 1 shows the overall flow chart of the KNN-QSAR method, which involves the following steps.

(1) Select a subset of n descriptors randomly (n is a number between 1 and the total number of available descriptors) as a hypothetical topological pharmacophore (HTP).
(2) Validate this HTP by a standard cross-validation procedure, which generates the cross-validated R^2 (or q^2) value for the KNN-QSAR model. The standard leave-one-out procedure has been implemented as follows. (i) Eliminate a compound in the training set. (ii) Calculate the activity of the eliminated compound, which is treated as an unknown, as the average activity of the K most similar compounds found in the remaining molecules (K is set to 1

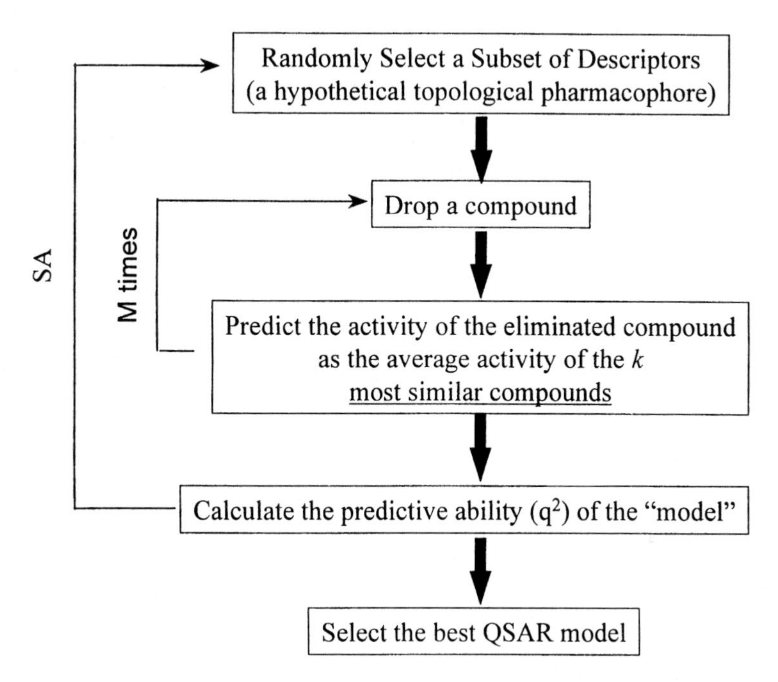

Figure 1. Flow chart of the algorithm for the construction of a KNN-QSAR model.

initially). The similarities between compounds are calculated using only the selected descriptors (i.e. the current trial HTP) instead of the whole set of descriptors. (iii) Repeat this procedure until every compound in the training set has been eliminated and predicted once. (iv). Calculate the cross-validated R^2 (or q^2) value using Eq. 1, where y_i and $\hat{y}_i$ are the actual activity and the predicted activity of the i-th compound, respectively; and $\bar{y}$ is the average activity of all the compounds in the training set. Both summations are over all the compounds in the training set.

$$q^2 = 1 - \frac{\sum (y_i - \hat{y}_i)^2}{\sum (y_i - \bar{y})^2} \quad (1)$$

Since the calculation of pair-wise molecular similarities, and hence the predictions, are based upon the current HTP, the derived q^2 value is indicative of the predictive power of the current KNN-QSAR model. (v) Repeat calculations for K = 2, 3, 4, ..., n. The upper limit of K is the total number of compounds in the dataset; however, the best value was found empirically to lie between 1 and 5. The K that leads to the best q^2 value is chosen for the current KNN-QSAR model.

(3) Repeat steps (1) - (2), i.e., the procedure of generating trial HTP's and calculating corresponding q^2 values. The goal is to find the best HTP that maximizes the q^2 value of the corresponding KNN-QSAR model. This process is driven by a generalized simulated annealing using q^2 as the objective function.

Library Design Using Pre-constructed QSAR Models. Figure 2 shows the schematic diagram of our approach called Focus-2D (5) to the targeted pentapeptide combinatorial library design. The algorithm includes the description, evaluation, and optimization steps. Pentapeptides are assembled from building blocks B_1, B_2, ... B_n (i.e., amino acids of 20 natural types). In order to identify potentially active compounds in the virtual library, Focus-2D employs stochastic optimization methods such as SA (13,14) and GA (15,16,17). The latter algorithm was implemented in this paper as follows.

Initially, a population of 100 peptides is randomly generated and encoded using topological indices or amino acid dependent physico-chemical descriptors. The fitness of each peptide is evaluated either by its chemical similarity to a biologically active probe or by its biological activity predicted from a pre-constructed QSAR equation (cf. Figure 2). Two parent peptides are chosen using the roulette wheel selection method (i. e., high fitting parents are more likely to be selected). Two offspring peptides are generated by a crossover (i. e., two randomly chosen peptides exchange their fragments) and mutations (i. e., a randomly chosen amino acid in an offspring is changed to any of the 19 remaining natural amino acids). The fitness of the offspring peptides is then evaluated and compared with those of the parent peptides, and two lowest scoring peptides are eliminated. This process is repeated for 2000 times to evolve the population. Finally, the frequency of each building block in the final population is calculated, and those with the highest frequency are proposed for the combinatorial synthesis of targeted library.

Database Mining. Similar procedure can be employed for database mining except that the actual rather than the virtual compounds are described in the first step. The protocol for the

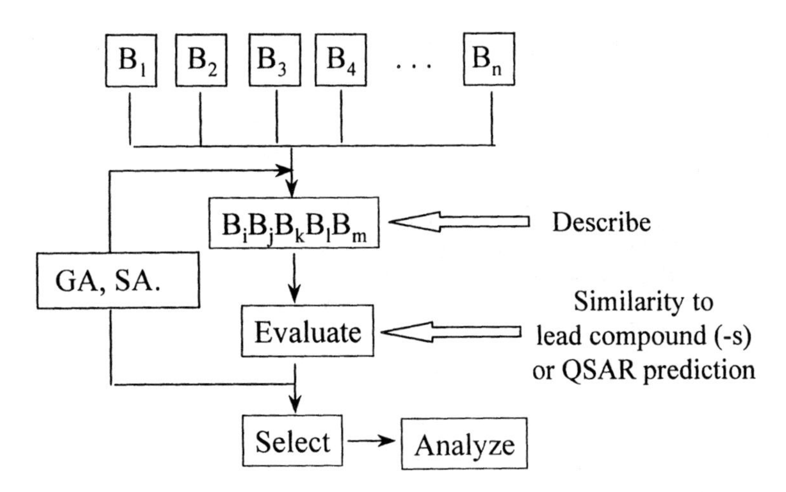

Figure 2. Flow chart of the FOCUS-2D algorithm for targeted library design.

similarity search is given in Figure 3. First, a similarity probe is chosen and both the probe and all compounds in the database are described using molecular connectivity indices. Then, the similarity of this probe molecule to every molecule in the database is calculated as the value of Euclidean distance between the two molecules in multidimensional descriptor space (cf. ref. 5) using either the whole set of descriptors or KNN-QSAR selected descriptors. All compounds in the database are sorted in the order of descending similarity to the probe molecule, and a certain number of the top ranking compounds is selected. The hit rate is evaluated as the percentage of known active molecules found among the selected compounds.

Results and Discussion

The main objective of this paper was to illustrate the principles of using QSAR models for database mining and rational design of targeted chemical libraries. Thus, we first show the results of QSAR modeling for several datasets using our original variable selection methods (i.e., GA-PLS and KNN) and then analyze the use of these models for selected examples of library design and database mining.

Targeted Bradykinin Library Design Using A Pre-Constructed QSAR Equation. We have applied our methodology to the rational design of a targeted library with bradykinin (BK) potentiating activity. Pentapeptide analogs of BK have been described either by topological indices or by a combination of physico-chemical descriptors, generated for each amino acid. The topological indices of virtual pentapeptides were calculated using the Molconn-X program (9).

We have also employed several amino acid based descriptors, including Z_1, Z_2, and Z_3 descriptors (related to hydrophilicity, bulk, and electronic properties of individual amino acids, respectively) reported by Hellberg *et al.* (18) and isotropic surface area (ISA) and electronic charge index (ECI) descriptors reported by Collantes and Dunn (19). In this case, virtual pentapeptides were encoded in the form of a string of descriptor values. Each string consisted of 15 descriptor values (five blocks of three descriptors per amino acid) when using Z descriptors or 10 descriptor values (five blocks of two) when using ISA-ECI descriptors. The following two sections describe two key steps of library design: (i) development of the QSAR model and (ii) the use of this model as fitness function for the evaluation of virtual library compounds.

Development of a QSAR Model. 28 BK potentiating pentapeptides (18,20) were used as a training set to develop a QSAR equation that was employed to predict the bioactivity of virtual library peptides. The log relative activity index (RAI) values of bradykinin potentiating pentapeptides were used as dependent variables. The detailed description of the assay as well as the calculation of relative activity index values were described in the original publications (18, 20).

The two most active compounds, VEWAK and VKWAP, were excluded from the training set. The calculated log RAI values for the training set compounds compared favorably with the experimental data (data not shown). Although the activities of the two excluded peptides were underestimated (the experimental values of log RAI were 2.73 and 2.35 for VEWAK and VKWAP, respectively), the QSAR equations correctly predicted them to have activities higher than those of compounds in the training set. Thus, the log RAI values of 1.79, 1.48, and 1.47 were obtained for VEWAK using ISA-ECI, Z_1-Z_2-Z_3, and topological indices,

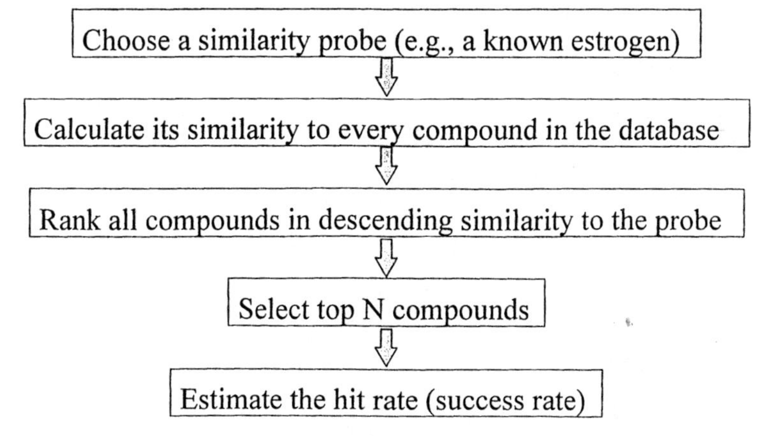

Figure 3. Flow chart of the algorithm for database mining, which is based on the similarity to a probe.

respectively, and the log RAI values of 1.80, 1.74, and 1.95 were obtained for VKWAP using ISA-ECI, Z_1-Z_2-Z_3, and topological indices as descriptors, respectively.

The statistics obtained from the PLS regression analyses and the GA-PLS method is shown in Table I. In order to test the reliability of the prediction using pre-constructed QSAR equations with these descriptors, we incorporated the modified "degree of fit" condition. According to this condition, if RSD of dependent variables of a virtual peptide is less than the RSD of the X matrix of the training set, the predicted values are considered to be reliable. If this condition is not met the log RAI of the virtual peptide is not predicted or set to a low log RAI number to avoid selecting it. The condition does not allow the Focus-2D program to over-extrapolate. Since the number of peptides in the training set is very small compared to theoretical number of different pentapeptides (3.2 million), the extrapolation of QSAR relationship should be done very carefully in small increments, and the "degree of fit" condition implemented here allows us to do this. The RSD values of the X matrix of the training set of 0.886, 0.818, and 0.195 were obtained for ISA-ECI, Z_1-Z_2-Z_3, and topological indices description methods, respectively and used to test the reliability of the prediction (Table 1).

Table I. Summary of the statistics for GA-PLS for 28 BK peptides.

	PLS		*GA-PLS*		
	ISA-ECI[a]	*Z_1-Z_2-Z_3*[b]	*Topological Indices*[c]		
# of crossovers	0	0	0	2000	10000
# of compounds	28	28	28	28	28
# of variables	10	15	160	45	23
ONC[d]	3	2	1	2	5
q^2 [e]	0.725	0.633	0.367	0.533	0.845
SDEP[f]	0.410	0.464	0.598	0.524	0.322
Fitness[g]	0.702	0.619	0.367	0.515	0.818
RSD of the X matrix[h]	0.886	0.818	0.381	0.134	0.195
SDEE[i]	0.313	0.315	0.544	0.466	0.260
R^2	0.840	0.831	0.476	0.630	0.899
F values	42.020	61.355	23.575	21.289	38.984

[a]ISA-ECI (n = 28, k = 3). [b]Z_1-Z_2-Z_3 (n = 28, k = 2). [c]Topological indices: (n = 28, k = 1) for 0 crossover; (n = 28, k = 2) for 2,000 crossovers; and (n = 28, k = 5) for 10,000 crossovers. [d]The optimal number of components. [e]Cross-validated R^2. [f]Standard error of prediction. [g]calculated as $[1 - (n - 1)(1 - q^2)/(n - c)]$. [h]The residual SD of the X matrix. [i]Standard error of estimate.

Focus-2D Using QSAR Equation. The results obtained with Focus-2D and a QSAR based prediction as the evaluation method, are shown in Figure 4 for Z_1-Z_2-Z_3 descriptors. The populations before (Figure 4a) and after (Figure 4b) Focus-2D as well as the population after the exhaustive search (Figure 4c) are shown. The populations after Focus-2D and after the exhaustive search were obviously very similar to each other. With Z_1-Z_2-Z_3 descriptors, Focus-2D analysis selected amino acids E, I, K, L, M, Q, R, V, and W. Interestingly, these selected amino acids include most of those found in two most active pentapeptides, i.e. VEWAK and VKWAP. Furthermore, the actual spatial positions of these amino acids were correctly

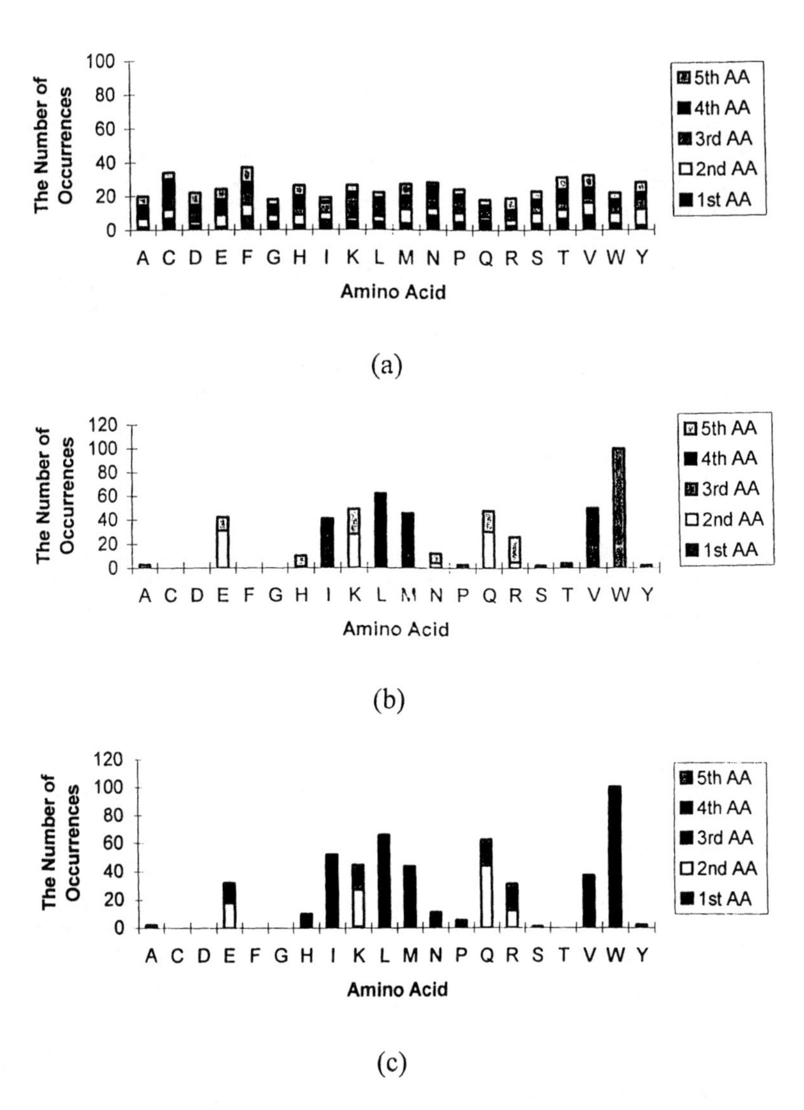

Figure 4. FOCUS-2D using Z_1-Z_2-Z_3 description method and a QSAR equation: (a) initial population; (b) final population after FOCUS-2D; and (c) final population after the exhaustive search.

identified: the first and fourth positions for V; the second and fifth positions for E; the third position for W; and the second and fifth positions for K.

Database Mining Using the Similarity to a Probe. In order to demonstrate the effectiveness of the method, a KNN-QSAR analysis of a set of estrogen receptor ligands was performed using molecular connectivity indices (MCI). The usefulness of KNN-QSAR based similarity search was then demonstrated in a database mining experiment, where a known estrogen receptor ligand was chosen as a probe molecule and molecular similarity was calculated using either the whole set of MCI descriptors, or a subset of 60 KNN-QSAR selected MCI descriptors.

Application of the KNN method to Estrogen Receptor Ligands. 58 estrogen receptor ligands were chosen as a comprehensive test case for the KNN-QSAR technique. This dataset was successfully analyzed earlier by Waller *et al.* (21) using Comparative Molecular Analysis (CoMFA) (22).

In KNN-QSAR method, *nvar* (the number of descriptors to be selected) can be set to any value that is less than the total number of descriptors that are generated by a molecular description method. Since the optimum value of *nvar* is not known *a priori*, several runs are usually needed to examine the relationship between the predictive power of a model (characterized by the q^2 value) and the number of descriptors selected (*nvar*). Figure 5 shows this relationship when MCI was used to describe each of the estrogen receptor ligands. When the real activity values for estrogen receptor ligands were used in the KNN-QSAR analysis, the q^2 values were 0.77, 0.63 and 0.48 for the 10-descriptor model, 60-descriptor model, and 120-descriptor model, respectively. In order to show the robustness of KNN-QSAR analysis, one needs to demonstrate that no comparable q^2 values can be obtained when randomly shuffled activity values or randomly assigned activity values (but within the same range as the real activity) are used in the KNN-QSAR analysis. Figure 5 also shows the q^2 vs. *nvar* relationships when three randomly assigned activity values were used in the KNN-QSAR analysis. Overall, these q^2 values are very low compared to those of the actual dataset. This suggests that the KNN-QSAR models obtained for the actual datasets are distinguishable from those for random datasets. One can also observe that the q^2 values decrease when the number of descriptors increases. On the surface, this may be counter-intuitive. The intuition may come from the fact that the more descriptors are used in multiple linear regression analysis, the higher regression coefficient is normally obtained. However, it should be kept in mind that the KNN-QSAR is not based on a regression method, but rather on the similarity principle. Theoretically, there should be no apparent trend in q^2 vs. *nvar* relationships, although in many practical situations, q^2 tends to decrease slightly when the number of descriptors increases. Conceivably, there should be one optimum number of descriptors, where either the q^2 is the highest or the separation between the q^2 for the real dataset and those for random datasets is the largest.

The plot of predicted vs. actual activity for a 10-descriptor model is shown in Figure 6. Apparently, the trend of the predicted values is similar to that of the real activity values. The results obtained in this work are better than those reported by Waller *et al.* (21) using CoMFA analysis in terms of the q^2 values (0.77 here vs. 0.59 obtained by Waller *et al.*)

Database Mining Using Estrogen Receptor Ligands as Probe Molecules. The database for this search was constructed artificially by putting together several known QSAR datasets with different pharmacological activity. This database contained 358 molecules

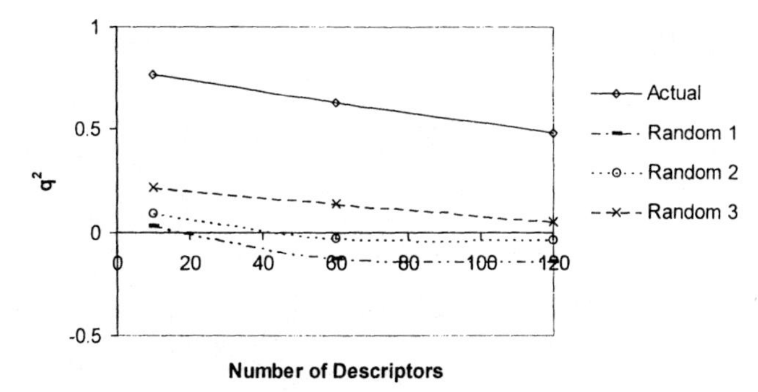

Figure 5. The relationship between q^2 and the number of variables (*nvar*) for estrogen receptor ligands. MCI were used as molecular descriptors. *nvar* is the number of descriptors selected for the final KNN-QSAR model. The results for both actual estrogen dataset and three datasets with random activity values are shown.

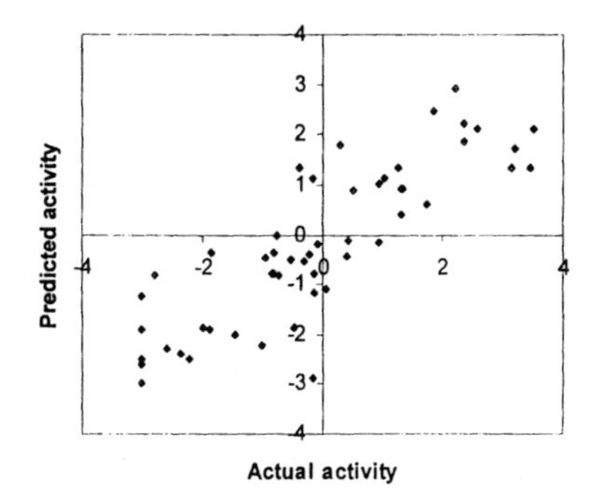

Figure 6. Predicted vs. actual activity obtained from a 10-descriptor KNN-QSAR model for estrogen receptor ligands using MCI as molecular descriptors.

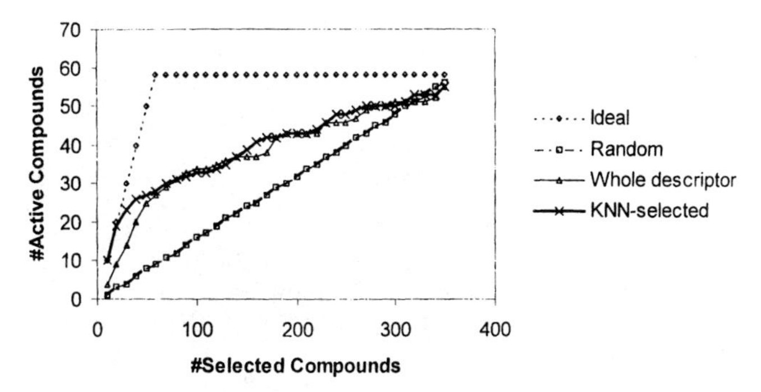

Figure 7. Comparison of the hit rates for known estrogen receptor ligands for ideal, random, or similarity based database mining using butylbenzylphthalate as a probe.

including 58 known estrogen ligands. Eight molecules were chosen arbitrarily from the known estrogen receptor ligands as the query molecules (or probes), one from each of the eight structural classes of these compounds described elsewhere (21). The protocol and the database for this similarity search experiment are described in Computational Details. The typical result of the similarity search is given in Figure 7, which shows the following four curves: (1) The curve of hit rates obtained in the ideal case, where every compound found among the top 58 structures is actually a known estrogen receptor ligand. This is the upper limit that anyone would like to reach. (2) The curve of hit rates obtained by a random search, i.e. by randomly selecting a certain number of compounds and then examining how many known estrogen receptor ligands are found. (3) The two curves of hit rates obtained from similarity searches based either upon the whole set of descriptors or upon the KNN-QSAR selected descriptors.

Our results show that, in most cases the hit rates obtained by similarity search for the known estrogen receptor ligands are more than two times higher than what would be expected from a random search. This demonstrates the effectiveness of the similarity search strategy. As mentioned above, we have used eight different estrogen receptor ligands as probes (all data not shown). In five out of eight cases the hit rates obtained using only descriptors selected by KNN-QSAR were better than those obtained using the whole set of descriptors (cf. Figure 7 as an example). It implies that the KNN-QSAR model has captured the critical structural features (descriptors) that are responsible for specific biological activities of the underlying compounds. In those three cases when the results obtained with KNN-QSAR selected descriptors were no better than those obtained using the whole set of descriptors, the probe molecules probably did not have many similar structures among known estrogen receptor ligands.

Conclusions and Prospectus.

One important aspect of any QSAR investigation is the potential application of the derived QSAR models. It is common to think that in the case of a 3D QSAR method such as CoMFA (22), the results should be used to predict the modifications of known compounds that may lead to more potent ligands. Such applications are not possible using GA-PLS or KNN-QSAR methods since the relationships between molecular descriptors used in these methods and the underlying chemical structures are not obvious. Thus, although these molecular descriptors can be calculated for any molecular structure, the prediction of molecular structure from descriptors is not straightforward if at all possible. However, as we demonstrate in this report, both QSAR methods could be used, in a fairly direct manner, to search for biologically active molecules, either in existing databases (i.e., for database mining) or virtual chemical libraries (i.e., for targeted library design). In this paper, we illustrate two practical applications of QSAR models for such searches. First, the pre-constructed QSAR equation can be used to directly predict biological activity of sampled chemical structures, and the selection of actual or virtual compounds can be based on the (high) value of predicted biological activity. Second, the selection of active compounds can be based on their similarity to a known active probe (lead) molecule. We showed that this similarity searches using the topological pharmacophore derived from the QSAR model, are in general more efficient than using all available descriptors. We believe that the integration of QSAR and rational library design or database mining provides a new exciting avenue for the development and application of QSAR methods that should be further explored in the future.

Literature Cited.

1. Gallop, M. A.; Barret, R. W.; Dower, W. J.; Fodor, S. P. A.; Gordon, E. M. *J. Med. Chem.* **1994**, *37*, 1233-1251.
2. Gordon, E. M.; Barret, R. W.; Dower, W. J.; Fodor, S. P. A.; Gallop, M. A. *J. Med. Chem.* **1994**, *37*, 1385-1401.
3. Johnson, M.; Maggiora, G. M. *Concepts and Applications of Molecular Similarity*; Wiley: New York, 1990.
4. Sheridan, R. P.; Kearsley, S. K. *J. Chem. Inf. Comput. Sci.* **1995**, 35, 310-320.
5. Zheng, W., Cho, S.J., and Tropsha, A. *J. Chem. Inf. Comp. Sci.* **1998** (in press).
6. Cho, S.J., Zheng, W., and Tropsha, A. *J. Chem. Inf. Comp. Sci.* **1998** (in press).
7. The program SYBYL 6.0 is available from Tripos Associates, 1699 South Hanley Road, St Louis, MO 63144.
8. Weininger, D. *J. Chem. Inf. Comput. Sci.* **1988**, 28, 31-36.
9. MOLCONN-X version 2.0, Hall Associates Consulting, Quincy, MA.
10. (a) Available from the authors WWW home page at http://mmlin1.pha.unc.edu/~jin/QSAR/. (b) Cho S. J., Cummins D., Bentley J., Tropsha A. J. Com.-Aided Molec. Design (under revision)
11. Wold, S., Ruhe, A., Wold, H. and Dunn, W. J., III. *J. Sci. Stat. Comput.* **1984**, 5, 735-743.
12 Carhart, R. E.; Smith, D. H.; Venkataraghavan, R. *J. Chem. Inf. Comput. Sci.* **1985**, 25, 64-73.
13. Bohachevsky, I. O.; Johnson, M. E.; Stein, M. L. *Technometrics* **1986**, *28*, 209-217.
14. Kalivas, J. H.; Sutter, J. M.; Roberts, N. *Anal. Chem.* **1989**, *61*, 2024-2030.
15. Goldberg, D. E. Genetic Algorithm in Search, Optimization, and Machine Learning. Addison-Wesley, Reading, MA, 1989.
16. Holland, J. H. *Scientific American* **1992**, *267*, 66-72.
17. Forrest, S. *Science* **1993**, *261*, 872-878.
18. Hellberg, S.; Sjöström, M.; Skagerberg, B.; Wold, S. *J. Med. Chem.* **1987**, *30*, 1126-1135.
19. Collantes, E. R.; Dunn, W. J. III *J. Med. Chem.* **1995**, *38*, 2705-2713.
20. Ufkes, J. G. R.; Visser, B. J.; Heuver, G.; Van Der Meer, C. *Eur. J. Pharm.* **1978**, *50*, 119-122.
21. Waller, C. L.; Minor, D. L.; McKinney, J. D. *Environmental Health Perspectives*. **1995**, 103, 702-707.
22. Cramer, R. D., III; Patterson, D. E.; Bunce, J. D. *J. Am. Chem. Soc.* **1988**, *110*, 5959-5967.

Chapter 14

Molecular Hologram QSAR

Trevor W. Heritage and David R. Lowis

Tripos Inc., 1699 S. Hanley Road, St. Louis, MO 63144

QSAR techniques have proven to be extremely valuable in pharmaceutical research, particularly 3D-QSAR. However, the complexity of descriptor calculation, conformer generation, and structural alignment renders the use of this type of QSAR non-trivial. Furthermore, demands for analysis of large data sets such as those generated by combinatorial chemistry and high throughput screening have compounded this problem. Molecular Hologram QSAR (HQSAR) is a new technique that employs specialized fragment fingerprints (molecular holograms) as predictive variables of biological activity. By eliminating the need for molecular alignment, HQSAR models can be obtained more rapidly than other techniques, rendering them applicable to both small and large data sets. HQSAR models are comparable in predictive ability to those derived from 3D-QSAR techniques and can readily be extended to support chemical database searching.

Since the works of Hansch and Fujita[1], and Free and Wilson[2] demonstrated the successful application of theoretical and computational methods to understanding and predicting biological activity, there has been considerable progress in the development of molecular descriptors and chemometric techniques. The entire field of Quantitative Structure-Activity Relationships (QSAR) has arisen, based upon the underlying assumption that the variations in biological activity within a series of molecules can be correlated with changes in measured or computed molecular features or properties of those molecules. In particular, the development of 3D QSAR techniques that attempt to correlate biological activity with the values of various types of molecular field, for example, steric, electronic, or hydrophobic has been of particular interest[3,4].

The most popular method of 3D QSAR in use today, Comparative Molecular Field Analysis[3] (CoMFA), uses steric and electrostatic field values computed at the intersections of a three-dimensional grid that surrounds the molecules in the data set. Although numerous successes in the use of 3D QSAR to predict biological activity have been reported, there remains the major limitation that the molecules in the data set must be mutually aligned based on some consistent rule or strategy[5,6]. Several approaches[7,8,9] to alleviate this problem have been attempted with only moderate success, and coupled with the conformational flexibility of the molecules in the data set, this problem remains the major barrier to 3D QSAR.

As a consequence, there is considerable interest in the development of alternative descriptions of molecular structure that do not require the alignment of molecules, such as autocorrelation vectors[10], molecular moment analysis[11], vibrational eigenvalue analysis[12] (EVA), and 3D WHIM descriptors[13]. In this chapter, we review a new descriptor of molecular structure, known as the Molecular Hologram, that is based solely on 2D connectivity information. As discussed later in this chapter, Molecular Holograms yield statistically robust QSAR models that are comparable, in statistical terms, to those derived using 3D QSAR techniques, with the key advantage that no 3D structure or molecular alignment is required.

Molecular Hologram QSAR Methodology

Molecular Hologram QSAR (HQSAR) involves the identification of those substructural features (fragments) in sets of molecules that are relevant to biological activity. A key differentiator of this method relative to other fragment based methods such as Free-Wilson[14], or CASE[15] analyses, is that the Molecular Holograms generated encode all possible fragments, including branched, cyclic, and overlapping fragments. Thus, each atom in a molecule will occur in multiple fragments and therefore increment several bins in the Molecular Hologram. Unlike maximal common subgraph algorithms and the Stigmata[16] approach which seek structural commonalities, HQSAR yields a predictive relationship between substructural features in the data set and biological activity using the Partial Least Squares[17] (PLS) technique.

Molecular Hologram Generation. A Molecular Hologram is a linear array of integers containing counts of molecular fragments, and originates from the traditional binary 2D fingerprints employed in database searching and molecular diversity applications. The process of hologram generation is depicted in Figure 1.

The input data set consists of the 2D chemical structures and the corresponding biological responses. The molecular structures are broken down into all possible linear, branched and cyclic combinations of connected atoms (fragments) containing between ***M*** and ***N*** atoms. Each unique fragment in the data set is assigned a pseudo-random, large positive integer value by means of a cyclic redundancy check (CRC) algorithm. Two key properties of the CRC algorithm are that:

(i). very few "collisions" between fragments are observed – that is, each and every unique fragment is assigned a unique integer value,

(ii). the integer value assigned to a particular fragment is always reproducible for that fragment – even between runs.

Each of these integers is then "folded" (or hashed[18]) into a bin (or position) in an integer array of fixed length ***L*** (***L*** is generally a prime number between 50 and 500). The occupancy values for each bin are then incremented according to the number of fragments hashing to their bin. Thus, all generated fragments are hashed into array bins in the range 1 to ***L***. This array is called a Molecular Hologram, and the associated bin occupancies are the descriptor variables.

The hashing function is used to reduce the dimensionality of the Molecular Hologram descriptor, but leads to a phenomenon known as "fragment collision". During fragment generation, identical fragments always hash to the same bin (since they have the same CRC number), and the corresponding occupancy for that bin is incremented. However, since the hologram length is, in most cases, considerably smaller than the total number of unique fragments encountered in the data set, different unique fragments will be hashed to the same bin, causing "collisions" between fragments. This is discussed further in the section on hologram length.

Hologram QSAR Model Building. Computation of the Molecular Holograms for a data set of structures yields a data matrix (***X***) of dimension ***R*** x ***L***, where ***R*** is the number of compounds in the data set and ***L*** is the length of the Molecular Hologram. For QSAR purposes, a matrix of target variables (biological activities) (***Y***) is also created. Standard PLS analysis is then applied to identify a set of orthogonal explanatory variables (components) that are linear combinations of the original ***L*** variables. Leave-one-out crossvalidation is applied to determine the number of components that yields an optimally predictive model.

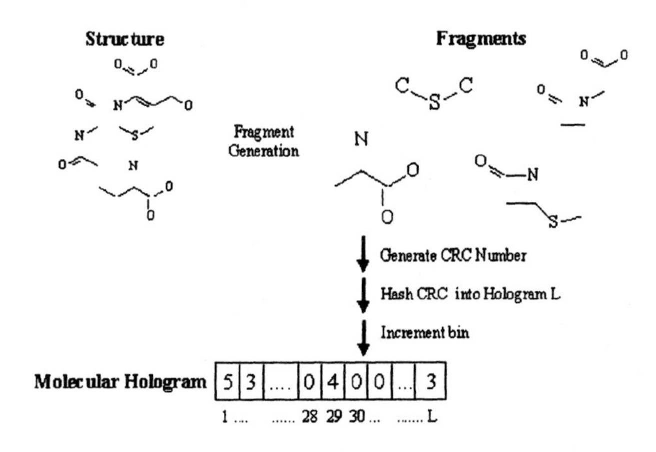

Figure 1. Generation of Molecular Holograms.

Once an optimal model is identified, PLS yields a mathematical equation that relates the Molecular Hologram bin values to the corresponding biological activity of each compound in the data set. The form of this equation for the generated QSAR model is shown by the following equation:

$$Activity_i = c_o + \sum_{l=1}^{L} x_{il} c_l$$

where x_{il} is the occupancy value of the Molecular Hologram of compound i at position or bin l, c_l is the coefficient for that bin derived from the PLS analysis, L is the length of the hologram, $Activity_i$ is the biological activity of compound i, and c_o is a constant.

HQSAR Parameters. As is the case with all other QSAR methods, careful selection of parameters is critical to the success of HQSAR. The key parameters involved in the generation of molecular holograms are hologram length (L), fragment size (M and N), and parameters that control how different fragments distinguished – *atoms, bonds, connections, hydrogens,* and *chirality*.

Hologram Length. The hologram length controls the number of bins in the hologram fingerprint. Since the hologram length is significantly less than the number of fragments in most compounds, alteration of this parameter causes the pattern of bin occupancies to change. The effect of this is to alter the distribution and frequency of fragment collisions. During HQSAR analyses it is important to compare and contrast models generated at several different hologram lengths to ensure that the result observed is not merely an artifact of fragment collisions – lack of consistency in the PLS results at several lengths is a good indication that this phenomenon is occuring. The use of prime number hologram lengths ensures that different fragment collision patterns are observed at each length.

Fragment Size. Fragment size controls the minimum (M) and maximum (N) number of atoms contained within any fragment. These parameters can be changed to bias the analysis toward smaller or larger fragments.

Fragment Distinction. Depending on the application and data set in question, HQSAR allows fragments to be distinguished based on *atoms*, *bonds*, *connections*, *hydrogens*, and *chirality* parameters. The *atoms* parameter enables fragments to be dstinguished based on the elemental atom types they contain, for example, allowing benzene be distinguished from pyridine. The *bonds* parameter enables fragments to be distinguished based on bond orders, for example, in the absence of hydrogen, allowing butane to be distinguished from 2-butene. The *connections* parameter provides a measure of atomic hybridization states within fragments. That is, with *connections* on, fragments are distinguished based on the number and type of bonds made to their constituent atoms

The *hydrogens* parameter enables the generation of fragments that include hydrogen atoms. A consequence of setting this option is that many more fragments are generated. The *chirality* parameter enables fragments to be distinguished based on atomic or bond stereochemistry. Thus, this parameter allows fragments containing a *cis* double bond to be distinguished from the *trans* counterpart, and R-enantiomers to be distinguished from S at chiral centers.

Application of Molecular Holograms in QSAR

One of the first demonstrations of the QSAR modeling power of HQSAR was obtained in a retrospective analysis of a data set endothelin inhibitors. The data set[19] consists of inhibition of endothelin-1 binding to A10 rat thoracic aorta smooth muscle cells for a series of 36 compounds containing an aryl sulfonamide moiety with an isoxazole analog bonded to the amide nitrogen. Analysis of the data set by the CoMFA[3] technique is not straight forward due to different charge computation schemes, structure optimization techniques, and structure orientation schemes, although a model with cross-validated-r^2 (i.e. q^2) of 0.70 and SE of 0.69 can be obtained[19]. Molecular Holograms were generated for each molecule in the data set using lengths in the range 53 to 201, and fragment sizes in the range 2 to 9 atoms. The model based based on holograms of length 53 gave cross-validated-r^2 of 0.59 and SE of 0.81 (see Figure 2). Figure 3 shows the outcome of randomization testing of the HQSAR model. Randomization testing involves randomly redistributing the activity data across the compounds and attempting to derive statistical models that correlate the scrambled data with the molecular descriptor. Figure 3 shows the distribution of randomized q^2 values relative to the observed q^2, and provides a means by which the liklihood that the observed correlation could have arisen by chance can be assessed.

One of the key advantages of the CoMFA and related techniques has been the capability to visualize, using 3D isocontour plots, those regions of space indicated by the PLS model to be highly correlated with the activity data. In HQSAR it is possible to identify, from their PLS coefficients, those bins of the molecular hologram that were most significant in explaining the variation in activity. The fragments in those bins can then be identified, and then each atom in the molecule is color coded based on the fragments that it occurred in. Figure 4 shows the color coding for four members of the sulfonamide endothelin data set described above. It is satisfying that the color coding observed in this set is consistent with the 3D isocontour maps derived from the CoMFA study. Thus, amino group substitution at the 5-position of the 1-naphthyl group is favourable in the most active compound (8), but shifting the substitution around the ring to the 6- or 7-position (compounds 11 and 14) leads to a decrease in activity as indicated by the color coding of the amino group nitrogen atom. A similar trend is seen in the 2-naphthyl series as indicated by compound 31, which has very poor activity.

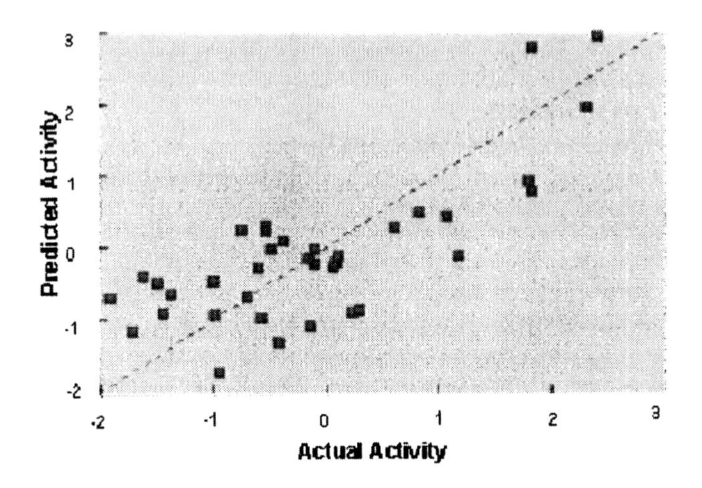

Figure 2. Cross-validated predicted activity *vs.* actual activity for the endothelin data set.

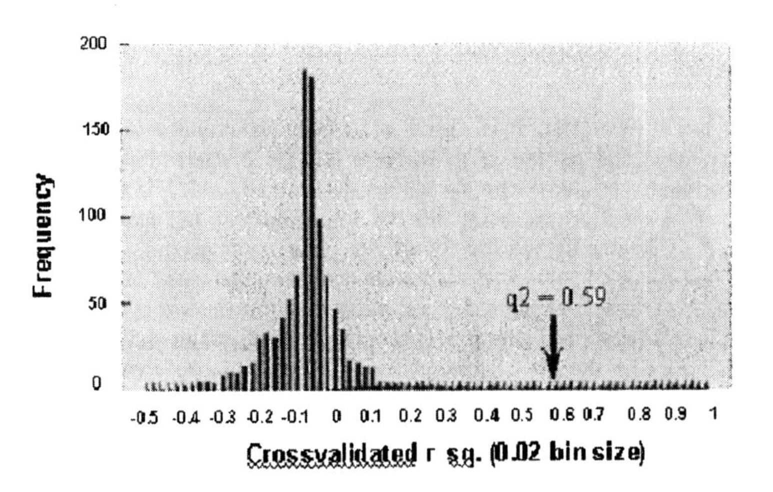

Figure 3. Histogram of cross-validated r^2 frequency of occurrence for 1,000 HQSAR runs with scrambled response data for the Endothelin data set.

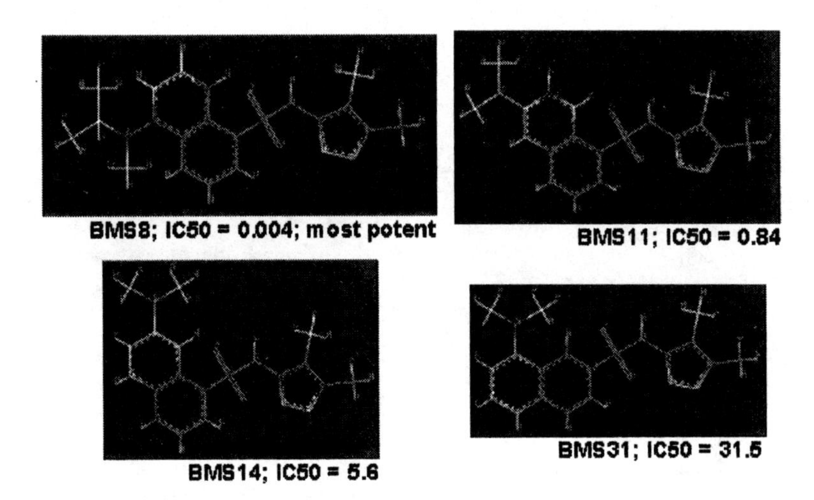

Figure 4. HQSAR model interpretation for four members of the Endothelin data set. (Figure is printed in color in color insert.)

In subsequent studies, the general applicability of the Molecular Hologram descriptor in QSAR studies has been investigated in detail using published data sets exhibiting a range of biological end-points. One key point to note, is that the HQSAR analyses shown in Tables A and B are extremely fast (60 to 120 seconds per data set on an SGI O2 R10K) and the data set preparation time is also minimal. In contrast, the 3D QSAR techniques may take several weeks of preparation in order to generate an appropriate conformation and mutual alignment of structures.

Comparison with 2D QSAR techniques. Table 1 shows a comparison between HQSAR and several 2D QSAR methods, including connectivity indices, clogP/cMr, and descriptors based on molecular formula attributes, for some published data sets. In every case, HQSAR outperforms the other 2D QSAR methods in terms of q^2 statistic – and in some cases by quite a significant margin. In those cases, where the other 2D QSAR techniques generated reasonable models, similar predictive performance as judged by the SE_{cv} statistic.

Comparision with 3D QSAR. Table 2 shows a comparison between HQSAR and 3D QSAR, primarily CoMFA, methods for some published data sets. Good PLS models (in terms of q^2) can be obtained for each of the eight data sets, that are comparable with the corresponding 3D QSAR model in most cases. The dependency of HQSAR on 2D molecular fragments does, however, reduce the generality of the method for *ab initio* predictions of activity for "unseen" compounds – particularly those that contain a large number of fragments that were not encountered in the training set. This is evidenced by the cross-validated standard error of prediction (*SEcv*) statistic shown in the table, which, in general is higher (worse) than the corresponding value obtained from the CoMFA study. In two cases leukotrienes[29] and triazines[26], HQSAR yields a significantly better QSAR model than the 3D technique, Apex-3D and CoMFA, respectively. In the case of the triazines, CoMFA yields a q^2 of 0.47, compared with $q^2 = 0.70$ for HQSAR. The CoMFA result can be significantly improved (to $q^2 = 0.61$) by explicit inclusion of lipophilicity parameters within the regression equation. This result indicates that the Molecular Holograms do incorporate a broad amount of information that has influence over biological activity. In the remaining case, where 3D QSAR performed less well, the angiotensins[30], HQSAR performed similarly.

Comparison of the HQSAR and CoMFA models shown in Table 2 indiates that, in general, CoMFA produces superior models in terms of predictive performance (SE_{cv}), but the similarity in the model statistics suggests that HQSAR may be used as a probe for preliminary SAR prior to spending significant amounts of time building a complex 3D QSAR model. In addition, the similar trend between HQSAR and CoMFA models gives confidence that HQSAR can be reliably applied in cases where CoMFA, or 3D QSAR, is inappropriate or awkward, for example to large data sets.

Table 1: Comparison of HQSAR and 2D QSAR techniques.

Data Set	N	HQSAR			2D QSAR		
		q^2	SE_{cv}	r^2	q^2	SE_{cv}	r^2
Triazolinones[20]	*42*	0.34	0.53	0.67	0.06	-	-
Phenyltrypamines[21]	*32*	0.56	1.13	0.85	0.46	-	-
Benzindoles[22]	*30*	0.69	0.55	0.93	0.53	-	-
MAO hydrazides[23]	*24*	0.80	0.26	0.91	0.80	-	-
Phenylthiothymines[24]	*40*	0.83	0.79	0.99	0.67	0.74	0.87
Bisamidines[25]	*37*	0.82	0.25	0.94	0.51	0.32	0.69

Table 2: Comparison of HQSAR with CoMFA 3D QSAR models.

Data Set	N	HQSAR			CoMFA		
		q^2	SE_{cv}	R^2	q^2	SE_{cv}	r^2
Sulfonamides[19]	*36*	0.59	0.81	0.90	0.70	0.69	-
Triazines[26]	*54*	0.66	0.63	0.84	0.47	-	0.66
Benzodiazepines *DS* [27 b]	*42*	0.65	0.62	0.90	0.70	0.59	0.99
Benzodiazepines *DI* [27 b]		0.62	0.55	0.92	0.73	0.45	0.96
Benzodiazepines *DI/DS*[27 b]		0.81	0.56	0.97	0.79	0.56	0.98
Steroids *CBG* [3]	*21*	0.71	0.81	0.85	0.75	0.66	0.96
Estrogenics[28]	*50*	0.78	0.70	0.89	0.67	-	0.89
Leukotriene antagonists[29]	*13*	0.80	0.64	0.99	0.51[a]	0.51[a]	0.87[a]
Angiotensins[30]	*28*	0.48	0.81	0.85	0.48	0.75	0.93

[a] 3D QSAR result obtained using Apex-3D QSAR instead of CoMFA.

[b] Data set published with two types of binding affinity data – DS is affinity for the diazepam sensitive subtype of the benzodiazepine receptor; DI is affinity for the diazepam insenstive receptor subtype; DI/DS is a measure of ligand selectivity between the two receptor subtypes.

Effect of Molecular Hologram Parameters

In a previous section it was indicated that Molecular Hologram QSAR performance is sensitive to the length selected, due to alteration of the distribution of fragments into hologram bins and changes to the pattern of so-called "fragment collisions". For several data sets, the variation in q^2 for a variety of Molecular Hologram lengths is shown in Figure 5. It is evident that there is no direct correlation between hologram length and the predictive quality of the QSAR model produced. This observation is reasonable since a change in hologram length leads to an unpredictable change in bin occupancies. Furthermore, the chart shows that for some hologram lengths, PLS is able to find much stronger (or weaker) correlations between the fragments and the biological activity. It should be noted that for all hologram lengths some kind of predictive model can be obtained. However, for the purpose of reporting, the most reasonable statistics to use might be those arising from the median model generated across a variety of molecular hologram lengths.

As would be expected, the predictive quality of an HQSAR model is also very dependent on the setting of the *atoms, bonds, connections, hydrogens*, and *chirality* parameters. The way in which HQSAR models will vary is, of course, highly dependent on the nature of the data set under investigation. For example, ignoring chirality during Molecular Hologram generation will have no effect on the final QSAR if the data set contains no enantiomers or bond stereomers – but will be critical for data sets where differences in biological activity are observed between stereoisomers.

Summary

QSAR techniques have proven extremely useful in the design of bioactive molecules Although classical QSAR techniques have provided useful correlations within essentially congeneric series of molecules, the major breakthrough has been with the advent of 3D QSAR. Although 3D QSAR techniques have been shown to have broader applicability and in general yield statistically more robust QSAR models, a major limitation has been the dependency on molecular fields to describe molecules. Inherent in the use of such field based descriptions is the need to identify a bioactive conformation and mutual alignment of the structures, where in most cases there is no a priori reason to select one conformation or alignment rule over another. The sensitivity of the resulting QSAR model to these, essentially arbitrary, choices has led researchers to investigate alternative 3D QSAR techniques that do not require molecular alignment, and also a resurgence of 2D methodologies such as that described herein.

Molecular Holograms provide an entirely empirical description of molecules that are based on traditional 2D fingerprint methodology. The significant advantage that this approach offers relative to 3D QSAR techniques is that it is only necessary to know the atomic connectivity information for the molecules, and thus, ambiguity due to 3D conformation and alignment decisions is removed.

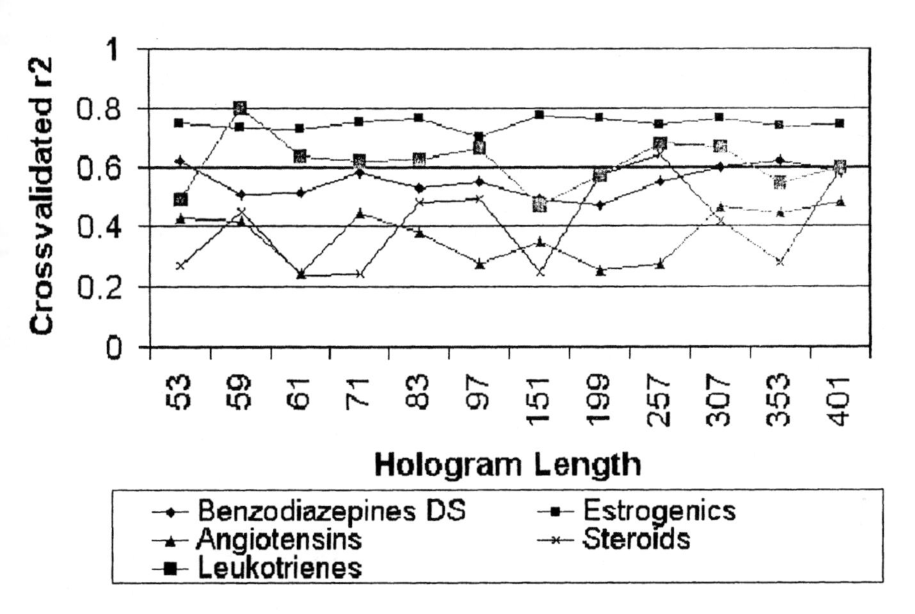

Figure 5. Variation in HQSAR cross-validated r^2 as a function of Hologram Length for several data sets.

The discussion of the QSAR modeling performance of HQSAR presented in this chapter illustrates the broad applicability of the descriptor and the statistical robustness of the resultant QSAR models in terms of cross-validation statistics. Extensive randomization testing of the PLS models discussed herein shows that the probability of obtaining similarly high correlations by chance to those actually obtained using Molecular Holograms is essentially zero. In most cases examined HQSAR outperforms the other 2D based methodologies investigated, in many cases producing results similar to CoMFA in terms of the ability to build statistically robust QSAR models.

The promising results presented herein may lead the reader to believe that HQSAR can be used as a replacement for 3D QSAR, but this is not the case. The superior speed of HQSAR permits its application to large data sets (of the order of thousands to tens of thousands) where 3D methods are simply inappropriate. Furthermore, since an HQSAR analysis of a traditional QSAR data set (of about 50 compounds) can be completed in under half an hour this is a useful technique to probe a data set for any preliminary SAR activity prior to investing more time in the derivation of a more complex model. In this way, HQSAR can be used to guide and monitor the construction of a QSAR data set, to which 3D QSAR can subsequently be applied to provide additional insight into the underlying structure-activity relationship and pharmacophoric information that is not so readily accessible from an HQSAR study.

References

1. Hansch, C.; Fujita, T. *J. Am. Chem. Soc.* **1964**, *86*, 1616-1626.
2. Free, S. M., Jr.; Wilson, J. W. *J. Med. Chem.* **1964**, *7*, 395-399.
3. Cramer, R. D.; Patterson, D. E.; Bunce, J. D. *J. Am. Chem. Soc.* **1988**, *110*, 5959-5967.
4. Kim, K. H.; Martin, Y. C.; J. Org. Chem. **1991**, *56*, 2723-2729.
5. Klebe, G.; Abraham, U.; Meitzner, T. *J. Med. Chem.* **1994**, *37*, 4130-4146.
6. Kellogg, G. E.; Semus, S. F.; Abraham, D. J. *J. Comput-Aided Mol. Design* **1991** *5*, 545-552.
7. Good, A. C. *J. Mol. Graph.* **1992**, *10*, 144-151.
8. Good, A. C.; Hodgkin, E. E.; Richards, W. G. *J. Chem. Inf. Comput. Sci.* **1992**, *32* 188-191.
9. Thorner, D. A.; Wild, D. J.; Willett, P.; Wright, P. M. *J. Chem. Inf. Comput. Sci* **1996**, *36*, 900-908.
10. Wagener, M.; Sadowski, J.; Gasteiger, J. *J. Am. Chem. Soc.* **1995**, *117*, 7769-7775.
11. Silverman, B. D.; Platt, D. E. *J. Med. Chem.* **1996**, *39*, 2129-2140.
12. Heritage, T. W.; Ferguson, A. M.; Turner, D. B.; Willett, P. *EVA: A Novel Theoretical Descriptor for QSAR Studies*; Book chapter in 3D QSAR in Drug Design, Volume 2, Kluwer Academic, (1997).

13. Clementi, S.; Cruciani, G.; Riganelli, D.; Valigi, R. *New Perspectives in Drug Design*; Academic Press, London, 1995, pp 285-310.
14. Free, S. M.; Wilson, J. W. *J. Med. Chem.* **1964**, *7*, 395.
15. Klopman, G. *J. Am. Chem. Soc.* **1984**, *106*, 7315.
16. Shemetulskis, H. E.; Weininger, D.; Blankley, C. J.; Yang, J. J.; Humblet, C. *J. Chem. Inf. Comput. Sci.* **1996**, *36*, 862.
17. Lindberg, W.; Persson, J-A.; Wold, S. Anal. Chem. **1983**, *55*, 643-648.
18. Knuth, D. E. *Sorting and Searching*; Adison-Wesley, Reading, MA, 1973.
19. Krystek, S. R. Jr.; Hunt, J. T.; Stein, P. D.; Stouch, T. R. *J. Med. Chem.* **1995**, *38*, 659-668.
20. Chang, L. L.; Ashton, W. T.; Flanagan, K. L.; Chen, T-B.; O'Malley, S. S.; Zingaro, G. J.; Siegl, P. K. S.; Kivlighn, S. D.; Lotti, V. J.; Chang, R. S. L.; Greenlee, W. J. *J. Med. Chem.* **1994**, *37*, 4464-4478.
21. Garratt, P. J.; Jones, R.; Tocher, D. A. *J. Med. Chem.* **1995**, *38*, 1132-1139.
22. Haadsma-Svensson, S. R.; Svensson, K.; Duncan, N.; Smith, M. W.; Lin, C-H. *J. Med. Chem.* **1995**, *38*, 725-734.
23. Hall, L. H.; Mohney, B. K.; Kier, L. B. *Quant. Struct.-Act. Relat.* **1993**, *12*, 44-48.
24. Hannongbua, S.; Lawtrakul, L.; Limtrakul, J. *J. Comput-Aided Mol. Design* **1996**, *10*, 145.
25. Montanari, C. A.; Tute, M. S.; Beezer, A. E.; Mitchell, J. C. *J. Comput-Aided Mol. Design* **1996**, *10*, 67.
26. McFarland, J. W. *J. Med. Chem.* **1992**, *35*, 2543-2550.
27. Wong, G.; Koehler, K. F.; Skolnick, P.; Gu, Z.; Ananthan, S.; Schonholzer, P.; Humkeler, W.; Zhang, W.; Cook, J. M. *J. Med. Chem.* **1993**, *36*, 1820-1830.
28. Oprea, T. I.; Garcia, A. E., *J. Comput-Aided Mol. Design* **1996**, *10*, 186.
29. Hariprasad, V.; Kulkarni, V. M. *J. Comput-Aided Mol. Design* **1996**, *10*, 284.
30. Belvisi, L.; Bravi, G.; Catalane, G.; Mabilia, M.; Salimbeni, A.; Scolastico, C.; *J. Comput-Aided Mol. Design* **1996**, *10*, 567.

Chapter 15

Adapting Structure-Based Drug Design in the Paradigm of Combinatorial Chemistry and High-Throughput Screening

An Overview and New Examples with Important Caveats for Newcomers to Combinatorial Library Design Using Pharmacophore Models or Multiple Copy Simultaneous Search Fragments

Arup K. Ghose, Veliarkad N. Viswanadhan, and John J. Wendoloski

Department of Molecular Structure and Design, Amgen Inc., One Amgen Center Drive, Thousand Oaks, CA 91320

Impressive advances in the fields of combinatorial chemistry and high-throughput screening have created a strong demand for computational methods for designing combinatorial libraries. Several computational and database tools aid in the design of such libraries. These include tools for reaction planning, scaffold selection, reagent searching, reagent diversification and virtual screening. Given a pharmacophore hypothesis we presented here a novel reaction based virtual library design method. It gave a systematic approach for reaction and reagent selection that will eventually lead to a combinatorial compound library satisfying the pharmacophoric geometry. Similarly, given the 3D structure of the protein-ligand complex (from X-ray or NMR), one can generate a pharmacophore hypothesis using multiple copy simultaneous search (MCSS) which can be used for a library design. An example of a focused library design using the second technique is presented.

1. Introduction

Tremendous advances in combinatorial chemistry[1,2] since its introduction in 1984, in the field of peptide synthesis[3], and the related fields of high-throughput screening[4], robotics and cheminformatics[5] have given us enormous opportunities to synthetically generate and biologically test diverse collections of organic compounds. A reduction in the time required to obtain 'hits' (active compounds which are not necessarily very potent) may be realized as a result. In addition, the advent of high-throughput screening has led to the generation of enormous SAR data that needs to be analyzed by computational chemistry methods, facilitating the goals of medicinal chemistry efforts *viz.*, generation, diversification and optimization of lead compounds.

The goal of generating a combinatorial library for a given biological target is to find active compounds or 'hits'. Obviously, the probability of finding such hits is enhanced if the library is specifically designed for the given target or the complementary pharmacophore. Thus, combinatorial library design should be an integral part of combinatorial chemistry. Two types of combinatorial libraries are used in high-throughput screening to generate SAR data. The first type is the *universal library*, which may be defined as a library designed or constructed such that it can provide hits against a wide range of biological targets. In such a library, there would be obviously no restrictions in favor of particular scaffolds or functional groups. It may include known drugs or ligands that conform the intuition of the medicinal chemist and compounds modeled computationally so that they form a diverse collection, accommodating a variety of pharmacophoric patterns. It will be important to profile the universal library so that, as the biological targets and the corresponding high-throughput screening assays are developed, these libraries can be prioritized for biological testing. The second library type, the *focused library*, is either designed around a scaffold with some R groups or without a common core, but designed to generate compounds likely to fit the pharmacophore in different ways. Once designed and synthesized, these libraries are registered in a database organized to correlate with their chemical, biological and physical data.

Clearly, the success in these efforts depends on the coordination among synthetic, computational and database aspects of combinatorial chemistry and selective application of structure determination techniques (X-ray crystallography and NMR). In this article a general discussion of computational and database aspects of combinatorial chemistry is initially presented. This is followed by a discussion of how to improve the utility of combinatorial chemistry by developing focused libraries. Design of focused libraries is discussed both when the 3D structure of the target protein is available and when such information is unavailable. Particularly, the recently developed technique of multiple copy simultaneous search[6] (MCSS) is shown to be useful in selecting scaffolds and corresponding reagents when 3D structural information on the target protein is available.

2. Computational and Database Aspects of Combinatorial Chemistry.

2.1 Information Technology in Combinatorial Chemistry. Information technology provides the means to create and access databases of compounds, both corporate (in-house) and commercial (3D or 2D structure, chemical, physical or biochemical data, etc.). The environment for storage, retrieval and manipulation of chemical databases can be created by commercially available software such as MDL-ISIS[7]. However, an information system for combichem is necessarily more complex, involving continual construction and management of *virtual libraries* (any computer generated library, universal or focused, that is amenable to high throughput synthesis) of compounds for planning and production stages and for analyzing the libraries for emerging targets. The *project library-central library* construct of MDL-ISIS[7] addresses some of these needs. An example of a compound library registered using that construct is shown in Figure 1. An enumerated version of the same library, depicting individual molecules, usually holds more information such as biological and physicochemical data, plate-id, well location, *etc.* pertaining to each compound in the library. The planning of further 'datamining' experiments (computational, synthetic or biochemical) is facilitated by access to these databases. A smooth communication between a high throughput chemical synthesis robot and the high throughput biological screening robot is another important aspect of combinatorial chemistry. Thus, information technology is critically important in ensuring the success of combinatorial chemistry.

2.2 Small Molecule Lead Development and Combinatorial Chemistry. Figure 2 shows an overview of the lead generation and optimization process in drug discovery and the role of combinatorial chemistry in the process. It is assumed that some SAR information or target protein 3D structure is available, which is used to formulate a plausible pharmacophore hypothesis. Such information often comes from screening of the corporate compound libraries. Database searches along with modeling and visualization would then generate computational 'hits' which may lead to candidate ligands that can be accessed either from a vendor or from an established synthetic route. These 'computational hits' may then be acquired and screened for generating new SAR data. This data provides the crucial information for designing focused

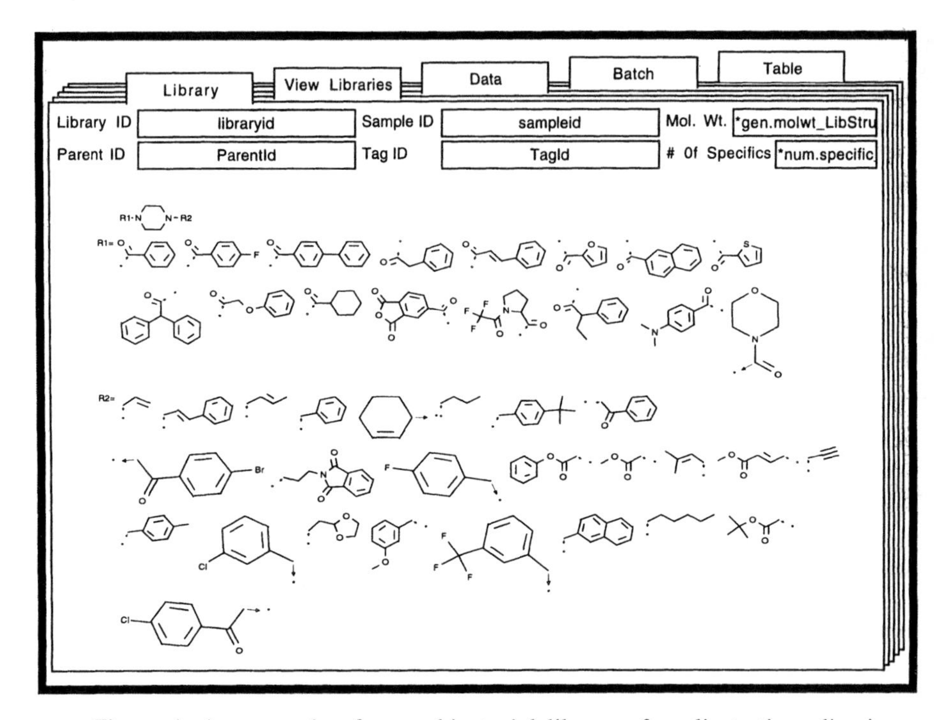

Figure 1. An example of a combinatorial library of cyclic tertiary diamines registered using MDL Project library-Central library construct.

libraries and finally for obtaining optimized leads, by an iterative process involving medicinal chemistry, X-ray crystallography, NMR and computational chemistry in addition to further biological testing. This process culminates in the selection of candidates for clinical development as attested by recent successes[8]. The design aspect of combinatorial libraries (focused and universal) is most critical to the success of lead development. Careful reaction planning, compound diversification and iterative experimentation may be considered integral to library design.

2.3 Reaction Planning, Reagent Searching and Screening. The first challenge in combinatorial chemistry efforts is to plan a combinatorial reaction scheme that can generate a large number of compounds with a central core (generally, but not always) in high yields. Searching the medicinal chemistry literature for common central cores is a popular route to scaffold selection. This may also be dictated in part by the types of reactions typically used by a group (for example, linear *vs.* multi-component condensation (MCC) syntheses). Figure 3 shows a hypothetical reaction scheme for a combinatorial library of tertiary amines. A search of the Available Chemical Directory[9] (ACD) for the reagents, compatible with the reaction scheme, generated 5.2 K tertiary amines, 1.6 K aldehydes and over 20 K alkyl halides, leading to

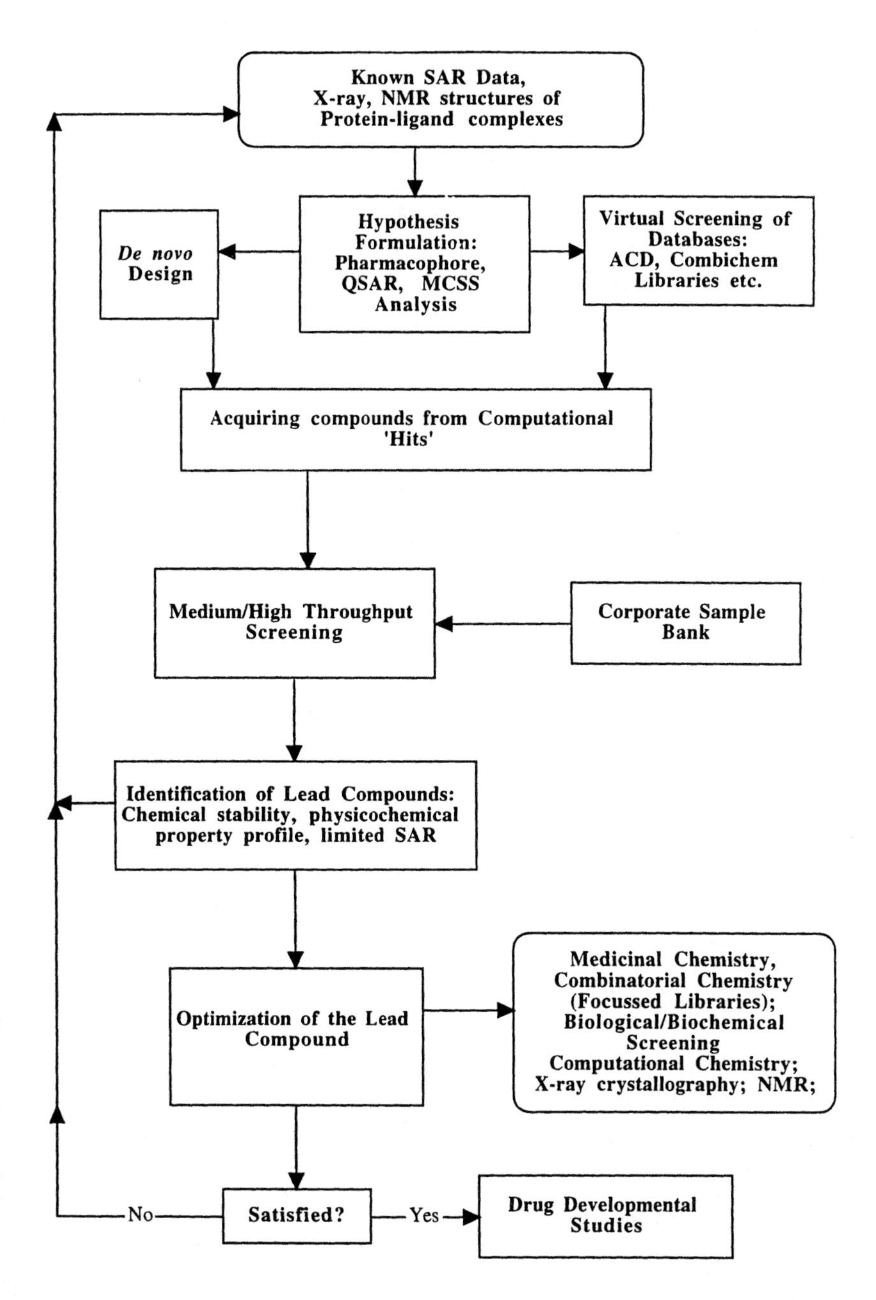

Figure 2 Lead generation and optimization process in the context of combinatorial chemistry.

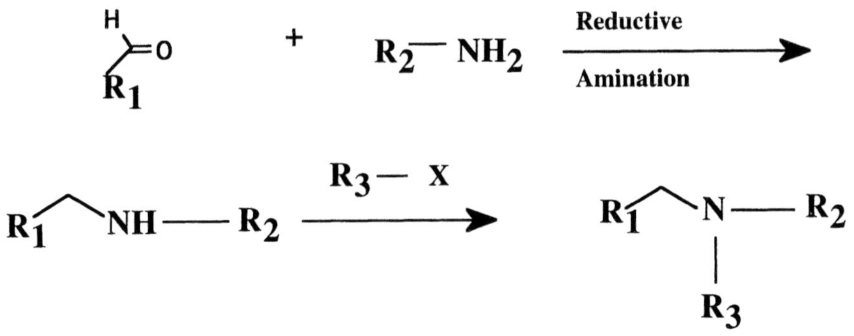

$$R_1CHO + R_2\text{—}NH_2 \xrightarrow[\text{Amination}]{\text{Reductive}}$$

$$R_1CH_2\text{—}NH\text{—}R_2 \xrightarrow{R_3\text{— X}} R_1CH_2\text{—}N(R_3)\text{—}R_2$$

Figure 3 A hypothetical reaction scheme for a combinatorial library of tertiary amines.

approximately 166 billion possible combinatorial compounds. This list is then pruned by a series of screening process that involves removal of polyfunctional, high. molecular weight compounds, etc. Such screenings still kept a list of compounds that is too big to be practicable.

2.4 Reagent Diversification. Since only a limited number of reagents is practical for any library, it is essential to use a diverse collection of reagents at each R group position. Many properties that are relevant to drug action are easily estimated by commercially available softwares. These include fingerprint descriptors (2D, 3D, atom pair, etc.), log *P*, dipole moment, molar refractivity, topological information contents, etc. While the chemical descriptor space can not be uniquely or completely defined, it is possible to select a subset of properties that represent others implicitly. For example, 2D fingerprints supplied by Tripos or MDL software implicitly contains information about hydrogen bonding donors and acceptors. Our clustering experiments with several different properties and combinations showed that 2D fingerprints are quite good general descriptors. However, a greater range of physicochemical properties is obtained when a diverse set of physicochemical descriptors is used. Table 1 shows the range of calculated log *P*'s (CLOGP) and molar refractivities (CMR) when 2D fingerprints are used alone (Set1) and when they are used in combination with calculated CLOGP and CMR values (Set2) for different structural classes. In these experiments a hierarchical clustering algorithm, as supplied by Tripos in their *Selector* module, was used and the compound nearest to the center of a cluster was selected.

3. Improvement of Combinatorial Libraries using Other Technologies.

Improvement of combinatorial libraries using other technologies may have two goals: (i) improvement of universal libraries so that the process of lead generation is accelerated; (ii) improvement of the focused libraries so that the optimization of the biological activity, bioavailability, blood stability, toxicity, etc. is accelerated. Computer technology and computational chemistry may be of great help in making such improvements by analyzing protein and ligand structures, physicochemical or quantum chemical properties. The analysis may be either towards *verification* or *construction*. In the first approach a possible *virtual library* is analyzed and only the members that satisfy the current requirements are made for biological testing. In the

Table 1. Examples showing the effect of using different physicochemical descriptors on the range of calculated properties in selecting compounds in a clustering algorithm.

Structural Class	no. of cpd's sel. / total	CLOGP range		CMR range	
		Set 1	Set 2	Set 1	Set2
Isothiocyanates	50/325	0.38 5.44	-0.15 7.90	2.65 7.40	2.20 7.60
Sulfonyl halides	25/149	0.55 2.65	0.55 3.83	1.59 5.51	1.59 5.81
Epoxides	25/285	-2.24 3.47	-2.24 7.05	1.70 6.86	2.00 7.60

second approach, a virtual library is designed after an analysis of the reagents or the final products. Once designed, all its members may be synthesized for biological evaluation. The ligand structural and physicochemical properties may be used to design the libraries when no protein structure is available; when the protein or the protein-ligand complex structure is available, such information will be used.

3.1 Virtual Libraries and Virtual Screening. As stated earlier, the number of possible compounds is too large to be synthesized due to storage and time factors. Even with robotics automation and combinatorial chemistry, we can make only a minute fraction of the total possibilities. Currently there are many computer programs available that can build libraries from a limited number of instructions. There are also programs that can produce reliable 3D structures for these molecules. Many physicochemical properties relevant for biological activity can be calculated within experimental errors. The physical space necessary to keep such virtual libraries is negligible compared to the real libraries. Under the circumstances we may build a large virtual library provided we can develop a fast and reliable *virtual screening* (computational screening) program.

Unlike physicochemical property prediction, the virtual screening for biochemical or biological affinity may not be very reliable or even possible, since the requirement for biological activity most often is not precisely defined. Even when the drug discovery research starts using a target protein, we may not have any structural clue of the protein binding site until we have a small molecule ligand with a reasonable affinity. The situation may be even worse when the drug discovery process starts with cell or animal assay.

The protein-ligand complex structure is by far the most valuable information to use in developing a virtual screen for a protein binding assay. In its absence the homology model may help to identify suitable ligands by giving an approximate nature of the binding site. The structure of the apoprotein, along with mutation data may also give important information for ligand design. Virtual screening may be developed using any such information.

After some hits are identified, analysis of the structural and physicochemical properties of the related ligands may lead to one or more of the following models: (i) A pharmacophore hypothesis, consisting of the chemical substructures necessary for the activity; (ii) A more precise 3D pharmacophoric geometry; (iii) 1-3D QSAR's; (iv) ligand-protein complex structure. Although a high throughput mass screening of the corporate compound library became the standard for identifying the initial lead compounds, any of the above mentioned available information can be utilized in a virtual screening for a new lead identification or lead optimization.

If we want to build a virtual library which is a few thousand fold bigger than what a high throughput screening assay can handle, virtual screening can be done in two steps: (i) a high throughput virtual screening (HTVS) that will screen out most unrealistic compounds; (ii) a medium throughput virtual screening (MTVS) to prioritize the hits of HTVS. The HTVS may be checking distance ranges between key pharmacophoric atoms, applying 1D-QSAR (a whole molecular property based QSAR, available in many QSAR packages) or 2D-QSAR (scaffold and substituent property based QSAR, the traditional Hansch type QSAR[10,11]). The MTVS may include flexible pharmacophore search, 3D-QSAR (three dimensional structure and spatial property based QSAR[12-14]), molecular mechanics energy minimization of protein-ligand complex, etc. More elaborate calculations like conformational analysis of the ligands, quantum chemical calculations, molecular dynamics, free energy perturbation may be applied mainly to exceptions and difficult problems. In any case a subset of the virtual library will be selected in the virtual screening process and the real library will be generated accordingly.

3.2 Focused Libraries using a Pharmacophore Hypothesis. We will discuss here a novel reaction based combinatorial library design approach using a 3D pharmacophore (hypothesis). This systematic approach may help us to decide reaction strategies and reagents that will lead to one or more combinatorial libraries satisfying the 3D pharmacophore geometry. From the standard medicinal chemistry structure activity relationships (SAR) data it is often possible to make a hypothesis regarding the substructures necessary for the biological activity of interest [15,16]. Comparison of the geometries of a few conformationally diverse compounds may even suggest the 3D geometrical arrangements of these groups. One approach for generating a combinatorial library may be to consider various bio-isosteric groups and attaching them to a suitable scaffold. Let us, for example, consider the pharmacophoric groups necessary for the matrix metalloproteinase (MMP's) activity [15], Figure 4. The 3D geometrical orientation of these pharmacophoric groups is

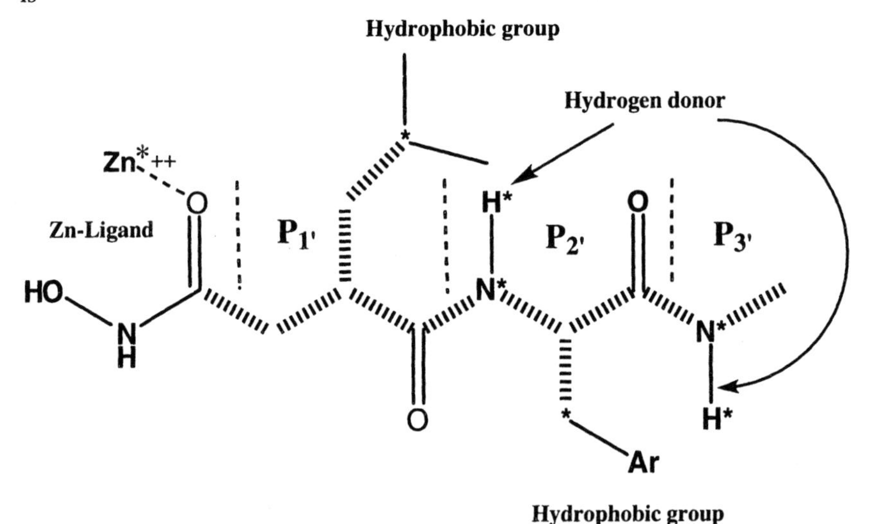

Figure 4. The five pharmacophoric groups that helped to bind with the MMP's are indicated by asterisk mark.

available from the X-ray structure of human fibroblast collagenase. If one wants to find a scaffold [17,18] where all the pharmacophoric groups can be put together he may easily identify scaffolds as shown in Figure 5. One can even build virtual combinatorial structures by altering the *R* groups. This approach may be useful in

designing a virtual library. However, the major problem of this approach is that unless there are suitable reactions and reagents available to attach these groups in a combinatorial fashion, such a library will remain a virtual library only! The method that was found to be very useful in our hands, consists of the following steps:
(i) Start from the most important pharmacophoric group. In the case of the MMP inhibitors it is the hydroxamate group. Keep in mind that this group may not be a good functionality for searching the reagent database like ACD, if it is not very common and if it can be made easily from a common reagent. For example, a carboxylic acid may be a better choice for hydroxamate in the present case.
(ii) Consider one or two other pharmacophoric groups close to the starting one to cover in the initial reagent search. Keep at least one functionality in this reagent that can be used for adding the rest of the pharmacophoric groups. Use pharmacophoric distance constraints to screen reagents. The nature of this *growing* functionality will dictate the reaction and the next reagent searching. The distance of the *growing* functionality from the rest of the pharmacophoric groups can be used to select the next set of reagents. A reagent having only the starting pharmacophoric group with several easy linking functionalities may also be a good choice.

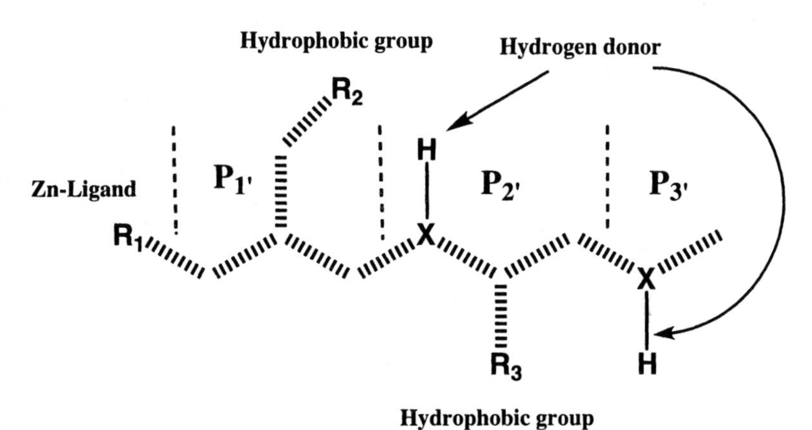

Figure 5. An example of a chemically unplanned combinatorial library. Without appropriate reactions and reagents the virtual combinatorial structures may be unattainable in the laboratory.

(iii) Where there are many choices for regents, diversification of the physicochemical properties of the ligands keeping the pharmacophoric features within a tighter range may be a good strategy.

In the case of the MMP inhibitors, one can start from a carboxylic group as a pharmacophoric group that may eventually be converted to a hydroxamate. The carboxylic acid should have a hydrophobic group to satisfy the distance and orientation requirement of the $P_{1'}$ hydrophobic group. It should have some functionalities to add the other pharmacophoric groups. Major advantages of this type of chemistry aided computational approach are: many structures that will be generated are already known inhibitors of the MMP's (which may be considered as a validation of the procedure); it will be easy to convince a chemist to perform the synthesis (an even better approach in this respect is to discuss the general idea to a chemist and get his input about the reaction and general nature of the reagent during this computer design).

3.3 Focused Libraries using a Protein Structure and the MCSS Computational Method. The major problem in designing compounds or libraries using pharmacophoric information is that the addition of currently nonexistent pharmacophoric groups is arbitrary and based on diversification of the physicochemical properties of the reagent. The process can be more direct if we have the target protein structure and an approximate knowledge of the binding pocket. Theoretical identification of the hot spots of the protein binding pocket was initiated by Goodford. In his GRID program[19], a rectangular box is created in the binding pocket and it is divided into smaller grids. At each grid point different types of representative probe atoms are placed and interaction with the protein is calculated. One major problem of this approach was that atoms in a molecule are always accompanied by dipolar bonds from the neighboring atoms. Both eletrostatic and van der Waals interactions of the neighboring atoms often invalidated the results based on the core atom alone. One immediate solution was to use small molecules like water instead of oxygen, methane instead of neutral carbon, etc. However, it was not a complete solution either since due to multiple minima problem one may get very different results depending on the initial orientation of the molecule. Studying the ligand in many different orientations was also slow. Karplus *et al.* [6] suggested a better approach, the Multiple Copy Simultaneous Search (MCSS). In this method each molecule was docked randomly at the active site in many different orientations. The energy of the whole system (protein and multiple ligands) was minimized keeping the protein fixed. The energy expression did not have interligand interactions, thereby allowing each ligand to move to the nearest minimum. The visual analysis of the MCSS results is available through Quanta software[20], Figure 6. Here the MCSS calculation was done on the N-terminus SH2 domain of the SHPTP2 protein. This intracellular protein binds various phosphotyrosine proteins in the cytoplasmic signaling pathways [21]. Methylphosphonic acid was used as the ligand in this calculation. In the analysis module, one can manipulate ligand display by changing the interaction energy boxes.

A few factors that may be difficult to handle in this type of approach are: (i) solvation of the protein; (ii) structural changes of the protein as a consequence of ligand binding. Solvation energy of the ligand may be computed and used in an *ad hoc* basis to increase the reliability of the interaction energy. In any case the interaction energy should be compared only for very similar groups. In general the interaction energies of the charged groups are much higher than the neutral groups. A recent publication by Joseph-Mccarthy *et al.* [22] claimed that earlier version of MCSS did not have the internal energy of the ligands, thereby giving bent benzene in some of their calculations. In other words, one should be aware of the MCSS version that he is using. Nevertheless, this method can be a very useful method for combinatorial library design. Most of the previous efforts in using MCSS fragments to design combinatorial libraries [22-24] concentrated in building structures that will satisfy the pharmacophoric geometry without taking into account the practical aspects of combinatorial chemistry. A chemistry driven strategy, as discussed in the previous section, once again may give structures that are synthetically tractable as well as a few may be very close to the known inhibitors.

Let us watch more closely the whole process using the SHPTP2 phosphotyrosine binding pocket as an example. When the MCSS calculation was done using a large variety of ligand at the N-terminus SH2 domain, one may identify at least four interesting binding regions, shown in Figure 7. Among these, two are obvious even by analyzing the peptide ligands in the X-ray crystallographic structures: (i) a cationic phosphate binding pocket; (ii) a hydrophobic pocket. Two other binding regions that may be utilized are the Lys-91 and Glu-17 side chains. MCSS interaction energy showed that the phosphate binding pocket is probably the best binding region. This is consistent with the fact that dephosphorylated peptides had very low binding affinity for SHPTP2. We will therefore start searching the reagent from the phosphate binding

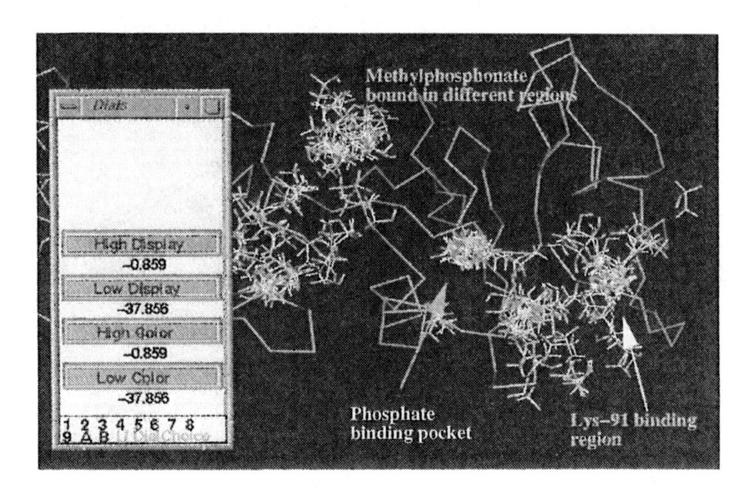

Figure 6. Analysis of the MCSS results of the N-terminus SH_2 domain of SHPTP2 with methanephosphonic acid as the ligand.
(Figure is printed in color in color insert.)

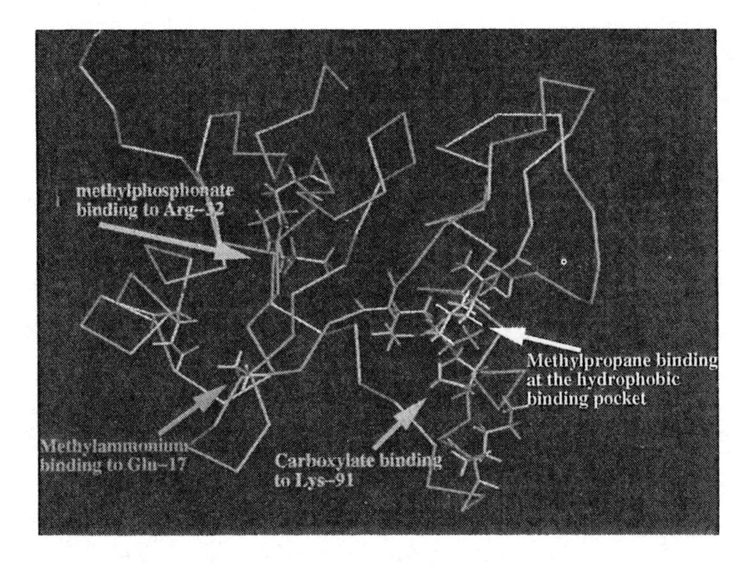

Figure 7. An MCSS deduced pharmacophore model for the SH_2 domain phosphotyrosine binding pocket of SHPTP2 protein.
(Figure is printed in color in color insert.)

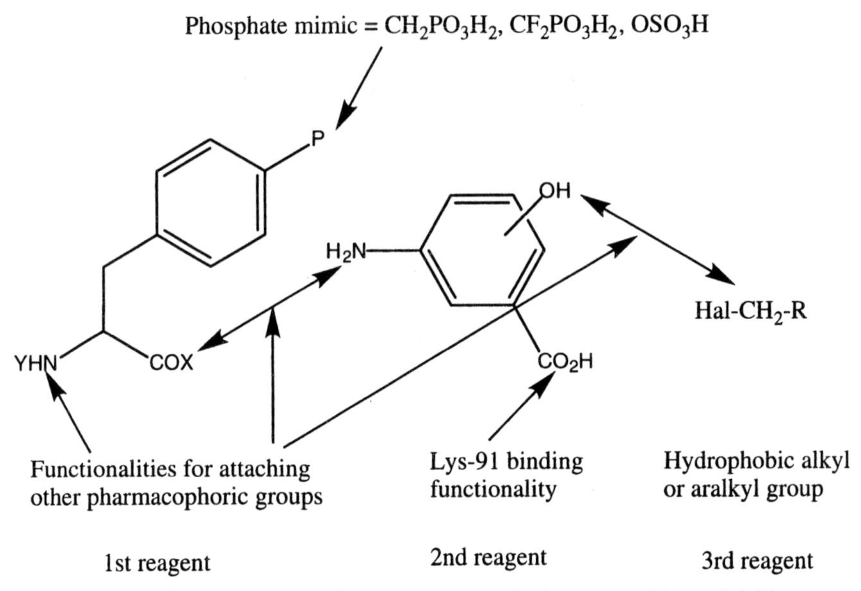

Figure 8. Successive search of reagents to design a combinatorial library that can attain the pharmacophoric geometry.

pocket. Phosphotyrosine type compounds with different phosphate mimicking groups may be the first choice if we do not want to make any major change here, Figure 8. We want to use the carboxy functionality to add reagents that can reach the hydrophobic pocket and the Lys-91 pocket. Here we will search for reagents that can be coupled with the carboxyl group, for example, amines or alcohols. It should have a hydrogen accepting functionality at a desired distance and orientation that will bind with the lysine side chain and so on. If the first reagent is not very flexible such a distance and orientation may be obtained either by model building and minimization or even from the structure of the X-ray ligand and the MCSS functional group. A few immediate hits in this search were the *m*-aminobenzoic acids. To add the hydrophobic groups one can easily think of a phenolic OH group which can be coupled with different alkyl halides.

These binding features of the protein were consistent with the relative binding affinities of several synthetic peptides, made here, whose IC50 value ranged between submicromolar to micromolar level. Interestingly enough, an independent effort by Lunney *et al.* [25] showed that several closely related compounds, Figure 9, showed a moderate binding affinity for PP60 Src SH2 domain. Although these two SH2 domains are structurally very similar there is one obvious difference between the Src SH2 domain and the SHPTP2 SH2 domain. The former has a much shorter loop where SHPTP2 Lys-91 is located. The CONH2 group in the Src SH2 inhibitors interacted with Lys-60 backbone.

4. Concluding Remarks

Combinatorial chemistry is definitely a major blessing in the field of drug design. However, one should not be carried away by the possible number of compounds that can be made, in theory, by this technology. It will be an extremely useful technology if we make smart libraries with all the knowledge of medicinal chemistry. Use of pharmacophore modeling, 3D-QSAR and especially target protein structure in the design of combinatorial libraries may be of major help in designing such smart libraries.

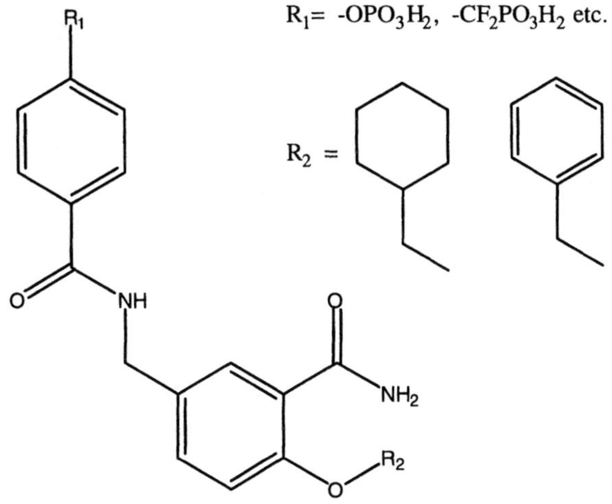

R_1= $-OPO_3H_2$, $-CF_2PO_3H_2$ etc.

R_2 =

Figure 9. A few closely related compounds as proposed for SHPTP2 SH2 domain, showed moderate activity for a closely related protein, PP60 Src SH2 domain.

Acknowledgment

The authors want to thank Drs. Martin Murphy, Gilbert Rishton, Alex Virgilio, Alex Kiselyov, Steven Jordan, Rashid Syed, David Sawutz, Keith Westcott and Wei Zheng for helpful discussions.

References

1. (a) Wilson, S. R. in *Combinatorial Chemistry: Synthesis and Application;* eds. Wilson, S. R.; Czarnik, A. W., John Wiley: New York, NY, 1997; Chapter 1, pp 1–23.
   (b) Armstrong, R. W.; David Brown, S.; Keating, T. A.; Tempest, P. A., in *Combinatorial Chemistry: Synthesis and Application;* eds: Wilson, S. R.; Czarnik, A. W., John Wiley: New York, NY, 1997; Chapter 8, pp 153–190.
2. Chaiken, I. M. and Janda, K. D.; eds. *Molecular Diversity and Combinatorial Chemistry*, American Chemical Society, Washington, DC.
3. Geyson, H. M.; Meloen, R. H. and Barteling, S. J. *Proc. Natl. Acad. Sci. (USA)* , **1984,** *91,* 3998-4002.
4. Bauer, B. E. in *Molecular Diversity and Combinatorial Chemistry*, Chaiken, I. M. and Janda, K. D.; eds. American Chemical Society, Washington, DC, 1996, p233.
5. Weinstein, J. N., Myers, T. G., O'connor, P. M., Friend, S. H., Fornace Jr., A. J., Kohn, K. W., Fojo, T., Bates, S. E., Rubinstein, L. V., Anderson, N. L., Buolamwini, J. K., van Osdol, W. W., Monks, A. P., Scudiero, D. A., Sausville,

E. A., Zaharevitz, D. W., Bunow, B., Viswanadhan, V. N., Johnson, G. S., Wittes, R. E., and Paull, K. D. *Science*, **1997**, 275, 343-349.
6. Miranker, A., Karplus, M., *Proteins,* **1991**, *11,* 29.
7. Integrated Scientific Information System (ISIS), MDL Information Systems, Inc., San Leandro, CA.
8. Borman, S. *Chem. and Engg. News*, **1996**, Feb 12, 29-54.
9. Available Chemicals Directory, MDL Information Inc., San Leandro, CA 94577.
10. Hansch, C., Maloney, P. P., Fujita,T., and Muir, R. M. *Nature,* **1962,** *194,* 178.
11. Hansch, C., Fujita,T., and Muir, R. M. *J. Am. Chem. Soc.,* **1964,** *86,* 1616.
12. Ghose, A. K., Crippen, G.M. *J. Med. Chem.,* **1985,** *28,* 333.
13. Cramer, R. D., Patterson, D. E., Bunce, J. D., *J. Am. Chem. Soc.,* **1988,** *110,* 5959.
14. Ghose, A. K., Crippen, G.M., Revankar, G. R., McKernan, P. A., Smee, D. F., Robins, R. K., *J. Med. Chem.,* **1989,** *32,* 746.
15. Ghose, A. K., Logan, M. E., Treasurywala, A. T., Wang, H., Wahl, R. C., Tomczuk, B. E., Gowravaram, M. R., E. P. Jaeger, Wendoloski, J. J. *J. Am. Chem. Soc.,* **1995,** *117,* 4671.
16. Ghose, A. K., Wendoloski, J. J. in *3D-QSAR in Drug Design,* Kubinyi, H., Martin, Y. C., Folker, G. Eds., Kluware Academic, 1997, in press.
17. Bartlett, P.A., Shea, J.T., Telfer, S.J. and Waterman, S. *Spec. Publ.-R. Soc. Chem., (Molecular Recognition in Chemical and Biological Problems*), **1989**,*78,* 182.
18. Eisen, M. B., Wiley, D. C., Karplus, M. and Hubbard, R. E., *Protein: Structure, Function and Genetics,* **1994**, *19,* 199.
19. Goodford, P. J. *J. Med. Chem.,* **1985,** *28,* 849.
20. Quanta-97, Molecular Simulations Inc., San Diego, CA 92121-3752.
21. Marengere, L.E., Waterhouse, P., Duncan, G. S. Mittrucker, H. W., Feng, G. S., Mak, T. W., *Science,* **1996,** *272,* 1170.
22. Joseph-McCarthy, D., Hogle, J. M., Karplus, M., *Proteins,* **1997**, *29,* 32.
23. Caflisch, A., *J. Comp.-Aided Mol. Des.,* **1996,** *10,* 372.
24. Caflisch, A. and Karplus, M. *Perspec. Drug Dis. Des.,* **1995,** *3,* 51.
25. Lunney, E. A., Para, K. S., Rubin, J. R., Humblet, C., Fergus, J. H., Marks, J. S., and Sawyer, T. K. *J. Am. Chem. Soc.* **1997,** *119,* 12471.

Chapter 16

The Basic Shape Topology of Protein Interfaces

John Lawton [1], Melanie Tudor [2], and W. Todd Wipke [3]

[1] **Molecular Engineering Laboratories, Department of Chemistry and Biochemistry, University of California at Santa Cruz, Santa Cruz, CA 95064**
[2] **Department of Chemistry, Mercer University, Macon, GA 31207**

Basic Shapes are a set of eight differential geometric shape descriptors that capture domain-independent local surface information. This paper describes the use of these shapes to study the surface complementarity of interactions regions in three classes of complexes: protein inhibitor-protein, protein oligomer, anf protein DNA. We derive a shape-shape association plot and a shape parameter affinity model (**SPAM**) that helps in analyzing the degree of shape complementarity.

Shape is an integral part of chemistry, particularly in the area of molecular recognition. At the lowest level, shape dictates the possible orientations that can occur between molecules, which influences their physical properties such as reactivity [1, 2], solubility [2], and associativity [3,4]. In this paper, we have chosen to study the nature of shape associations at protein interfaces.

Previously, we have demonstrated that with a program called QSDock (Quadratic Shape Descriptor Docking Algorithm) [5, 6], it is possible to align accurately and efficiently similar or complementary molecules using only shape. QSDock uses *local* surface properties that include surface normals and principal curvatures to determine transformations intended to optimize either the shape similarity or the shape complementarity between two molecules. The method was found to be both fast and accurate. It docks molecules two orders of magnitude faster than other docking algorithms, and the docked molecule positions have average root mean square deviations (**rmsd**) from crystal ligand orientations less than 1.0 Å. A key feature of the QSDock program was the *explicit* use of shape which reduced the computational complexity of docking.

Shape information can also be used *explicitly* to improve the accuracy of molecular shape comparisons. This can be done by directly comparing the local shapes of two

[3] Corresponding author.

aligned molecular surfaces. Currently shape comparisons are done *implicitly*, by calculating the common surface area or common volume between two molecules, using a three dimensional (**3D**) grid based approach [7]. We were particularly interested in using shape complementarity as a metric for ranking our plausible dockings.

Before one can attempt to quantify shape associations, it is important to experimentally observe the shapes and their interactions in analogous known systems. This raises the obvious question, "How complementary in shape are the interface surfaces of molecules that are found to bind in nature?". A second question then follows, "Is the shape complementarity in nature constant, or is it sensitive to the size of the molecules interacting?". The **3D** structures of complexes determined from crystal and NMR experiments are valuable resources that have made it possible to survey the types of shapes and the types of shape associations that occur at the interface regions. This information should provide a better understanding of the degree to which shape complementarity exists at protein interfaces and the importance of shape complementarity for molecular recognition.

Background. In 1986, Connolly used the shape features of protein surfaces as a basis for docking proteins [8]. His reasoning was that proteins could be docked by associating a set of complementary features of each surface. He successfully docked the *alpha* and *beta* subunits of hemoglobin by matching a set of three or four peaks to a set of pits. While he was not able to elucidate all the docked conformations in his test set, the novel use of shape gave us a glimpse of the potential of explicit shape representation for molecular shape comparison. Since then, several other groups have presented work based on the perception of local and global shape properties of molecules [9–16].

In this paper we present a topological survey of the *basic shapes* [17,18] of protein interfaces for three classes of protein complexes. The *basic shapes* are a set of differential geometric shape descriptors that capture domain-independent local surface information. The intent of this work is to lay the computational groundwork for quantitative measure of molecular shape complementarity or similarity.

Shape: Local Versus Global. We choose to focus on the local aspect of molecular shape. Hence, the shape of each molecule is broken down to a set of shapes distributed over the whole surface. This choice underscores our interest in sub-shape comparisons, and ultimately in developing methodology for quantitatively measuring the shape complementarity between two objects that may differ in size.

Experimental

The **3D** atom coordinates used in this work were taken from the Brookhaven Protein Databank. [19] The dataset consists of three classes: protein inhibitor–protein complexes (**PI-PR**), protein oligomer complexes (**P-OLI**), and protein–DNA complexes (**P-DNA**). A complete listing of all PDB files used in this work is included in Table 1.

Protein Complex Classes. The **PI-PR** class features proteins that present a portion of their backbone as the ligand to a receptor on another protein. The ligand region of

Table 1: Protein complexes used in this study

PDB	Protein inhibitor-Proteins	R (Å)
1cho	α-Chymotrypsin/turkey ovomucoid (3rd domain) [20]	1.8
2ptc	Trypsin/PTI [21]	1.9
2sec	Subtilisin/elgin-c [22]	1.8
2sni	Subtilisin novo/chymotrypsin inhibitor 2 [22]	2.1
	Protein Oligomers	
1bov	Verotoxin-1 [23]	2.2
1lyn	Sperm lysin [24]	2.75
2pab	Phosphofruktokinase [25]	1.8
4hvp	HIV-1 protease [26]	2.3
	Protein-DNA	
1cgp	Catabolite gene activator protein(CAP)-DNA complex [27]	3.0
1gat	Erythroid transcription factor GATA-1 [28]	N/A
1tsr	P53 core domain protein-DNA complex [29]	2.2

the backbone interacts with the other protein in a fashion analogous to a peptidic ligand-receptor complex. These interactions typically have a greater percentage of electrostatic character than typical protein interfaces. [3] In terms of shape associations, it is expected that at short distances, the protein inhibitor associations with the receptor should be primarily between convex shapes on the inhibitor and concave shapes on the receptor. As the distance increases we expect the shape associations to be analogous to protein oligomers, which associate mainly through non-specific Van der Waals interactions.

The **P-OLI** complexes used in this study consisted of identical subunits. The shape distributions for each subunit are expected to be highly similar, and the shape associations should be symmetrical.

The proteins in the **P-DNA** class were constrained to have interactions with the major groove of DNA, although they were not precluded from interacting with the minor groove. The major groove of DNA appears as a smooth surface that flows along the base pair trajectory. The local shape of these surfaces is not expected to be high in information. However, the recognition of DNA is a chemical phenomenon, where the protein interacts with a specific sequence of base pairs. [30] If chemical recognition occurs at the base pairs, shape profiles for the protein may be similar to the shape profiles of the **P-OLI** protein inhibitor.

Shape Perception. The perception of molecular shape is a two step process that first involves the generation of a molecular surface, followed by the characterization of the *local* shape for each point on the molecular surface. We have chosen to use the Connolly Ms-Dot program [31] using a probe radius of 1.4 Å and a surface point density of 4 pts/Å^2 for generating molecular surfaces. The Connolly solvent accessible surface [32,33] provides a continuous surface which is required by our subsequent shape characterization methodology. We then smoothed the solvent accessible surface using a simple convolution algorithm to remove local surface undulations. The convolution algorithm works by averaging the **3D** coordinates of each point and its neighbors within a distance $d = 2.0$ Å. This gives a smoothed molecular surface (**SMS**) which is used as the basis for shape characterization. The Connolly solvent accessible surface and our smoothed molecular surface of methotrexate are pictured in Figure 1 for comparison.

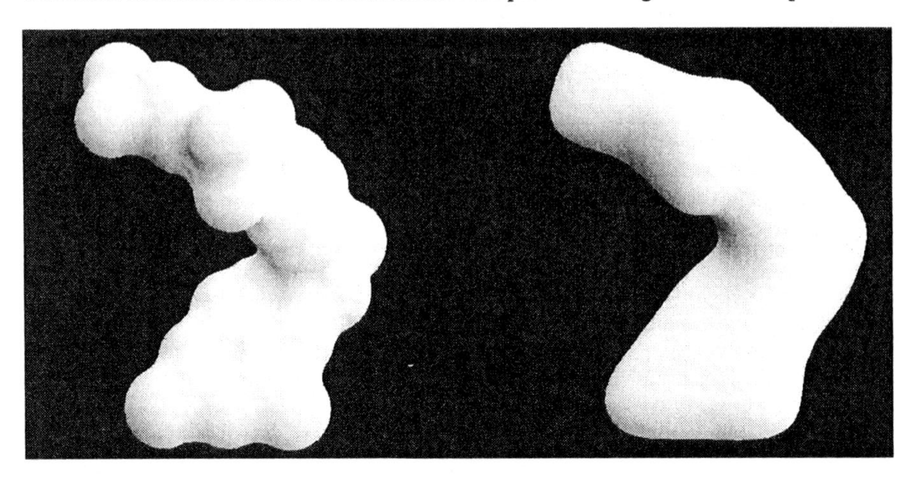

Figure 1: Rendered Connolly Molecular surface (left) and a rendered smoothed molecular surface (right) of methotrexate.

The emphlocal range curvatures [5,6] were calculated for each point on the **SMS**, by fitting a polynomial of the form $ax^2 + bxy + cy^2$ to a circular surface patch of radius $r = 2.0$ Å. Determination of the least squares estimators β [34] for parameters a, b, and c, gave the coefficients of the second fundamental form II or the *Hessian* matrix. For each point, the *principal directions*, the direction of the minimum curvature (k_{min}) and maximum curvature (k_{max}), were calculated by determining the eigenvectors of II. [35]

Gaussian and Mean Curvature. The Gaussian curvature (K) and the mean curvature (H) were used to classify shapes on the molecular surface. Gaussian and mean curvature represent the local second-order surface characteristics that possess the necessary invariance properties for this work. [17] The values for the Gaussian curvature and the mean curvature were computed from the principle curvatures k_{min} and k_{max} using Equations 1 and 2, respectively.

$$K = k_{min} k_{max} \tag{1}$$

$$H = (k_{min} + k_{max})/2 \tag{2}$$

The Gaussian curvature is the product of the principal curvatures and the mean curvature is the average of the principal curvatures. Compared to the principal curvatures, the Gaussian curvature is more sensitive to noise and the mean curvature is less sensitive to noise in the surface points. Additionally, Gaussian curvature is an *intrinsic* property of the surface which makes it insensitive to its orientation. The Gaussian curvature represents the continuity of the curvature of a surface region; it is positive for both concave and convex regions, negative for saddle regions and zero for flat regions. The mean curvature is an *extrinsic* property of the surface and is sensitive to its orientation. Its sign is negative for convex regions and positive for concave regions on the surface.

The *Basic Shapes*. The basic shapes are derived from the signs of the Gaussian and mean curvature of the surface, which yields eight basic surface types: peak, ridge, saddle ridge, flat, minimal saddle, saddle valley, valley, and pit. The use of Gaussian and mean curvature, as opposed to the principal curvatures, allows saddle shapes to be resolved into saddle ridge, saddle valley and minimal saddle. Table 2 depicts the mapping of Gaussian and mean curvature to surface type. The signs of the Gaussian and mean curvature are computed using Equation 3

$$K = \begin{cases} + & \text{if } K > z \\ 0 & \text{if } -z \leq K \leq z \\ - & \text{if } K < -z \end{cases} \tag{3}$$

where $\boldsymbol{z} \in [-0.05, 0.05]$ is a zero threshold .

Table 2: Basic shapes derived from the signs of the Gaussian curvature $K = k_{min}k_{max}$, and the mean curvature $H = (k_{min} + k_{max})/2$. For values of zero in the table, a zero threshold $\boldsymbol{z} \in [-0.05, 0.05]$ was used.

	K		
H	+	z	-
-	Peak	Ridge	Saddle ridge
z	None	Flat	Minimal saddle
+	Pit	Valley	Saddle valley

Surface Interpenetration. In comparing the shape complementarity of molecular surfaces, surface overlap or surface *interpenetration* is inevitable. It is the result of the approximate nature of molecular surfaces in combination with situations where there are electrostatic attractions at the molecular interface. The Algorithm for determining inter-surface associations must be generalized to handle surface interpenetration as well as non-overlapping conditions.

Inter-surface Distances. Surface associations were determined (see algorithm 1) by projecting points from the destination surface $\boldsymbol{S'}$, onto a surface normal $\boldsymbol{n}$ emanating from the source surface $\boldsymbol{S}$. The shortest distance $\boldsymbol{t}$ along the normal, subject to a projection distance threshold $\boldsymbol{d_t}$, was taken as the surface association distance. Additionally, the index to the *closest* point S'_j on the destination surface, was also returned. (see Figure 2)

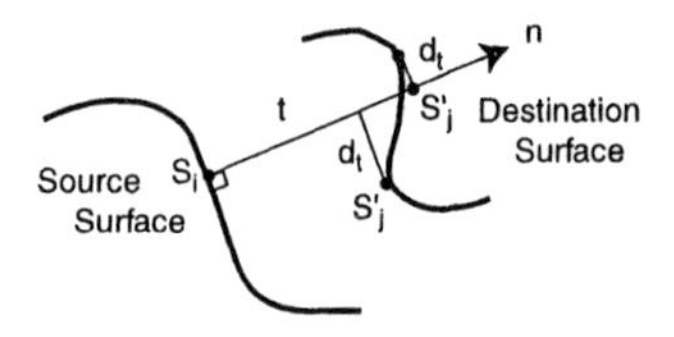

Figure 2: Inter-surface association is determined by finding the shortest distance t along the surface normal that is within a distance threshold d_t to points on the destination surface.

The projection distance threshold d_t was derived empirically from the approximate point dispersal as a function of the point density used to generate the Connolly Molecular surface. For a surface generated at 4 pts/Å^2 the projection distance threshold was set to $d_t = 0.35$ Å.

A grid based representation was used in the function *GetNeighbors* (see Algorithm 1) to minimize the computational overhead when determining the set of neighboring points S'_n. The functions *PointLineDist* and *PointOntoLine* are simple geometric algorithms that have been described elsewhere. [36]

Inter-surface Associations. The procedure for determining surface associations (Algorithm 2) loops twice, using each surface as both the source and the destination surface. It is similar to an approach used by Lawrence [37] to quantify surface complementarity based on the angle and inter-surface distance between associated surface normals of opposing molecular surfaces. It has been found experimentally that surface associations from points on the source surface with lower curvature had better inter-surface alignments. Thus, surface associations were only considered if the absolute value of the shape index on the source surface was lower than the shape index of the destination surface. The shape index [35] of a surface $S \in [-1, 1]$ was derived from the local range curvatures k_{min} and k_{max} using the following equation:

$$S = -\frac{2}{\pi} atan \left(\frac{k_{max} + k_{min}}{k_{max} - k_{min}} \right) \tag{4}$$

This had the effect of limiting the number of surface normals emanating from surfaces with a high degree of curvature, thus reducing a source of spurious associations, and more closely followed human perception of surface-surface associated points.

Results

In fourteen complexes used in this study, the **PI-PR** class had maximum surface interpenetration $t_{min} \in [-1.50, -1.25]$ Å, while the **P-OLI** and **P-DNA** classes had maximum surface interpenetration distances over the range $t_{min} \in [-0.75 - 0.50]$ Å (Table 3). (No surface interpenetration less than -1.5 Å was detected in the datasets used

Input: s, S', d_{max};
Output: N, t_{min}, j;

1: $S'_n \leftarrow$ GetNeighbors(s, d_{max}, S');
2: $N \leftarrow$ Count(S'_n)
3: **if** N is 0 **then**
4:    return 0 {No neighbors were found}
5: **end if**
6: **for** $i \in S'_n$ **do**
7:    $d' \leftarrow$ PointLineDist($s.n, S'_{n,i}$);
8:    **if** $d' > d_t$ **then**
9:       goto step 4;
10:    **end if**
11:    $p' \leftarrow$ PointOntoLine($s.n, S'_{n,i}$);
12:    $t \leftarrow$ Dist(s, p');
13:    **if** $t < t_{min}$ **then**
14:       $t_{min} \leftarrow t$;
15:       $j \leftarrow i$;
16:    **end if**
17: **end for**

Algorithm 1: **InterSurfDist:** Given a set of points on the destination surface $\boldsymbol{S'}$ a maximum association distance $\boldsymbol{d_{max}}$, and a point $\boldsymbol{s = S_i}$ on the source surface, the inter-surface distance is calculated by projecting each neighboring point onto the surface normal, and ranking the distance $\boldsymbol{t}$ along the surface normal. The number of neighbors $\boldsymbol{N}$, the shortest distance $\boldsymbol{t_{min}}$, and the associated index $\boldsymbol{j}$ are returned.

Input: S, S';
Output: M;

1: **for** $i \in S$ **do**
2:    $N, j, t \leftarrow$ InterSurfDist(S_i, S'); {See Algorithm 1}
3:    **if** N is 0 **then**
4:       goto step 1 {No point was found}
5:    **end if**
6:    **if** Ang(S_i, S'_j) $< 90°$ **then**
7:       goto step 1 {Invalid surface association ($inside \Leftrightarrow outside$)}
8:    **end if**
9:    **if** |ShapeIndex(S_i)| $\leq$ |ShapeIndex(S'_j)| **then**
10:       goto step 1 {source S_i must be lower than dest S'_j}
11:    **end if**
12:    $M[S_i, S'_j] \leftarrow M[S_i, S'_j] + 1$ {Increment Shape Association Profile}
13: **end for**
14: $S \leftrightarrow S'$ {Swap surfaces}
15: goto step 1 {Do a second pass}

Algorithm 2: **CreateShapeAssocProfile:** Given two sets of **3D** coordinates $\boldsymbol{S}$ and $\boldsymbol{S'}$ representing a source and destination surface, create a Shape Association Profile by tabulating all shape pair associations. (The *ShapeIndex* function is defined in Equation 4)

Table 3: Results from the analysis of surface associations, where t_{min} is the shortest inter-surface distance, and t_{maxN} is the inter-surface distance where the maximum number of associations were found. A negative distance means that the surfaces are overlapping, implying electrostatic attractions.

complex	t_{min} (Å)	N	t_{maxN} (Å)	N
PI-PR	-1.50	(4)	0.25	(1164)
P-OLI	-0.75	(7)	0.50	(1932)
P-DNA	-0.75	(4)	0.50	(642)

in this analysis.) The inter-surface association distance ranges for t_{maxN}, was shorter for the **PI-PR** complexes $t_{maxN} \in [0.00, 0.25]$ Å, relative to the **P-OLI** and **P-DNA** complexes which had a distance range of $t_{maxN} \in [0.25, 0.50]$ Å.

Shape Information. Shape association information is presented in two forms, shape subunit profiles and shape association profiles. Shape subunit profiles show the type and relative number of shapes, as a function of distance, present at the molecular interface. Shape association profiles are used to convey information about the relative degree that two shapes are associated for a given distance range. For each shape association profile, forty shape association matrices were used to characterize shape associations over a distance the range $t \in [-5.0, 5.0]$ Å, where each matrix held shape association frequencies over a 0.25 Å distance range. Shape profiles for subunits of a given class were determined by summing up the rows or the columns of a shape association matrix at each distance increment. Since shape subunit profiles are more fundamental, we will present our results first using that analytical method.

Shape Subunit Profiles. The shape subunit profiles for the **PI-PR**, **P-OLI**, and **P-DNA** complexes depicted in Figure 3 facilitate analysis of the number and the types of shapes that occur at protein interfaces, as a function of inter-surface distance. The frequency of surface associations is at a maximum at the inter-surface distance range $t \in [0.00, 0.50]$ Å for all three classes of proteins. The number of associations decreases rapidly as the inter-surface distance increases. There is one notable exception to this trend in the **P-OLI** dataset, where the frequency of valleys increases to a second smaller maximum at $t = 4.25$ Å.

At shorter distances, the shape subunit profiles consist mainly of peaks, ridges valleys, and pits. As the distance increases, the histograms tend to flatten out and show significantly less bias towards convex and concave shapes.

Shape Association Profiles. The shape association profiles at distance ranges $t \in [0.25, 0.50]$ Å and and $t \in [4.75, 5.0]$ Å are depicted in Figures 4 and 5, respectively. The plot on the left hand side of the figures are the shape association counts normalized to give the joint probability for a given association between two shapes. The plots on the right hand side of figures 4 and 5 are log odds plots where the shape as-

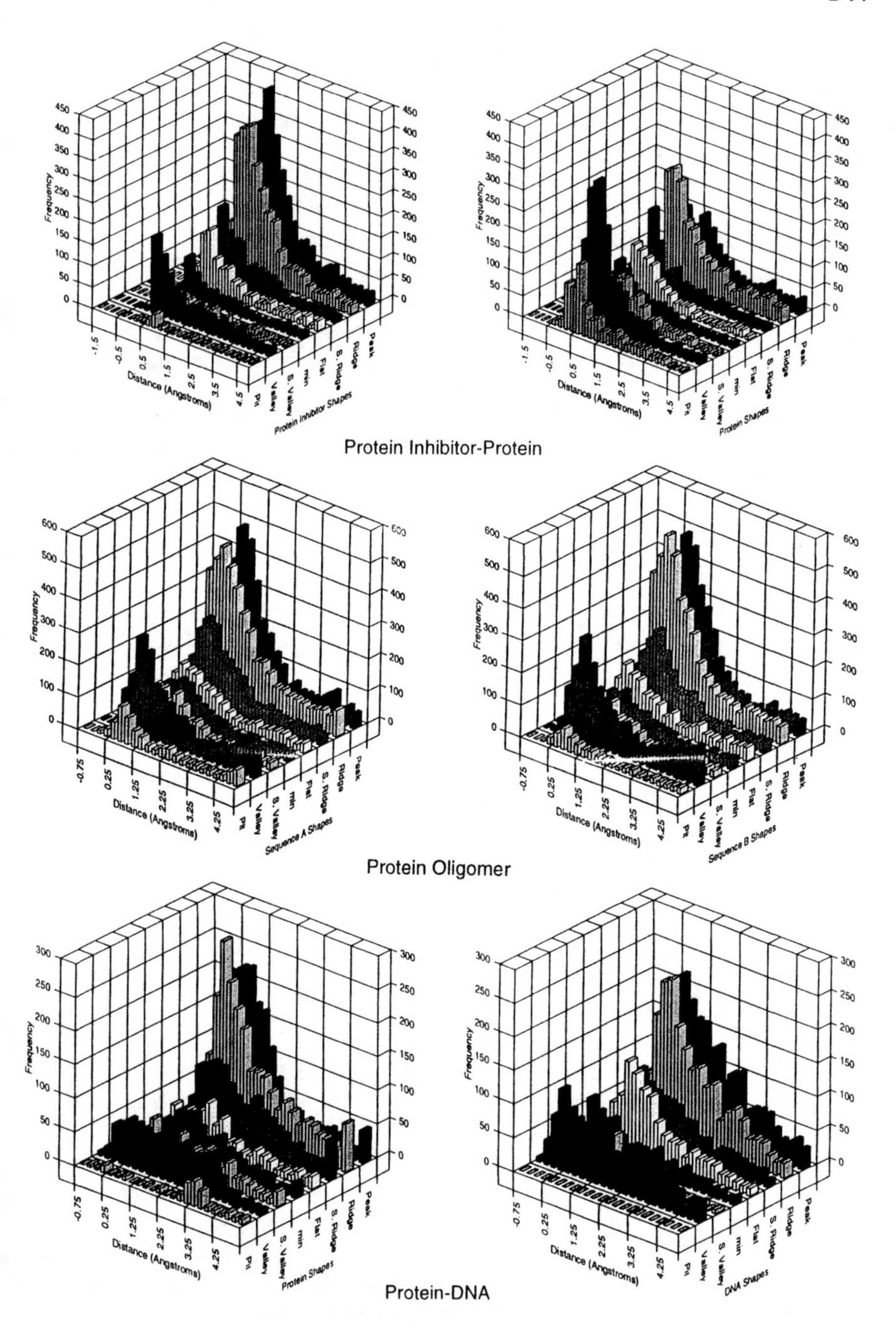

Figure 3: Shape subunit profiles derived from the contact pairs of points at the protein interface, for protein inhibitor-protein complexes (top), protein oligomer complexes (middle), and protein-DNA complexes (bottom).

sociation probabilities are normalized by the independent probabilities for each shape. A positive log odds score indicates that the association has a higher probability of occurrence than random and is therefore statistically significant. A log odds score that is negative indicates that the association has a lower probability of occurrence than random and is statistically significant.

At shorter distances shape associations were observed in three out of four quadrants of the shape matrix, which indicates a propensity for convex-concave, concave-convex and convex-convex interactions. The absence of concave-concave shape associations, corresponding to the *fourth* quadrant, was expected. At the longer inter-surface distance range of $t \in [4.75, 5.0]$ Å, the preferences for convex-concave and convex-convex interactions were still apparent in the **PI-PR** and **P-DNA** datasets. However, convex-concave associations were no longer dominant in all datasets.

We feel it is necessary to comment on the data derived from the **P-OLI** dataset. The similarity of the shape subunit profile and and symmetry of the shape association profile of the **P-OLI** dataset was consistent with the fact that all complexes were homodimers, with each subunit possessing essentially identical shape.

Discussion

The graphs in Figures 3-5 summarize the qualitative results of both shape distributions and shape associations for three classes of protein interactions. Most surface associations occur at distances between 0.0-0.5 Å, decrease rapidly, then level off as the surface distance increases. (Figure 3) At short inter-surface distances, the shape distribution favors highly curved shapes such as peaks, pits, ridges, and valleys.

In the **PI-PR** class, the protein inhibitor shows more peak and ridge with a little pit and valley, whereas the protein receptor shows a high proportion of valley and pit interactions, as one might expect, since the inhibitor is behaving as a small molecule fitting into the protein receptor. The **P-OLI** class shows a complete symmetry, each partner has the same shape distribution showing that the mode of interaction is symmetrical. In the **P-DNA** class the protein shows more peak and ridge than the than valley, whereas the DNA shows significant amount of valley shapes as we would expect for interactions in the *groove*.

At $t \in [0.0, 2.5]$Å, the shape association profile for the **PI-PR** class (Figure 4, left) shows a high probability for inhibitor peaks and ridges interacting with receptor pits and valleys. In the **P-OLI** class, we see a similar pattern, but now its symmetrical. Surprisingly there is a high probability for peak-peak interactions, but when we observe the log odds plot for peak-peak (Figure 4, right) we see that there is no significance to this association, but there is significance to the peak-pit and peak-valley associations. The large negative log odds peak for valley-valley means that that association occurs less frequently than the random probability of the valley-valley association. Moving to the **P-DNA** class, the shape associations (Figure 4, bottom left) are similar to **P-OLI**, except we see less peak-pit. Again we see a high probability for peak-peak associations which in the log odds plot (Figure 4, right) has no significance, i.e. does not occur more frequently than would be expected on a random basis. The log odds plot does show peak-valley interactions to be significant as well as protein ridge-DNA saddle valley, but

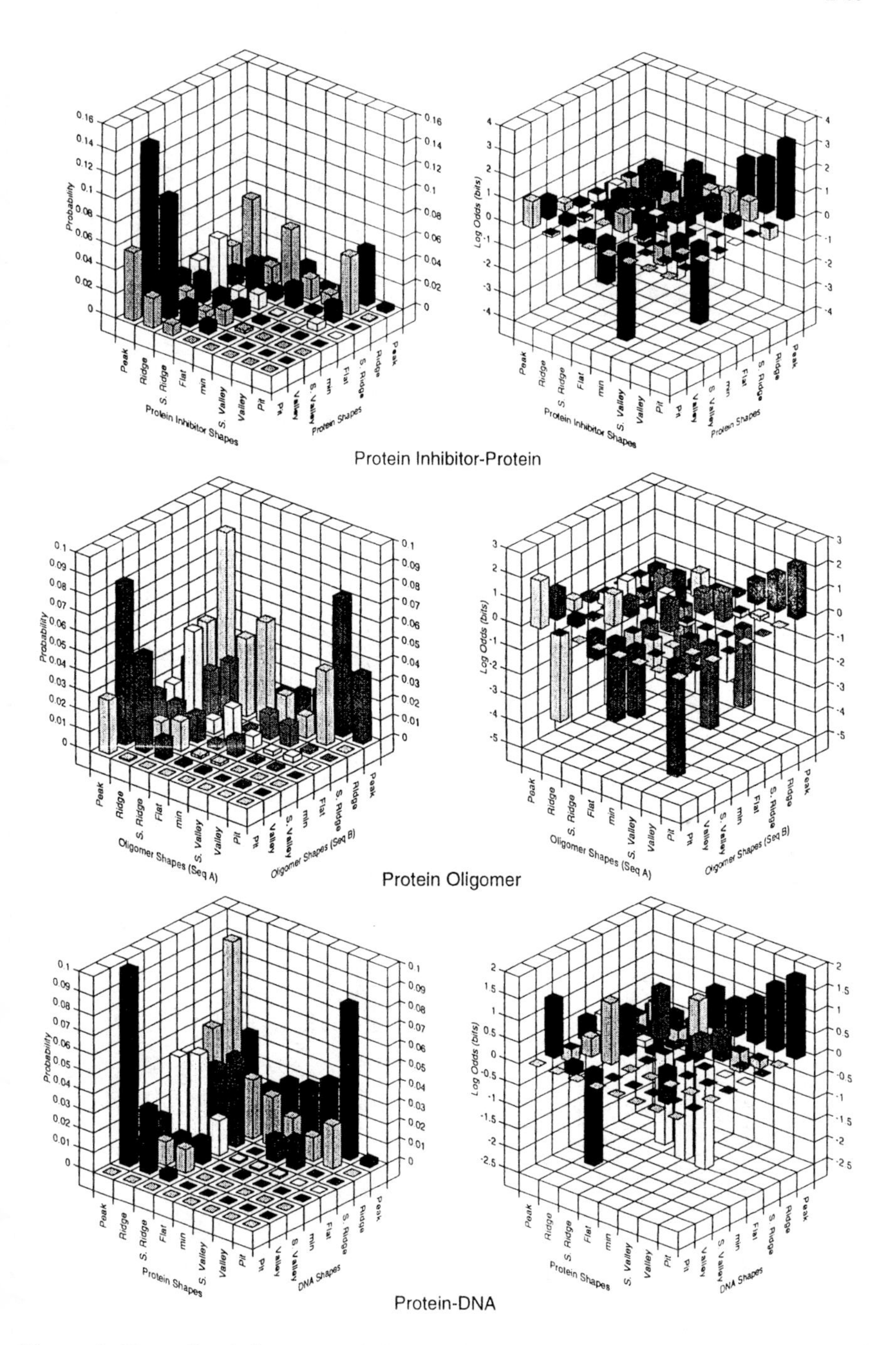

Figure 4: Normalized shape association profiles (left) and log odds shape association profiles (right) over the range $t \in [0.0, 0.25]$ Å.

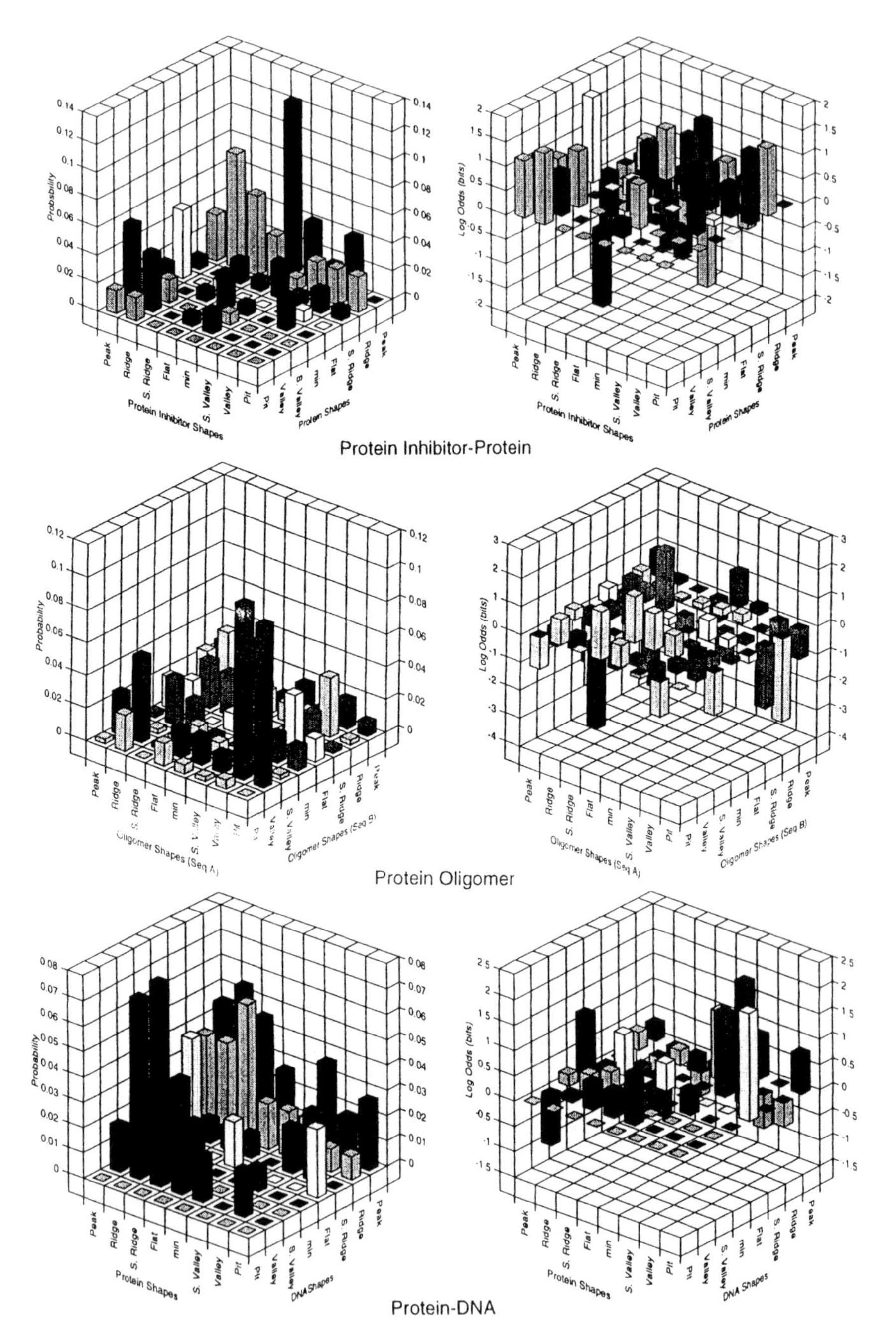

Figure 5: Normalized shape association profiles (left) and log odds shape association profiles (right) over the range $t \in [4.75, 5.0]$ Å.

protein saddle ridge-DNA valley interaction is significantly absent, for they cannot fit together. Similarly, flat-flat interactions have a low probability, but occur even less often than expected. Peaks on the DNA interacting with pits on the protein are infrequent, but significant. Saddle ridge-saddle ridge interactions show up as significant interactions in **PI-PR** and **P-DNA**, although their frequency is only moderate.

At longer interaction distances $t \in [4.75, 5.00]$Å (Figure 5) there is an interesting inhibitor flat-protein peak interaction that is significant in the **PI-PR** class. In the **P-OLI** class there is a surprisingly high probability of pit-valley and valley-valley interaction of reasonable significance. Similarly in the **P-DNA** class protein pits are correlated to DNA flats and protein ridge to DNA flats. The general trend we observe is that at these greater inter-surface distances there is poor and sometimes inverse shape complementarity. The concave-concave relationship shows there is a cavity between the surfaces. There are no concave-concave shape associations at $t \in [4.75, 5.00]$Å, for the close surface separation precludes the surfaces bending away from each other.

Shape Complementarity. With our representation of surface interfaces, it is possible to qualitatively assess the degree of shape complementarity of an interface for a given 0.25 Å distance range. Assuming that complementary interactions involve surface associations between shape pairs of opposite degrees of curvature, i.e., pit-peak, ridge-valley, ... peak-pit, we would expect the diagonal elements of the matrix to have high shape association probabilities as well as a high degree of correlation (positive log odds scores). This is not observed in the normalized shape association matrices but is observed in the log odds plots as a general trend at short distances (Figure 4, right), but not at larger distances (Figure 5, right). We see the highest correlations at each corner of the shape matrix for complementary interactions between surfaces of high curvature, but with weaker complementary correlations between the flatter surfaces. As the inter-surface distance increases, the shape complementarity of the interactions is further reduced, and significant negative complementarity appears. It is difficult to see any diagonal trend in the probabilities plots at the larger distance, but there is still a diagonal bias in the log odds plot (Figure 5, right), especially in the **PI-PR** class.

We conclude that the molecular shape parameter affinity model (**SPAM**) is quite helpful in analyzing interacting surfaces of complexes for shape complementarity. We will report elsewhere other applications of the SPAM model such as scoring molecular dockings.

Acknowledgments

This work was supported by NIH grant EY06914 and NSF REU grant CHE-9619961. We wish to acknowledge discussions with Brian Goldman and Ed Dratz, and the use of the Quadratic Shape Descriptor (QSD) program developed by Brian Goldman.

References

[1] Abraham, W. H. "Scales of Solute Hydrogen Bonding – Their Construction and Application to Physicochemical and Biochemical Processes". *Chem. Soc. Rev* **1993**, *22*, 73–83.

[2] Abraham, W. H. "Applications of Solvation Equations to Chemical and Biochemical Processes". *Pure and Appl. Chem.* **1993**, *65*, (12), 2503–2512.

[3] Chothia, C.; Janin, J. "Principles of Protein-Protein Recognition". *Nature* **1975**, *256*, 705–708.

[4] Janin, J.; Chothia, C. "The Structure of Protein-Protein Recognition Sites". *J. Biol. Chem.* **1990**, *256*, (7), 16027–16030.

[5] Goldman, B. B. *Applications for Ligand Docking and Similarity Searching*, Ph.D. Dissertation, University of California at Santa Cruz, 1997.

[6] Goldman, B. B.; Wipke, W. T. "QSD Quadratic Shape Descriptors .1. Rapid Superposition of Dissimilar Molecules Using Geometrically Invariant Surface Descriptors". *J. Chem. Inf. Comput. Sci.* **1998**, submitted.

[7] Stouch, T. R.; Jurs, P. "A Simple Method for the Representation, Quantification, and Comparison of the Volumes and Shapes of Chemical Compounds". *J. Chem. Inf. Comput. Sci.* **1986**, 4–12.

[8] Connolly, M. L. "Shape Complementarity at the Hemoglobin $\alpha_1\beta_1$ Subunit Interface". *Biopolymers* **1986**, *25*, 1229–1247.

[9] Mezey, P. G. "The Shape of Molecular Charge Distributions: Group Theory without Symmetry". *J. Comp Chem.* **1986**, *8*, (4), 462–469.

[10] Leicester, S. E.; Finney, J. L.; Bywater, R. "Description of Molecular Surface Shape Using Fourier Descriptors". *J. Mol. Graphics* **1988**, *6*, 104–108.

[11] Zachman, C.; Heiden, W.; Schlenkrich, M.; Brickmann, J. "Topological Analysis of Complex Molecular Surfaces". *J. Comp Chem.* **1991**, *13*, (1), 76–84.

[12] Connolly, M. L. "Shape Distributions of Protein Topography". *Biopolymers* **1992**, *32*, 1215–1236.

[13] Mezey, P. G., *Shape in Chemistry, An Introduction to Molecular Shape and Topology*, VCH Publishers: New York, NY, 1993.

[14] Duncan, B. S.; Olson, A. J. "Approximation and Characterization of Molecular Surfaces". *Biopolymers* **1993**, *33*, 219–229.

[15] Grant, J. A.; Pickup, B. T. "A Gaussian Description of Molecular Shape". *J. Phys. Chem* **1995**, *99*, 3503–3510.

[16] Leherte, L.; Latour, T.; Vercauteren, D. P. "Similarity and Complementarity of Molecular Shapes: Applicibility of a Topological Analysis Approach". *J. Comput-Aided Mol. Design* **1996**, *10*, 55–66.

[17] Besl, P. J.; Jain, R. C. "Invariant Surface Characteristics for 3D Object Recognition in Range Images". *Computer Vision Graphics and Processing* **1986**, *33*, 33–80.

[18] Besl, P. J.; Jain, R. C. "Segmentation Through Variable-Order Surface Fitting". *IEEE trans. Pattern Anal. Machine Intell.* **1988**, *10*, (2), 167–198.

[19] Bernstein, F. C.; Koetzle, T.; Williams, G.; Meyer, E.; Brice, M.; Rodgers, J.; Kennard, O.; Shimanouchi, T.; Tasumi, M. "The Protein Data Bank: A Computer-Based Archival File for Macromolecular Structures". *J. Mol. Biol.* **1977**, *112*, 535–542.

[20] Fujinaga, A.; Sielecki, R.; Read, R. J.; Ardlett, W. "Crystal and Molecular Structures of the Complex of α-Chymotrypsin with its Inhibitor Turkey Ovomucoid Third Domain at 1.8 Å Resolution". *J. Mol. Biol.* **1987**, *195*, 397–418.

[21] Marquart, M.; Walter, J.; Deisenhofer, J.; Bode, W.; Huber, R. "The Geometry of the Reactive Site and of the Peptide Groups in Trypsin, Trypsinogen and Its Complexes with Inhibitors". *Acta Cryst.* **1983**, *B39*, 480–490.

[22] McPhalen, C. A.; James, M. N. G. "Structural Comparison of Two Serine Proteinase-Protein Inhibitor Complexes: Eglin-C-Subtilisn Carlsberg and CI-2-Subtilisn Novo". *Biochemistry* **1988**, *27*, 6582–6598.

[23] Stein, P. E.; Boodhoo, A.; Tyrell, G. J.; Brunton, J. L.; Read, R. J. "Crystal Structure of the Cell Binding β-Oligomer of Verotoxin-1 from *E-coli*". *Nature* **1992**, *355*, (6362), 748–750.

[24] Shaw, A.; Mcree, D. E.; Vacquier, V. D.; Stout, C. D. "The Crystal Structure of Lysin, A Fertilization Protein". *Science* **1993**, *262*, (5141), 1864–1867.

[25] Blake, C. C. F.; Geisow, M. J.; Oately, S. J.; Rérat, B.; Rérat, C. "Structure of Pre-albumin: Secondary, Tertiary and Quaternary Interactions Determined by Fourier Refinement at 1.8 Å". *J. Mol. Biol.* **1978**, *121*, 339–356.

[26] Miller, M.; Sathdanarayana, J. S. B. K.; Thoth, M. V.; Marshall, G. R.; Clawson, L.; Selk, L.; Kent, S. B. H.; Wlodawer "Structure of Complex of Synthetic HIV-1 Protease with a Substrate-Based Inhibitor at 2.3 Å Resolution". *Science* **1989**, *246*, 1149–1152.

[27] Schultz, S. A.; Steitz, G. C. S. T. A. "Crystal Structure of a CAP-DNA Complex -The DNA is Bent by 90-Degrees". *Science* **1991**, *253*, 1001–1007.

[28] Clore, G. M.; Gronenborn, A. M. "Structures of Larger Proteins, Protein-Ligand and Protein-DNA Complexes by Multidimensional Heteronuclear NMR". *Protein Science* **1994**, *3*, 372–390.

[29] Cho, Y. J.; Gorina, S.; Jeffery, P. D.; Pavletich, N. P. "Crystal Structure of a P53 Tumor Supressor DNA Complex -Understanding Tumorigenic Mutations". *Science* **1994**, *265*, (5170), 346–355.

[30] Jeffrey, G. A.; Saenger, W., *Hydrogen Bonding in Biological Structures*, Springer-Verlag: Berlin, Heidelberg, Germany, 1991.

[31] Connolly, M. "Molecular Surface-Dot (ms, qcpe 429)" 1981.

[32] Connolly, M. L. "Solvent Accessible Surfaces of Proteins and Nucleic Acids". *Science* **1983**, *221*, 709–713.

[33] Connolly, M. L. "Analytical Molecular Surface Calculation". *J. Appl. Cryst.* **1983**, *16*, 548–558.

[34] Rawlings, J. O., *Applied Regression Analysis*, Wadsworth and Brooks/Cole: Pacific Grove, California, 1988.

[35] Koenderink, J. J., *Solid Shape*, The MIT Press: Cambridge Ma, 1990.

[36] Bower, A.; Woodwark, J., *A Programmers Geometry*, Butterworths: London, England, 1983.

[37] Lawrence, M. C.; Colman, P. M. "Shape Complementarity at Protein-Protein Interfaces". *J. Mol. Biol.* **1993**, *234*, (4), 946–950.

Chapter 17

Evolutionary Algorithms in Computer-Aided Molecular Design: A Review of Current Applications and a Look to the Future

David E. Clark

Computer-Aided Drug Design, Rhone-Poulenc Rorer Ltd., Dagenham Research Centre, Rainham Road South, Essex RM10 7XS, United Kingdom

In the last few years, evolutionary algorithms have proved to be very powerful search and optimization procedures in many rational drug design applications. This chapter begins with a brief introduction to the two main classes of evolutionary algorithm in use at present, genetic algorithms and evolutionary programming. Following some comments concerning the implementation of evolutionary algorithms, a number of selected applications of these algorithms will be reviewed. The application areas covered include protein-ligand docking, *de novo* molecular design, QSAR and combinatorial libraries. After examining some pros and cons of evolutionary algorithms, some future directions for their application will be postulated.

Many of the problems encountered in rational drug design are inherently difficult from a computational viewpoint. In many cases, computer-aided molecular design (CAMD) tasks involve the optimization of a function whose response surface is complex and/or a search through vast (combinatorial) solution spaces. This state of affairs is ameliorated somewhat by the fact that, in rational drug design, one is usually content to locate "good" solutions rather than the "best" (or global optimum). Consequently, the application of heuristic algorithms becomes attractive.

In the last decade, evolutionary algorithms have emerged as a class of heuristic algorithm and have found widespread application in the field of computer-aided molecular design. Milne recently calculated that, between 1989 and 1992, there were only 5 published papers with a chemical orientation that employed evolutionary algorithms. Since 1993, however, that figure has exploded to reach

210 (*1*). My own research corroborates these figures, as the graph shown in Figure 1 demonstrates. It is clear that the number of papers published in the field of computer-aided molecular design that have employed evolutionary algorithms has grown rapidly, and continues to do so. Note that the figure for 1997 only counts papers published up to the month of August.

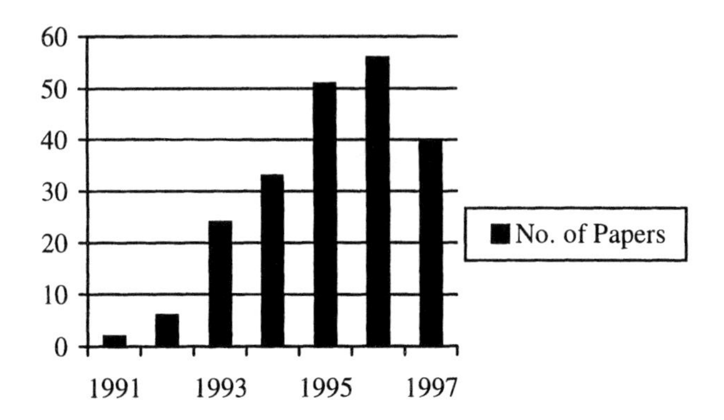

Figure 1: Number of publications in CAMD using evolutionary algorithms

It is the purpose of this chapter to introduce the most prominent classes of evolutionary algorithm and then survey some selected applications in the area of rational drug design. Various issues surrounding the implementation of evolutionary algorithms will be aired and the pros and cons of this type of algorithm discussed. Finally, I shall seek to delineate some possible future directions for the application of evolutionary algorithms to rational drug design.

Evolutionary Algorithms

Broadly speaking, evolutionary algorithms are a class of search and optimization algorithm inspired by the mechanisms observed in natural selection. Traditionally, three main groups of evolutionary algorithms have been distinguished:

- Genetic algorithms (with genetic programming as an offshoot)
- Evolutionary programming
- Evolution strategies

All of these types of evolutionary algorithm share key unifying features such as the use of reproduction, random variation, competition and selection, giving rise to the evolution of superior solutions over time. Consequently, the historical distinctions between them are rapidly blurring and researchers in the evolutionary algorithm community now prefer the umbrella term of "evolutionary computation" to cover the above classes as well as some others not mentioned here (*2*).

For the purposes of this chapter, the focus will be upon genetic algorithms and evolutionary programming because these have received the most attention from practitioners of rational drug design. In what follows, a brief description of these two types of algorithm will be given. For more details on genetic algorithms, evolutionary programming and other kinds of evolutionary computing, the reader is referred to the comprehensive work edited by Baeck, Fogel and Michalewicz (*3*) and a growing number of introductory texts (*4-10*).

Genetic Algorithms

The steps involved in a simple genetic algorithm are given below and explained in more detail thereafter:

1. Randomly initialise a starting population of N members
2. Assign each member a fitness score using a fitness function
3. Select a pair of parents for reproduction
4. Generate offspring using crossover and/or mutation
5. Assign each offspring a fitness score using a fitness function
6. Replace least fit members of population by the offspring if the offspring are superior in fitness
7. Go to 3 until termination or convergence

In common with other evolutionary algorithms, a standard implementation of a genetic algorithm works with a *population* of individuals, each of which represents, or is itself, an attempted solution of the problem under study. The quality of each of these attempted solutions is determined by a *fitness function* which assigns a score or *fitness* to each of the population members. The genetic algorithm proceeds by the repetitive application of *genetic operators* to the population; these operators being *selection*, *crossover* and *mutation*, in the simplest case.

The genetic algorithm uses a *selection* operator to choose a pair of parents from the population from which offspring will be produced. The selection operator generally embodies some kind of "survival of the fittest" mechanism in so far as parents are selected for breeding with a probability that is proportional to their fitness. Thus, fitter population members are more likely to pass on their characteristics to the next generation.

The generation of offspring from the parents is effected by the application of the *crossover* and/or *mutation* operators. Crossover involves the splicing of the parent solutions at some randomly chosen point and then combining the resulting pieces to form offspring that differ from the parents while maintaining some of their characteristics. Mutation simply applies a random perturbation to the offspring sometimes resulting in the creation of new genetic material and thereby helping to maintain the diversity of the population. These two operators are shown below for a simple case in which the population members are binary strings.

Crossover (at point marked "X"):

Parent #1: 100100001X1000011 ----------------> Offspring #1: 1001000010010001
Parent #2: 100010000X0010001 ----------------> Offspring #2: 1000100001000011

Mutation (at right-most bit):

Parent: 1001000011000011 ------------------> Offspring: 1001000011000010

With the repeated application of the selection, crossover and mutation operators, the average fitness of the population increases over time until, eventually, the population converges on what is hopefully a good, if not optimal, solution.

Evolutionary Programming

The steps constituting a simple evolutionary programming algorithm are as follows:

1. Randomly initialise a starting population of N members
2. Assign each member a fitness score using a fitness function
3. Generate one offspring per population member using only mutation
4. Assign the N offspring fitness scores using a fitness function
5. Allow parents and children to compete for survival
6. N survivors form next parent population
7. Go to 3 until termination or convergence

As with a genetic algorithm, the first steps are the creation of a starting population and the calculation of the fitnesses of the population members. At this point, however, the two algorithms diverge. In evolutionary programming, all of the population members give rise to offspring using only a mutation operator. The N parents and N children then compete for survival. A popular mechanism for this competition is known as *tournament selection*. This involves comparing the fitness of each population member against the fitness of a fixed number of randomly selected opponents from within the population. Every time the chosen population member's fitness exceeds that of an opponent, it receives a win. Once the whole population has been tested in this way, the members are ranked according to the number of wins they have achieved and the top N survive to form the next parent population. As with the genetic algorithm, this process of reproduction and selection continues until the population converges or some predetermined time limit is exceeded.

Implementation Issues

Having introduced the basic *modus operandi* of the most popular evolutionary algorithms, it is now necessary to examine some broader issues and questions surrounding their practical implementation.

The first, and most important, of these is to consider whether an evolutionary algorithm is the most appropriate choice for solving the problem in hand. As Luke points out (*11*), not all problems can be solved efficiently using genetic methods. There are problems that already possess very specialized solution strategies and it is unlikely that an evolutionary algorithm will improve greatly upon these. A good example of this in the CAMD context is the work of Brown *et al.* (*12*) in which a genetic algorithm compared unfavourably to Ullmann's subgraph isomorphism algorithm for the task of 2-D substructure searching.

Once it has been decided to pursue an evolutionary algorithm-based approach, the key maxim to bear in mind is "mould the algorithm to the application, not *vice versa*". There are two areas where this particularly applies. The first is in the choice of the encoding of the solutions, or indeed if the solutions should be encoded at all. It is **not** mandatory to represent solutions by strings, be they binary or otherwise! The work of Glen and Payne (*13*) and Westhead *et al.* (*14*) in the CAMD context illustrates this point nicely. In both these cases, genetic operators were applied directly to 3D structures during the course of *de novo* design. Rather than forcing an unnatural encoding upon solutions, it is preferable to use natural, problem-related representations (*2*).

Similar considerations apply to the choice and nature of genetic operators to be used. Here especially, the incorporation of problem-specific knowledge can lead to operators that are best suited to the application in hand. Again, the work of Glen and Payne (*13*) is a good example of this, as will be detailed later. The importance of this cannot be overemphasized, particularly in the light of the "No Free Lunch" theorems propounded by Wolpert and Macready (*15*) which state broadly that for any algorithm, any elevated performance on one class of problems is exactly countered by poorer performance on another class. Thus, a "black box" evolutionary algorithm applied to a problem is unlikely to be as successful as one in which problem-specific knowledge has been incorporated. In general, many practical applications have shown that the best solutions are obtained after making substantial modifications to standard algorithms (*2*). A quote from Michalewicz (*6*) suffices to summarize these points:

"It seems that a 'natural' representation of a potential solution for a given problem plus a family of applicable 'genetic' operators might be quite useful in the approximation of solutions of many problems, and this nature-modelled approach ... is a promising direction for problem solving in general", quoted in (*2*).

Finally, and obviously, the construction of the fitness function for any application is a key consideration. An inappropriate fitness function will fail to guide the evolutionary algorithm to the desired solutions - "garbage in, garbage out"! As fitness function evaluation is often the rate-limiting step in the operation of an evolutionary algorithm, it is worth seeking to optimize its computational efficiency, or choosing a more computationally efficient option where two equally accurate alternatives suggest themselves.

To reiterate, the key message of this section is that when implementing evolutionary algorithms, better results will be obtained if the problem in hand, rather than tradition or dogma, dictates the choices made.

Review of Current Applications

In what follows, selected applications of evolutionary algorithms in the field of rational drug design will be reviewed. As Figure 1 has illustrated, there is a large number of applications from which to choose and space considerations impose a strong constraint on the number that can be covered here. For more detailed coverage, the reader is directed to these recent reviews (*16-20*). The following applications have been chosen because they represent the state-of-the-art in terms of evolutionary algorithms applied to rational drug design and/or because they exemplify some interesting feature of evolutionary algorithms that I wish to underline. The application areas are the following: protein-ligand docking, de novo molecular design, QSAR, molecular diversity/combinatorial libraries and finally a miscellany of interesting applications will be listed briefly.

Protein-Ligand Docking. One of the areas where evolutionary algorithms have met with most success in rational drug design is in protein-ligand docking. Both of the examples chosen in this section are the subject of other chapters in this book - what follows may serve as a brief summary.

GOLD. The first of the docking programs to be considered here is GOLD (Genetic Optimization of Ligand Docking) developed by Jones and co-workers (*21,22*). In GOLD, the ligand is allowed full conformational flexibility, including "flipping" of alicyclic rings, and the receptor is permitted partial flexibility *via* the rotation of single bonds connected to terminal acceptor or donor groups in the amino acid residues within the active site. One of the most interesting features of GOLD, from the perspective of this review, is that it is a *parallel* genetic algorithm. In other words, rather than operating on a single population of individuals, the algorithm maintains several populations in what is known as an "island model". The only communication between the various populations consists of the periodic migration of selected individuals between the "islands".

Each individual in GOLD is represented by four strings. The first two are binary strings that encode the conformation of the ligand and the protein, respectively. The latter two are integer strings that contain mappings between donor groups in the ligand and acceptors in the protein and *vice versa.* The evaluation of an individual's fitness involves the generation of the conformations of the ligand and protein and then the least-squares fitting of acceptor to donor points dictated by the integer strings. Once the respective orientations and conformations of ligand and protein have been thus determined, the fitness score is calculated as the sum of terms involving hydrogen-bonding, steric and torsional energies. Parent selection is effected by a roulette wheel mechanism and offspring are bred by crossover and mutation of the binary and integer strings. A further operator is necessitated by the parallel nature of GOLD: a *migration* operator controls the migration of individuals between populations.

GOLD's performance was evaluated over a set of 100 protein-ligand complexes extracted from the Brookhaven Protein Databank (*21*). In 71% of the cases, GOLD was judged to have docked the ligand correctly - an extremely impressive result. More details on GOLD and results obtained with the program are presented in the chapter by Jones and co-workers.

EPDOCK. The second docking application to be reviewed here is EPDOCK, a program developed by Gehlhaar and co-workers (*23-25*). EPDOCK represents an interesting application for a number of reasons. First, as its name suggests, it employs evolutionary programming, rather than the more common (at present) genetic algorithm. Second, the work has involved collaboration with an expert in evolutionary computation, David Fogel (of Natural Selection Inc., San Diego), something which may become more common as rational drug designers seek to extract the maximum value from evolutionary algorithm applications.

In contrast to GOLD, EPDOCK permits only the ligand to be flexible. The conformation and orientation of the ligand are encoded in a real-valued string and multiple offspring are generated from each parent using mutation, which in this case involves the addition of a Gaussian random variable to elements of the string. Another key feature of interest is that a further level of sophistication is incorporated into EPDOCK - the mutation is what is termed "self-adaptive". In addition to the string containing the conformation and orientation information, each member of the population also has another string associated with it. This string contains, for each of the docking variables, a parameter that determines the size of the mutations available for that variable. These strings also evolve during the operation of EPDOCK so that the algorithm chooses for itself the most appropriate mutations for each docking variable as the search proceeds. The fitness of the individuals is determined using a fast molecular recognition potential (*24*), which is described in more detail in the chapter by Gehlhaar. Parent selection is carried out using a tournament selection scheme similar to that described earlier.

EPDOCK has been shown to successfully dock ligands in a number of challenging test cases (*24*) and considerable work has been invested in seeking optimal values for some of the operational parameters used by the program (*23*).

De Novo Molecular Design. There have been fewer applications of evolutionary algorithms to *de novo* molecular design than to docking; the one chosen for review here is the "Chemical Genesis" program of Glen and Payne (*13*). The program permits the fragment-based generation of (novel) 3D structures subject to a variety of constraints that may be scalar (e.g., log P), surface-derived (e.g., charge distribution) or grid-based (e.g., active site shape).

As was mentioned in an earlier section, one of the key features of the work of Glen and Payne is that the population members are the 3D structures themselves; there is no encoding. The genetic operators are applied to the evolving structures directly and, consequently, are able to embody many application-specific features. There are, in fact, two crossover operators, one allowing exchange of terminal portions of structures and the other permitting the excision and exchange of central portions of the solutions. Added to this are twelve mutation operators enabling, for instance, the variation of a structure's conformation and orientation and the addition or deletion of fragments. The fitness function applied to each member is simply the sum of the deviations from the applied constraints and parent selection uses the standard roulette-wheel method. A useful additional feature of Chemical Genesis is the production of a "fossil record" for each run of the program. This file records the evolution of each of the population members making possible a dynamic playback of the process of structure generation.

QSAR. As with docking, QSAR has been a fruitful area for the application of evolutionary algorithms. The instance that will be described here is the Genetic Function Approximation (GFA) method developed by Rogers (*26-28*) and now available as a module of the Cerius^2 software (*29*).

In GFA, each population member encodes a set of basis functions that will comprise a QSAR. These basis functions may be quantities such as logP, molecular weight and so forth. QSAR models are derived from these basis functions using a fitting technique, such as linear regression, to generate appropriate coefficients. The fitness of each of the individuals is then determined by evaluating the QSAR derived from it using Friedman's Lack-of-Fit (LOF) measure. This measure seeks to balance the error in the predictivity of the QSAR model against the number of terms used to derive it. Thus, GFA aims to select models that predict well without being overfitted. Parents are selected using a probabilistic mechanism such that their chance of selection is inversely proportional to their LOF score. New members are bred from the parents using a crossover operator which splices and recombines the strings of basis functions. An optional mutation operator can also be invoked to swap one basis function for another within a string.

In experiments with GFA using the standard Selwood dataset (*27*), the eight top-scoring models produced by the program were found to be as good as, or superior to, other published QSARs for that set. An advantage offered by an evolutionary algorithm for QSAR is that each run of the program generates a number of possible models, rather than just the single one generated by traditional techniques. Comparing various models output by the program can lead to added insights and by combining features from several models, superior QSARs may be produced. This benefit has also been noted by So and Karplus (*30*) when applying their Genetic Neural Network methodology. The advantages of multiple models will be discussed more later because it is generally relevant to CAMD applications. GFA can also plot out a graph of feature usage during the course of a program run; this helps to show the relative utility of the different features considered as basis functions. Other advantages of GFA compared to traditional QSAR methods are presented by Rogers (*27*).

Combinatorial Libraries and Molecular Diversity. The rapid rise in recent years in the number of computational techniques applied to the design and analysis of combinatorial libraries needs little introduction. Suffice it to say that evolutionary algorithms have figured large in this application area. Indeed, the first published paper applying computers to library design appears to have been the genetic algorithm developed by Sheridan and Kearsley (31). In this section two other GA applications will be briefly reviewed.

The papers of Singh *et al.* (*32*) and Weber *et al.* (*33*) share a number of common features of interest. Both employ simple genetic algorithms for the design of combinatorial libraries; the former for a peptide library targeted at stromelysin, the latter using the Ugi reaction to generate products aimed to be active against thrombin. The most significant aspect of the algorithms described by these groups is that they used the measured biological activity of the products at each generation as the fitness function, thus demonstrating that the fitness function used by an evolutionary algorithm need not be a computationally-derived measure!

Obviously, the use of experimental quantities will mean an appreciable time lag between the generation of a population of individuals and its fitness evaluation, but the results from both studies were compelling and make this kind of approach worthy of further investigation. In the case of Singh *et al.* (*32*), potent stromelysin substrates were found after making only 300 out of a possible 64 million hexapeptides. Similarly, Weber's group located Ugi products with low micromolar activity against thrombin after synthesizing only 400 out of a possible 160,000 molecules (*33*).

Miscellaneous. Some other rational drug design applications are worthy of a brief mention. One of these is the GASP program for pharmacophore elucidation and flexible molecular overlay, now a commercial product (*34,35*). Also of interest is the GERM program for the generation of receptor models from sets of overlaid structures (*36,37*). An exciting recent development is the CONSOLV program that seeks to predict water locations within enzyme active sites (*38*). CONSOLV is also of interest in that it represents the fruits of a collaborative effort between the Protein Structural Analysis and Design Laboratory and the GARAGe (Genetic Algorithms Research and Applications) Groups of Michigan State University. In the field of 3D database searching, Wild and Willett have employed a genetic algorithm for matching the molecular electrostatic potential field of a query molecule against those belonging to molecules in a compound collection (*39*). They also studied various parallel implementations of their genetic algorithm. A currently burgeoning application area for evolutionary algorithms is the assignment and refinement of NMR spectral data, for example in the FINGAR (*40*) and GARANT (*41*) programs. Finally, a field of potential future relevance to rational drug design, the search for effective protein folding algorithms, has also seen many evolutionary algorithm applications (*42*).

Obviously, this selection represents a small and somewhat subjective sample of the many evolutionary algorithm applications in the field of computer-aided molecular design. An extensive bibliography is available on request from the author (email david-e.clark@rp-rorer.co.uk) or on the World-Wide Web at http://panizzi.shef.ac.uk/cisrg/links/ea_bib.html.

Pros and Cons of Evolutionary Algorithms

Having reviewed some applications of evolutionary algorithms in rational drug design, what are some of the pros and cons of this class of algorithm compared to other types?

In their favour, evolutionary algorithms represent a methodological framework that is easy to understand and to work with. Most scientists have at least a passing familiarity with the biological mechanisms that underpin evolutionary algorithms and this aids the comprehension of their basic principles and operations. Evolutionary algorithms also allow a great deal of flexibility in their implementation and use. "Off-the-shelf" implementations are available (usually free of charge) and can often be readily adapted for the problem in hand. Alternatively, as has been shown, it is possible to create applications that are extensively customized for a particular situation. The latter choice is likely to yield improved performance for the specific instance for which it has been customized, while the former framework will be more transferable between applications.

Evolutionary algorithms are also noteworthy for their robustness. In general, reasonable performance can be obtained without the expenditure of great effort in parameter tuning, although obviously, optimal performance may require much work in this direction. A further beneficial aspect is that evolutionary algorithms are extremely amenable to parallelization and because the computationally demanding fitness evaluation step is decoupled from the rest of the algorithm, gains of speedups of almost 100% are to be expected when migrating to a parallel environment.

Of most relevance to rational drug design, perhaps, is that evolutionary algorithms are characterized by their ability to cope with difficult or unconventional response (energy) surfaces. This is a characteristic shared with other heuristic algorithms such as simulated annealing (*43*) or tabu search (*44*). An added advantage of evolutionary algorithms over these last two types of heuristic search, however, is the generation of a population of solutions from each run. This is particularly helpful in rational drug design problems where there is sometimes uncertainty as to the accuracy of the fitness function, in other words, how well it reflects the biological reality. In this instance, it is useful to see not just the top-scoring solution but a whole range.

The factors that weigh against evolutionary algorithms are two-fold. Firstly, they are inherently stochastic, which means that it is usually necessary to repeat runs several times with different random seeds to be sure of the results. This means that evolutionary algorithms will always be at a disadvantage if they are to be compared to a deterministic algorithm for the same purpose. Again, this underlines the need to consider carefully whether an evolutionary algorithm is an appropriate choice for the problem under consideration. However, in many cases, problems will only be amenable to solution by some class of heuristic algorithm; in this situation, the stochastic nature of evolutionary algorithms obviously becomes less of a problem. Secondly, as mentioned above, for optimal performance a good deal of parameter tuning may be necessary. Furthermore, it is nearly impossible to guarantee that one has ever arrived at a set of optimal parameters, particularly if they are coupled in some way. This is the so-called meta-optimization problem. However, again this is a difficulty shared with other heuristic search algorithms and so often it is not a large disincentive to the use of evolutionary algorithms. In addition, the use of self-adaptation, such as in EPDOCK, may help to ameliorate these parameter optimization problems.

Future Directions

This review has sought to present some of the current applications of evolutionary algorithms in rational drug design and to highlight strengths and weaknesses of the algorithms in general. In this section, a look to the future is attempted to see what directions might be taken in both the basic research into evolutionary computation and its application to CAMD problems.

It is to be expected that new application areas for evolutionary algorithms will arise within rational drug design. One possibility is the use of evolutionary algorithms for data mining within the ever-increasing databases of information arising from combinatorial chemistry and high-throughput screening experiments. Genetic algorithms are already in use for this kind of task in other application domains and so might be expected to be useful tools for seeking trends and relationships in biochemical data.

As mentioned previously, evolutionary algorithms lend themselves well to parallelization. While this will be most impressive on parallel architectures or distributed networks, the work of Jones *et al.* (*21*) demonstrates that using multiple subpopulations on a serial machine can lead to faster convergence and thus, greater efficiency compared to a single, monolithic population. As parallel machines become more widely available and techniques for distributed computing become more advanced, it is likely that more parallel applications of evolutionary algorithms will emerge.

The subject of self-adaptation is currently the subject of much interest in the field of evolutionary computation (*2*,*3*). The use of self-adaptive parameters is not limited to evolutionary programming and researchers are experimenting with self-adaptive operators in genetic algorithms too. A different approach aiming to solve the same problem, that of parameter optimization, is *meta-evolution*. Meta-evolution involves the use of one evolutionary algorithm to control a population of other evolutionary algorithms all with different parameter values. As these operate on a problem, it becomes possible over time to determine which are performing well and thus what are good sets of parameters. Meta-evolution clearly suggests the use of parallel architectures and this is being investigated in the DAGA-2 program (*45*). Without doubt, as these methods develop, they will begin to find their way increasingly into rational drug design applications.

Another area of activity within the evolutionary computation community is that of developing the basic theory of evolutionary algorithms. To quote from (*2*), "we know that they work, but we do not know why"! As the theoretical foundations of evolutionary computing become firmer, it is to be expected that this will help to provide better rules-of-thumb to guide implementations as well as improving the basic performance of the algorithms themselves.

Some workers are beginning to find that rather than using evolutionary algorithms in isolation, improved results can be obtained by hybridizing them with other types of search or optimization algorithm (*46-49*). Indeed, a recent paper comparing heuristic search algorithms for docking showed that the performance of all the heuristic algorithms could be improved by hybridization with a local optimizer and/or another heuristic algorithm (*50*). Other hybridizations of evolutionary algorithms are already in existence in rational drug design applications; for instance, EPDOCK employs a local minimizer to refine the solutions generated by the evolutionary programming algorithm (*24*), and So and

Karplus's Genetic Neural Networks combines evolutionary programming with a neural network to develop QSARs (*30*).

It is possible too that the future will see more applications of the two other types of evolutionary algorithm mentioned in the introduction - evolution strategies and genetic programming. The latter is already starting to be applied to problems of interest to rational drug designers. For instance, the group working on CONSOLV found that they obtained better results with genetic programming than with a genetic algorithm (*51*). Other genetic programming applications are given in (*52,53*).

As suggested earlier, it seems likely that the future will see more collaboration between researchers in evolutionary computation and those seeking to apply the methods to rational drug design problems. This can only be of benefit to both parties, giving the former experience of challenging "real-world" problems which may perhaps stimulate new techniques or approaches while the latter gain access to the technical expertise of researchers steeped in the theory and practice of evolutionary algorithms.

Finally, as evolutionary algorithms gain more widespread acceptance and validation, it seems inevitable that more commercial software products for rational drug design will embrace them thereby adding to the list of those currently available (*29,35,54*).

Conclusions

This review has sought to survey the current status of evolutionary algorithms as applied to rational drug design. In general, the applications have met with considerable success, although not in all cases, underlining the importance of choosing the right algorithm for the problem and of being aware of the implications of the No Free Lunch theorems (*15*). With these caveats in mind, it seems likely that as the field of evolutionary computation advances, it will continue to find ready and successful application to the considerable problems encountered in the process of rational drug design.

Acknowledgments

I should like to thank my colleagues in the Computer-Aided Drug Design group at Rhone-Poulenc Rorer (Drs. Richard A. Lewis, Stephen D. Pickett and Paul Bamborough) for their help and advice in the preparation of this review. Drs. David B. Fogel and Thomas Baeck are also thanked for providing reprints of their work and for helpful discussions. In addition, I am very grateful to Prof. Peter Willett and Dr. Val Gillet of Sheffield University for allowing me to post the EA bibliography on their departmental web pages.

Literature Cited

1. Milne, G. W. A. *J. Chem. Inf. Comput. Sci.* **1997**, *37*, 639-644.
2. Baeck, T.; Hammel, U.; Schwefel, H. -P. *IEEE Trans. Evol. Comput.* **1997**, *1*, 3-17.
3. Baeck, T.; Fogel, D. B.; Michalewicz, Z., Eds.; *Handbook of Evolutionary Computation*; Oxford University Press/Institute of Physics: New York, NY, 1997.
4. Mitchell, M. *An Introduction to Genetic Algorithms*; MIT Press: Cambridge, MA, 1996.
5. Baeck, T. *Evolutionary Algorithms in Theory and Practice*; Oxford University Press: New York, NY, 1996.
6. Michalewicz, Z. *Genetic Algorithms + Data Structures = Evolution Programs*; Springer: Berlin, 1996.
7. Fogel, D. B. *Evolutionary Computation: Toward a New Philosophy of Machine Intelligence*; IEEE Press: Piscataway, NJ, 1995.
8. Holland, J. H. *Adaptation in Natural and Artificial Systems*; 2nd Ed.; MIT Press: Cambridge, MA, 1992.
9. Davis, L., Ed.; *Handbook of Genetic Algorithms*; Van Nostrand Reinhold: New York, NY, 1991.
10. Goldberg, D. E. *Genetic Algorithms in Search, Optimization and Machine Learning*; Addison-Wesley: Reading, MA, 1989.
11. Luke, B. T. In *Genetic Algorithms in Molecular Modelling*; Devillers, J., Ed.; Academic Press: New York, NY, 1996; pp 35-66.
12. Brown, R. D.; Jones, G.; Willett, P.; Glen, R. C. *J. Chem. Inf. Comput. Sci.* **1994**, *34*, 63-70.
13. Glen, R. C.; Payne, A. W. R. *J. Comput.-Aided Mol. Des.* **1995**, *9*, 181-202.
14. Westhead, D. R.; Clark, D. E.; Frenkel, D.; Li, J.; Murray, C.W.; Robson, B.; Waszkowycz, B. *J. Comput.-Aided Mol. Des.* **1995**, *9*, 139-148.
15. Wolpert, E. D. H.; Macready, W. G. *IEEE Trans. Evol. Comput.* **1997**, *1*, 67-82.
16. Judson, R. In *Reviews in Computational Chemistry;* Lipkowitz, K. B.; Boyd, D. B., Eds.; VCH: Weinheim, 1997, Vol. 10; pp 1-73.
17. Maddalena, D. J.; Snowdon, G. M. *Exp. Opin. Ther. Pat.* **1997**, *7*, 247-254.
18. Parrill, A. *Drug Discov. Today* **1996**, *1*, 514-521.
19. Devillers, J. In *Genetic Algorithms in Molecular Modelling*; Devillers, J., Ed.; Academic Press: New York, NY, 1996; pp 1-34.
20. Clark, D. E.; Westhead, D. R. *J. Comput.-Aided Mol. Des.* **1996**, *10*, 337-358.
21. Jones, G.; Willett, P.; Glen, R. C.; Leach, A. R.; Taylor, R. *J. Mol. Biol.* **1997**, *267*, 727-748.
22. Jones, G.; Willett, P.; Glen, R. C. *J. Mol. Biol.* **1995**, *245*, 43-53.

23. Gehlhaar, D. K.; Fogel, D. B. In *Evolutionary Programming V: Proceedings of the Fifth Annual Conference on Evolutionary Programming*, Fogel, L. J.; Angeline, P. J.; Baeck, T., Eds.; MIT Press: Cambridge, MA, 1996; pp 419-429.
24. Gehlhaar, D. K.; Verkhivker, G. M.; Rejto, P. A.; Sherman, C. J.; Fogel, D. B.; Fogel, L. J.; Freer, S. T. *Chem. Biol.* **1995**, *2*, 317-324.
25. Gehlhaar, D. K.; Verkhivker, G. M.; Rejto, P. A.; Fogel, D. B.; Fogel, L. J.; Freer, S. T. In *Evolutionary Programming IV: Proceedings of the Fourth Annual Conference on Evolutionary Programming*; McDonnell, J. R.; Reynolds, R. G.; Fogel, D. B., Eds.; MIT Press: Cambridge, MA, 1995; pp 615-627.
26. Rogers, D. In *Genetic Algorithms in Molecular Modelling*; Devillers, J., Ed.; Academic Press: New York, NY, 1996; pp 87-108.
27. Rogers, D. In *Proceedings of the Sixth International Conference on Genetic Algorithms*; Eshelman, L. J., Ed.; Morgan Kaufmann, San Mateo, CA, 1995; pp 589-596.
28. Rogers, D.; Hopfinger, A. J. *J. Chem. Inf. Comput. Sci.* **1994**, *34*, 854-866.
29. Cerius^2.GFA. Molecular Simulations Inc., 9685 Scranton Road, San Diego, CA, USA.
30. So, S. S.; Karplus, M. *J. Med. Chem.* **1996**, *39*, 1521-1530.
31. Sheridan, R. P.; Kearsley, S. K. *J. Chem. Inf. Comput. Sci.* **1995**, *35*, 310-320.
32. Singh, J.; Ator, M. A.; Jaeger, E. P.; Allen, M. P.; Whipple, D. A.; Soloweij, J. E.; Chowdhary, S.; Treasurywala, A. M. *J. Am. Chem. Soc.* **1996**, *118*, 1669-1676.
33. Weber, L.; Wallbaum, S.; Broger, C.; Gubernator, K. *Angew. Chem. Int. Ed. Eng.* **1995**, *34*, 2280-2282.
34. Jones, G.; Willett, P.; Glen, R. C. *J. Comput.-Aided Mol. Des.* **1995**, *9*, 532-549.
35. GASP. Tripos Inc., 1699 S. Hanley Road, St. Louis, MO, USA.
36. Walters, D. E.; Muhammad, T. D. In *Genetic Algorithms in Molecular Modelling*; Devillers, J., Ed.; Academic Press, New York, NY, 1996; pp 193-210.
37. Walters, D. E.; Hinds, R. M. *J. Med. Chem.* **1994**, *37*, 2527-2536.
38. Raymer, M. L.; Sanschagrin, P. C.; Punch, W. F.; Venkataraman, S.; Goodman, E. D.; Kuhn, L. A. *J. Mol. Biol.* **1997**, *265*, 445-464.
39. Wild, D. J.; Willett, P. *J. Chem. Inf. Comput. Sci.* **1996**, *36*, 159-167.
40. Pearlman, D. A. *J. Biomol. NMR* **1996**, *8*, 49-66.
41. Bartels, C.; Guentert, P.; Billeter, M.; Wuethrich, K. *J. Comput. Chem.* **1997**, *18*, 139-149.
42. Pedersen, J. T.; Moult, J. *Curr. Opin. Struct. Biol.* **1996**, *6*, 227-231.
43. Kirkpatrick, S.; Gelatt, C. D.; Vecchi, M. P. *Science* **1983**, *220*, 671-680.
44. Cvijovic, D.; Klinowski, J. *Science* **1995**, *267*, 664-666.
45. Wang, G.; Goodman, E. D.; Punch, W. F. *Simultaneous Multi-Level Evolution*; GARAGe Technical Report 96-03-01; Michigan State University: East Lansing, MI, 1996.
46. Myung, H.; Kim, J. H. *Biosystems* **1996**, *38*, 29-43.

47. Mahfoud, S. W.; Goldberg, D. E. *Parallel Comput.* **1995**, *21*, 1-28.
48. Varanelli, J. M.; Cohoon, J. P. In *Proceedings of the Sixth International Conference on Genetic Algorithms*; Eshelman, L. J., Ed.; Morgan Kaufmann: San Mateo, CA, 1995; pp 174-181.
49. Pal, K. F. *Biol. Cybernetics* **1995**, *73*, 335-341.
50. Westhead, D. R.; Clark, D. E.; Murray, C. W. *J. Comput.-Aided Mol. Des.* **1997**, *11*, 209-228.
51. Raymer, M. L.; Punch, W. F.; Goodman, E. D.; Kuhn, L. A. In *Genetic Programming 1996: Proceedings of the First Annual Conference*; Koza, J. R; Goldberg, D. E.; Fogel, D. B.; Riolo, R. L., Eds.; MIT Press: Cambridge, MA, 1996; pp 375-380.
52. Koza, J. R. In *Handbook of Evolutionary Computation*; Baeck, T.; Fogel, D. B.; Michalewicz, Z., Eds.; Oxford University Press/Institute of Physics: New York, NY, 1997; Section G6.1.
53. Handley, S. In *Proceedings of the Fifth International Conference on Genetic Algorithms*; Forrest, S., Ed.; Morgan Kaufmann: San Mateo, CA, 1993; pp 271-278.
54. LeapFrog, FlexiDock. Tripos Inc., 1699 S. Hanley Road, St. Louis, MO, USA.

Chapter 18

Further Development of a Genetic Algorithm for Ligand Docking and Its Application to Screening Combinatorial Libraries

Gareth Jones [1,5], Peter Willett [1], Robert C. Glen [2], Andrew R. Leach [3], and Robin Taylor [4]

[1] Krebs Institute for Biomolecular Research and Department of Information Studies, University of Sheffield, Western Bank, Sheffield S10 2TN, United Kingdom
[2] Tripos Inc., 1699 South Hanley Road, St. Louis, MO 63144
[3] Glaxo Wellcome Medicines Research Centre, Gunnells Wood Road, Stevenage SG1 2NY, United Kingdom
[4] Cambridge Crystallographic Data Centre, 12 Union Road, Cambridge CB2 1EZ, United Kingdom

We have previously reported the development and validation of the program GOLD (Genetic Optimization for Ligand Docking), an automated ligand docking program that uses a genetic algorithm to explore the full range of ligand conformational flexibility with partial flexibility of the protein. The validation of the algorithm exposed certain weaknesses, particularly in its handling of hydrophobic ligands. We describe here a number of modifications designed to overcome these shortcomings. Using the improved algorithm we have investigated the relationship between the program's scoring function and ligand activity and applied the algorithm to screening combinatorial libraries.

Using lipase as a target, GOLD docked a library of carboxylic acids and amines extracted from the Fine Chemicals Database. By predicting the binding modes of product from the predicted dockings of reactants the problem is considerably simplified. The success of this approach was verified by comparing the prediction obtained by docking the component reactants with the result obtained by docking the product. The procedure was repeated using a second library of sulphonyl chlorides and amines.

Prediction of small-molecule binding modes to macromolecules of known three-dimensional structure is a problem of paramount importance in rational drug design (the "docking" problem). There have been many different approaches to solving the docking problem (*1,2*), including: deterministic approaches (*3,4*); simulated annealing

[5] Current address: Arena Pharmaceuticals, 6166 Nancy Ridge Drive, San Diego, CA 92121

(*5*); genetic algorithms (GAs, *6,7*) and evolutionary programming (*8*). GAs provide an evolutionary search paradigm that enables the rapid identification of good, though not necessarily optimal, solutions to combinatorial optimization problems (*9,10,11*). This technique has proved remarkably successful in almost all areas of computational chemistry and molecular modeling, to which it has been applied (*12*).

GOLD (*6,13*) is an automated ligand docking program that uses a GA to explore the full range of ligand conformational flexibility. Conformational flexibility is encoded in the GA in binary bitstrings where the angle of rotation about each rotatable bond is encoded in a byte. In order that GOLD efficiently elucidates binding modes, hydrogen bonds have been directly encoded into the GA, using integer strings, such that each position on the string ecoded a particular ligand donor-hydrogen or acceptor while the value at that position encoded a complementary protein acceptor or donor-hydrogen. When decoding a GA chromosome least-squares fitting is employed to try and form some of these encoded hydrogen bonds. A simple scoring function is used to rank generated binding modes. This comprised a term for hydrogen bonding (which took account of the fundamental requirement that water must be displaced from both donor and acceptor before a bond is formed); a pairwise dispersion potential between the protein and the ligand and a molecular mechanics term for the internal energy for the ligand. The program has been extensively evaluated on a large number of protein-ligand complexes (*13*).

The validation studies of GOLD revealed weaknesses in predicting the binding mode of hydrophobic ligands (*13*). Here, we describe a number of improvements, designed primarily to overcome this limitation. Using the improved algorithm a number of activity studies have been carried out and results on sets of influenza A neuraminidase and α-chymotrypsin ligands are presented. Finally, GOLD has been used to screen combinatorial libraries. Given the large size of combinatorial libraries, the selection of a small number of reactants or products for further study against a known target is both a demanding and important problem in rational drug discovery. Rather than enumerating a library and docking every library compound, we simplify the problem by docking only library monomers. The binding modes of high affinity products can then be elucidated from the docked monomers.

A knowledge of how GOLD works would be helpful for the first section of this chapter which deals with improvements to the program (see references *6* and *13*). However, the sections on activity prediction and library screening can be read separately without any great understanding of GOLD.

Extending the Test Suite. In (*13*) we reported the validation of GOLD on 100 test complexes extracted from the Protein Data Bank (PDB) (*14*). Shortly afterwards, one of us (RT) expanded this test suite to 134 complexes. Further expansion without including complexes very similar to those selected, or biasing the dataset towards peptidic ligands did not seem possible. Using the same subjective classification of *Good* (where all protein ligand interactions were correctly predicted), *Close* (the binding mode and all important interactions were elucidated), *Errors* (the prediction was partially correct but had significant errors) and *Wrong* (where the predictioun was completely wrong) we obtained the results shown in Table I. Overall we achieved prediction rate of 72%, with 96 complexes being in the *Good* and *Close* categories. All results can be viewed at http://www.ccdc.cam.ac.uk/prods/gold.html.

Table I. Results of docking predictions on a dataset of 134 complexes. The missing complex is 1ACL, for which no binding mode could be elucidated.

Subjective Result	*Number*	*PDB identification codes*
Good	53	1ABE 1ACM 1ACO 1CBX 1COY 1CPS 1DBB 1DBJ 1FKG 1FKI 1HDY 1HEF 1HYT 1LST 1MDR 1MRK 1PBD 1PHD 1POC 1SRJ 1STP 1TPP 1ULB 1XIE 2ADA 2CGR 2CHT 2CTC 2PHH 2SIM 3AAH 3PTB 3TPI 4DFR 4PHV 7TIM 8GCH 1AEC 1AHA 1ASE 1HSL 1BMA 1CIL 1FRP 2GBP 1GLP 1LAH 1LPM 1MMQ 1MRG 1TRK 1TNL 1WAP
Close	43	1BLH 1DIE 1DR1 1DWD 1EPB 1GHB 1GLQ 1IDA 1IVE 1LDM 1PHA 1PHG 1RNE 1SLT 1TKA 1TMN 1XID 2DBL 2PK4 2YHX 3CPA 3GCH 3HVT 4CTS 5P2P 6ABP 6RNT 1APT 1AZM 4EST 1ATL 1BBP 1BYB 1CBS 1COM 1FEN 1HFC 1IMB 1LCP 1NCO 1TNG 1TNI 1TPH
Errors	15	1BAF 1EAP 1ETR 1HDC 1LIC 1RDS 1ROB 6RSA 1ACK 2CMD 1CTR 2LGS 1LNA 1SNC 1UKZ
Wrong	22	1AAQ 1ACJ 1DID 1EED 1ETA 1HRI 1ICN 1IGJ 1MCR 1MUP 2R07 1NIS 1TDB 2AK3 2MTH 2PLV 3CLA 4FAB 2MCP 1CDG 1LMO 1TYL

In (*13*) several flaws were identified in the algorithm. This extended verification confirmed these shortcomings, which were mainly related to the recognition of hydrophobic interactions, and we now describe a number of improvements that have been made to correct the algorithm.

Improving the Algorithm

Increased VDW Weighting. The fitness function used by GOLD comprised three main terms: a hydrogen bonding energy; a pairwise van der Waals steric energy between the protein and the ligand; and an internal energy term for the ligand. The total fitness score was the linear sum of these terms, with each term being equally weighted. During testing it was observed that there was insufficient weight given to large areas of hydrophobic contact and that there was insufficient probability of charged ligand groups being solvent accessible. This was attributed to an imbalance in the fitness function, with hydrophobic contact being undervalued relative to hydrogen bonding, since, while the van der Waals interaction between the protein and the ligand accounts for dispersive interactions, it fails to score for the favorable entropic effects of water displacement. It was felt that this imbalace might be easily corrected by scaling the contribution the van der Waals energy between the protein and the ligand made to the fitness function by a weight. Using the results from the experiments in (*13*) each test system was examined in turn. The fitness scores of the 20 solutions were recalculated using a new trail weighting for the the van der Waals energy between the protein and the ligand. The binding mode with the highest fitness score was then selected as a prediction. This procedure was repeated using a number of trial weights and a factor of 1.375 was found to give the most correct predictions over the 100 complexes. This new weight was then applied within the algorithm and proved successful at correcting the imbalance between hydrophobic and polar interactions within the fitness function. Several complexes such as 4FAB and 1ACJ were now correctly docked by GOLD. The ability of GOLD to dock polar ligands was unaffected.

As an alternative to increasing the van der Waals weighting an approximate surface area calculation was used to determine desolvation energies (*15*). However, this proved no more discriminatory than increasing the van der Waals energy weighting and was very time consuming.

Shape Fitting. Another problem observed during testing was a systematic problem in placing hydrophobic groups in deep cavities. The chromosome encoding used in GOLD meant that the GA searched patterns of hydrogen bonding motifs: the algorithm was thus directed to discover hydrogen bonding interactions, but found hydrophobic interactions by "accident". Moreover, given the tendency for bad bumps to occur as the ligand approaches deep cavities, cavity binding may actually have been discouraged.

The chromosome encoding has now been extended to encompass hydrophobic interactions. A grid with inter-point spacing of 0.25Å was placed across the active site. An augmented carbon probe atom was placed at each grid point and the van der Waals interaction between the probe and the active site measured. This probe atom had a van der Waals radius of 2.5Å and the physical properties of an sp^3 carbon. If the interaction energy was less than -2.5kcal/mol then the grid point was considered to be a favorable position for ligand hydrophobic atoms and the point was labeled. Ligand carbons bonded to at least one hydrogen were considered to be hydrophobic and were also labeled. A new integer string was added to the chromosome which encoded possible hydrophobic interactions, such that the position on the string was a ligand hydrophobic atom label and the value on the string was a label of a hydrophobic grid point. During least-squares fitting and the application of genetic operators this string was handled in an analogous fashion to the integer strings that encode for hydrogen bonding (as described in detail by Jones *et al.* (*6*)).

This extension of the chromosome encoding allowed GOLD to dock apolar ligands and ligands which did not form any hydrogen bonds to the protein, and removed the need for the small ligand model (*13*). The binding modes of complexes such as 1HRI and 2R07, which were originally incorrectly predicted, were now successfully elucidated, without any adverse effects being observed when docking hydrophilic ligands.

Solvent Exposed Charged Residues. Another problem observed in the testing of GOLD was the binding of ligand groups to charged solvated residues. The exterior of a protein generally contains many charged solvated residues, such as lysines, arginines and glutamic and aspartic acids. These residues, being highly solvated and mobile, are not normally favorable for ligand binding. However, with the GOLD algorithm such groups, being charged, are very attractive. An approximate surface area algorithm was used to identify such residues. Using this algorithm the ratio of surface areas of the residue in the absence and presence of the rest of the protein was determined. If this ratio was greater than 0.5 then the residue was considered solvent exposed. Such residues had their hydrogen bonding energies recalculated using a water dielectric of 78.5, which made such residues much less attractive to ligand binding. For example, the binding mode of the ligand in 6RSA was originally incorrectly predicted, with the ligand binding to a solvent exposed residue, but is now predicted correctly.

Torsional Distributions. As currently documented GOLD searches all single acyclic torsions over 360°. However, searches of small molecule crystallographic databases, such as the CSD (*16*), show that many common torsions are restricted. Figure 1 shows two common torsions and their distribution in the CSD. GOLD can now read such torsional distributions and search only those ligand torsional angles which are seen in small molecule crystals. By default a small torsional library of 24 common torsions is available, but the user can also use the library of over 200 torsions from the MIMUMBA program (*17*).

Re-mapping the Chromosome. In the docking procedure described in (*13*) the least-squares fitting process is applied in two passes. The chromosome is decoded to give a mapping in 3D space between points in the ligand and points in the protein active site. The first pass of least-square fitting is applied to all points, while the second pass is applied to close contacts from the first pass. Thus a significant portion of the chromosome will not be used in performing the final least-squares fitting. In the improved docking algorithm the unused chromosome mappings are reset to the dummy value of '-1' (that is, they are un-mapped). The crossover operator has been altered so that if, at a particular position, one parent has a dummy value and the other parent does not, the child will always inherit the non-dummy value. These changes have the effect that a child is much more likely to inherit meaningful chromosome mappings than previously and this was reflected in improved fitness scores.

Current Results on 100 Test Systems. The continued improvement of the algorithm has resulted in a superior rate of prediction over the 100 test systems previously used to test our algorithm (*13*). Table II compares the root mean squared deviations (RMSD) of GOLD predictions to the crystallographically observed binding mode obtained using the current algorithm with those previously reported (*13*). As before, GOLD was run a number of times and the solution with the highest score was retained as a prediction. Up to 20 dockings were performed per test system. Due to the increased reliability and reproducibility of the algorithm it was felt that it was not necessary to generate 20 GA solutions, as, in many cases all 20 predictions would be identical. So if, at any time, the best three solutions were all within 1.5Å RMSD of each other docking was terminated. However, in at least two cases, if the algorithm had run to 20 dockings a much better RMSD to the observed binding mode would have been obtained (previously the GA was always run 20 times). Examination of the

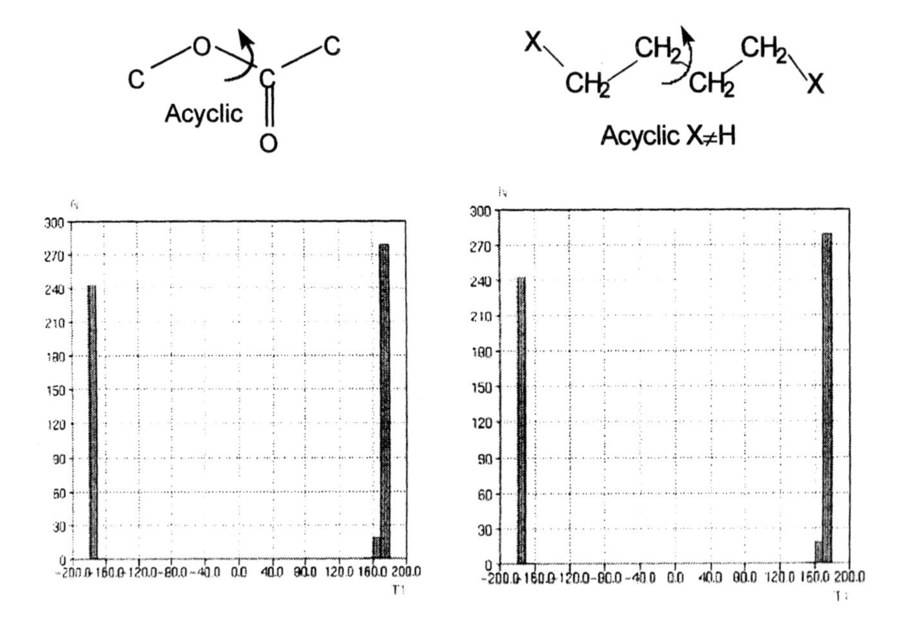

Figure 1. Distributions for ester (on left) and alkyl torsions.

table shows that the new algorithm shows a clear improvement with 72% of test systems having an RMSD of 2.0Å or less and 81% having an RMSD of 3.0Å or less.

Table II. Current and previously reported results for GOLD on 100 PDB complexes listed in (*13*).

RMSD (Å)	*Total*	*Sum Total*	*Previous Sum Total*
<=0.5	11	11	8
>0.5,≤1.0	34	45	35
>1.0, ≤1.5	15	60	55
>1.5, ≤2.0	12	72	66
>2.0, ≤2.5	7	79	68
>2.5, ≤3.0	2	81	71
>3.0	19	100	99

Comparison of Fitness Score With Binding Data

As we have shown in validation experiments, GOLD is clearly capable of reproducing the binding mode of many ligands with considerable accuracy. However, it would also be desirable if the program were able to predict the activity of ligands. For this reason the relationship between GOLD fitness scores and the activity of known ligands was explored for two test systems.

Influenza A Neuraminidase. Activity and, where available, structural data for influenza A neuraminidase (NA) ligands was obtained (*18-24*). For many of these compounds crystal structures were available (either publicly from the PDB, or as proprietary Glaxo Wellcome structures). The 34 compounds included 10 inactive molecules, which exhibited high structural similarity to some of the active compounds (Neil Taylor, Personal Communication). Using GOLD, each ligand was docked 10 times (each docking comprised 100000 GA operations) into the active site of NA and the solution with the highest GA fitness score retained. The results of these experiments are shown in Table III. From the table it can be seen that, with the exception of the inactive BANA ligands (*18*), GOLD always reproduced the observed binding mode.

Figure 2 plots the relationship between observed IC_{50} values (logarithmic scale) and GOLD fitness scores. While the relationship is clearly not ideal there does appear to be a marked correlation and this is borne out by non-parametric statistical tests which show strong evidence of rank correlation (Spearman test r_s=-0.65, p<0.001; Kendall test, τ=-0.48, p<0.001). If we consider 10μM to be a cutoff for activity then there are 15 actives and 19 inactives in the dataset. The question arises as to whether the GOLD score can be used to predict activity. Using a score of 74 or greater to indicate activity we obtain Table IV. From the table it seems that the GOLD

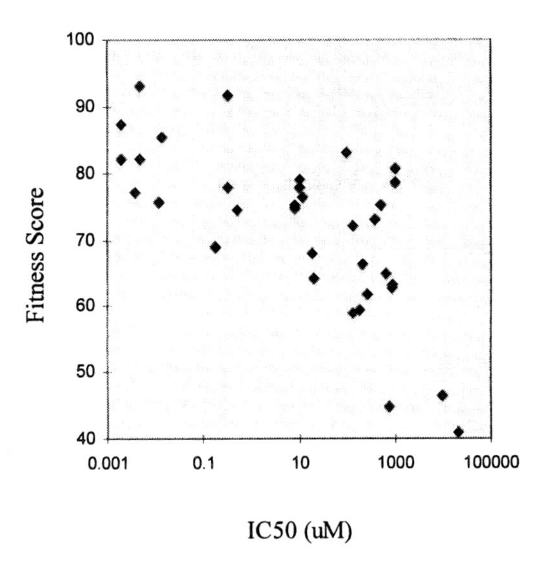

Figure 2. Plot of GOLD fitness score against IC_{50} values for NA ligands.

Table III. Compound activities and Fitness scores for NA inhibitors. Where the crystal structure of the complex is available the PDB code (or Glaxo Wellcome structure id) and the RMSD between the GOLD prediction and observed binding mode are shown. Where no reference is given for a compound, the structure is proprietary. The compound id is that used in the reference.

Compound Id	*$IC_{50}(\mu m)$*	*Fitness Score*	*Crystal Structure Id*	*RMSD*	*Notes*
3a (*21*)	0.002	82.02			4
4l (*20*)	0.002	87.39			4
2a (*21*)	0.004	76.99	gs0023	1.13	1
2 (*21*)	0.005	92.96	gs0015	0.38	1
4j (*20*)	0.005	81.96			4
3b (*21*)	0.012	75.52			4
5e	0.014	85.34			4
2b (*21*)	0.18	68.93			4
4 (*21*)	0.32	91.59	gs00154	0.32	1
4g (*20*)	0.32	77.78			4
1a (*21*)	0.5	74.40			4
DANA (*19*)	8.6	74.73	1nnb	1.13	1
DANA (*18*)	8.6	75.20	1ivf	1.18	2
EPANA (*23*)	~10	78.98	1iny	1.25	1
EPANA (*23*)	~10	77.78	1inx	1.54	2
4d (*20*)	12	76.24			4
1a (*21*)	19	67.97			4
1 (*24*)	20	64.14			4
2 (*24*)	>100	83.10			5
inactive4	>130	58.86			4,6
inactive8	>130	71.94			4,6
inactive10	>180	59.31			4,6
inactive3	>210	66.34			4,6
inactive5	>270	61.80			4,6
inactive7	>390	72.91			4,6
inactive2	>500	75.18			4,6
inactive1	>640	64.80			4,6
BANA105 (*18*)	750	44.85	1ivd	2.55	3
inactive6	>880	62.74			4,6
inactive9	>900	63.13			4,6
APANA (*23*)	~1000	78.43	1inw	1.28	2
NANA (*22*)	~1000	80.66	2bat	1.21	2
BANA106 (*18*)	10000	46.39	1ivc	2.30	2,3
BANA108 (*18*)	>20000	41.03	1ive	2.84	2,3

1. Subtype N9 crystal structure.
2. Subtype N2 crystal structure.
3. GOLD failed to predict BANA geometries correctly as these compounds contain non-planar amide bonds. However, the position of the benzene ring and acid group were correctly predicted.
4. Docked into the protein crystal structure of gs00023.
5. Docked into the protein crystal structure of gs00155.
6. Glaxo Wellcome inactives from assay data. These inactives have a high structural similarity to the actives.

score is a good indicator of activity and it is most unlikely that this distribution could have arisen through chance (χ^2=15.27, p<0.001, ν=1).

Table IV. Counts of predicted and observed activities for NA inhibitors.

	GOLD predicted activity	*GOLD predicted inactivity*
Active	14	1
Inactive	5	14

While this experiment appears successful, several shortcomings were exposed. The active conformations of the NANA (*22*) and PANA (*23*) ligands differ in their ring conformations, which GOLD failed to discriminate. Additionally, inspection of the crystal structures shows that Glu276 is mobile, forming interactions with Arg224 or with the ligand (*20*); account needs to be taken of protein flexibility for GOLD to score properly for this effect.

α-Chymotrypsin. The NA test suite contains a set of polar ligands which GOLD is most adept at handling. In order to see if the program was as successful with hydrophobic ligands, GOLD was used to dock a series of 103 α-chymotrypsin inhibitors (*25*) using the PDB protein structure in 6CHA (*26*). Since the ligands were very simple (most were near-rigid tricyclic hydrophobic compounds) each GA docking run only comprised 50000 operations. This allowed a 100% speed-up of the algorithm, though at the expense of the quality of predicted binding modes. This reduced parameterisation was tested on the 100 systems used to verify GOLD (*13*) and good results were obtained (70 predictions being within 2.0Å RMSD of the crystal structure and 79 predictions being within 3.0Å, though some loss of predictive quality was noted, especially for large and flexible ligands). Each ligand was docked three times into the active site (if the first two dockings were within 1.5Å RMSD of each other then a third solution was not generated) and the solution with the highest fitness score retained.

Figure 3 plots the observed K_i against the calculated fitness scores. Here the relationship is much poorer than that observed for the NA inhibitors. It is not clear if this is due to failures in the fitness function or the inability to reproduce the correct binding mode (for which data is not available). The Kendall test indicates that there is a meaningful relationship between the rankings of fitness scores and K_i values (τ=-0.1495, p=0.033), while the relationship just fails to show statistical significance in the Spearman test (r_s=-.1909, p=0.065). While these results are poor, it is worth emphasizing that, due to its hydrophobic nature, this dataset represents a worst case scenario for GOLD.

Comments on Activity Studies. It is worth noting that if GOLD can successfully predict binding modes then there are a number of alternative approaches available to attempt to predict binding free energies accurately (e.g. *27*). Finally, activity prediction requires an estimate of both the entropic components to binding and the free energy

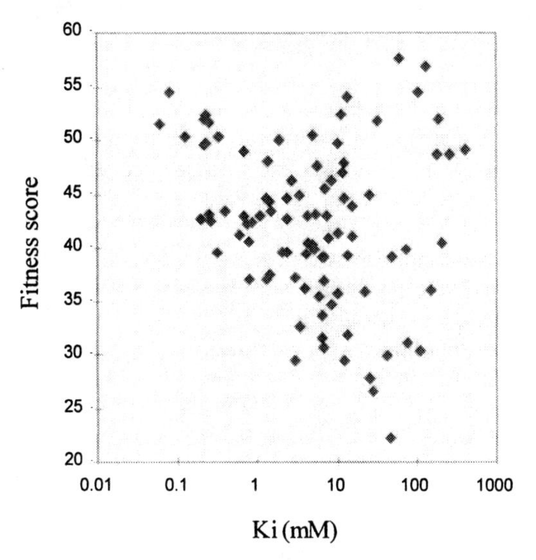

Figure 3. Plot of GOLD fitness score against K_i values for α-chymotrypsin ligands. Two inactive outliers with very low fitness scores are not shown.

difference between different molecules and these components are missing from the GOLD fitness function, though it would not be too difficult to account for the loss of torsional freedom in the ligand and protein (by including an appropriate term for each rotatable bond fixed on binding).

Docking Combinatorial Libraries

Combinatorial chemistry methods (*28*) are playing an increasingly important role in rational drug discovery. This section describes the use of GOLD to dock combinatorial libraries. However, rather than enumerate a library and dock in all products, we dock in all monomers and elucidate the binding modes of high affinity products from the binding modes of the monomers. The attraction of such is method is obvious: given a library of two monomer pools, one of size N and the other of size M, we need only perform $N+M$ dockings, rather than $N \times M$ dockings.

For this method to be applicable it is assumed that the binding mode of high affinity ligands can, in general, be reproduced by partitioning the ligand into fragments and docking the fragments. This approach is illustrated using the crystal structure of methotrexate bound to dihydrofolate reductase (DHFR) (PDB code 4DFR, *29*), as shown in Figure 4. The ligand was split in two and both parts terminated with a methyl group. The two fragments were docked into DHFR using GOLD. 25 GA runs of 100000 operations were performed and the solutions for each fragment with the highest fitness scores are shown in Figure 4. The upper dark structure is the GOLD prediction for the pyridine moiety only, while the lower dark structure is the GOLD prediction for the binding mode of the remainder of the ligand. Over all heavy atoms, the RMSD between the predictions for the two fragments and the crystallographically observed binding mode is 1.5Å. Thus, in this case, it is clear that by docking fragments of the ligand we can reproduce the observed binding mode of the whole ligand to a fair degree of approximation. Further experimental evidence is provided by the SAR by NMR methodology where the binding modes of fragment compounds were obtained by NMR and the fragments were linked to produce high affinity ligands (*30*).

Some fragment docking methods rely upon this hypothesis that docking portions of a ligand can be used to determine the binding mode of the whole ligand (e.g. *31*). However, it cannot in general be assumed that the binding mode of a ligand can be reproduced by fragment docking and such an approach would be undesirable for an accurate docking algorithm. However, the assumption is necessary in this case, where we are attempting to select a small number of products from a large library.

Experiments using a library of acids and amines. A set of carboxylic acids and amines were extracted from the Available Chemicals Directory (ACD, *32*). This library comprised 426 acids and 105 amines. From these, three-dimensional structures in SYBYL mol2 file format (*33*) were generated using CONCORD (*34*). Atom-type checking software developed for use with the GOLD docking program was then used to ensure that all structures were correctly typed. Where appropriate, structures were manually corrected within SYBYL. The acid and amine reactants will form products

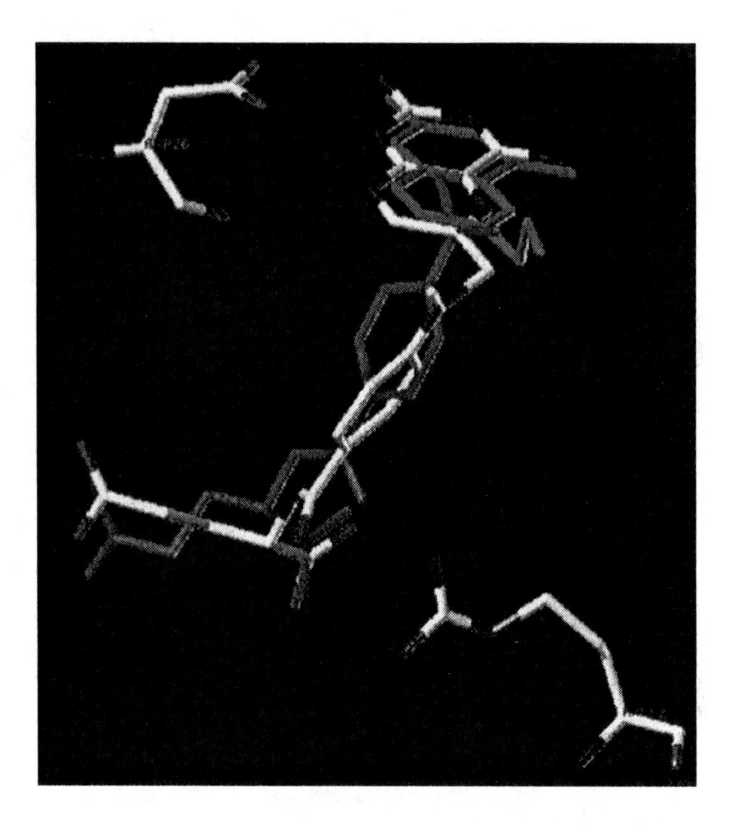

Figure 4. The crystal structure of methotrexate bound in DHFR is shown shaded by atom type. The upper dark structure is the GOLD prediction for the pyridine moiety only, while the lower dark structure is the GOLD prediction for the binding mode of the remainder of the ligand.

with an amide bond. Since the amide bond will be present in every product, this group was formed in both the acids and amines and terminated with a methyl (i.e. each amine, RNH_2, was converted to the corresponding acetylamino compound, RNHCOMe, and each acid, $R'CO_2H$, to the corresponding methylamide, R′CONHMe). SYBYL scripts were used to automate this process (*33*).

In order to test the method, lipase was chosen as an example target (active site coordinates were taken from the PDB structure 1CRL, *35*). Each reactant was docked into lipase three times to give three possible GOLD solutions. However, if the first two GOLD solutions were within 1.0Å RMSD of each other then the third solution was not generated. In general, most monomers were simple ligands and, for this reason, each GA run comprised only 50000 operations.

Following this docking procedure there were 1076 GOLD solutions for the acid monomers and 270 GOLD solutions for the amine monomers. This docking stage took just under one CPU-week on an SGI R4400 processor. Each GOLD solution obtained for an acid monomer was compared with every GOLD solution for the amines. If the RMSD of all equivalent atoms in the amide linkage were within 1.5Å then the two monomers were merged to form a new structure. The merging involved deleting the amide linkage in the acid and the terminating methyl group in the amine, then connecting the acid to the amine. The van der Waals energy of this new structure was calculated using the Tripos force field. If this energy exceeded 500kcal/mol then the structure was rejected; otherwise, the new structure was considered to be a good prediction of the binding mode of the product formed from the two monomers. The comparison stage took 1.5 hours CPU on an SGI R4400 processor.

Following the comparison stage a number of predictions were obtained for the product binding modes. In many cases there were more than one prediction for the same product, since multiple binding modes were present for each monomer. If this were the case a fitness score for each product was obtained by summing the fitness scores of the component monomers. The product with the highest fitness score was then retained, while all other predictions were rejected. In practice, it was found that, when there were many predictions for the same product those predictions were usually very similar (resulting from very similar GOLD predictions for the monomers).

Table V summarizes the results of this experiment. From approximately 300000 possible superpositions of GOLD solutions, the binding modes of 129 unique products with reasonable energy were determined. It is to be hoped that these products will exhibit high affinity for the target. Since the enumerated library contains 44730 products, this experiment resulted in considerable data reduction, providing a small number of compounds for further study via modeling experiments or experimental assay. If required, further data reduction can be achieved by ranking the products using the GOLD fitness scores for the component monomers. In addition to the data reduction achieved in the products, data reduction has also been achieved in the number of monomers.

Table V. Summary of acid and amine library experiments. All CPU times are for an R4400 SGI processor.

No. Acids	426
No. Amines	105
No. of available products	44730
No. GOLD solutions for acids	1088
No. GOLD solutions for amines	280
Total no. of possible superpositions	304640
No. of product superpositions found	311
No. of products following energy screen	237
No. of unique products	129
No. of acids in products	34
No. of amines in products	49
CPU time to dock acids	124h
CPU time to dock amines	33.7h
CPU time for superposition and energy screen	1.5h
Total CPU time	159h

In this experiment the underlying assumption is that the binding mode of a high affinity product can be predicted from the binding modes of the component monomers. While this assumption can only be strictly verified through experimental assays and crystallographic studies, some limited verification may be obtained by using GOLD to dock in a product. The predicted binding mode obtained using GOLD to dock the product directly can then be compared with the binding mode obtained by superimposing the predicted binding modes of the monomers.

This verification procedure was applied to the 129 products described above. Firstly, good coordinates were produced for each product. These were obtained from the two component monomers with attached amide-bond linkages. The amide bond in one monomer was least-squares fitted to the amide bond in the other. The amide bond in one monomer and the terminating methyl groups in both monomers were deleted and the addition of a single bond then completed the product. Prior to docking the product was minimized using the MAXIMIN2 module in SYBYL.

The docking was done in two passes. Firstly, each product was docked into lipase using GA runs of 50000 operations to generate three GOLD solutions (if the first two solutions were within 1.0Å RMSD then only two solutions were generated). The GOLD solution with the highest fitness score was taken as a predicted binding mode for the product. This prediction was then compared with the binding mode previously obtained by docking in the two monomers and an RMSD obtained for the deviation of heavy atoms. For those few cases where the RMSD was greater that 3.0Å a second docking procedure was performed. The second docking comprised 10 GA runs of 100000 operations. Again, the GOLD solution with the highest fitness score was taken as a predicted binding mode for the product.

This two-pass docking was performed in order to reduce CPU usage; ideally, all products would have been docked using 10 GA runs of 100000 operations. The RMSD values obtained are shown in Table VI. It can be seen that in the vast majority of cases the binding mode obtained by docking the product is very similar to that obtained by docking the component monomers and superimposing them. Only 13% of products (16 compounds) have an RMSD in excess of 3.0Å, indicating a significant difference in the binding modes obtained using the two methods.

Table VI. Acid and amine library. RMSD of direct predictions for product binding mode against product binding mode predictions from docked monomers.

RMSD	*Total*	*Percent*
Under 1.0Å	40	31
Under 2.0Å	92	71
Under 3.0Å	113	88

This verification stage was relatively time consuming, requiring 209 CPU hours (8 days and 17 hours) on an R4400 SGI processor.

An example of the results obtained for one product are shown in Figure 5. For clarity, only a few residues in the active site of lipase are displayed (colored by atom type) and only heavy atoms are shown. The predicted binding modes for acid monomer 2531 and amine monomer 70017 are shown as dark colored structures. It can be seen that the amide bond linkage occupies the same area of 3-D space. The predicted binding mode obtained by GOLD when the product (also present in the ACD, no. 248910) is docked into the active site is shown in the third structure (colored by atom type). It can be seen that the predicted binding mode of the product corresponds closely to that obtained by docking the two monomers.

Experiments using a library of sulphonyl chlorides and amines. The procedure described above was repeated for a second library of sulphonyl chlorides and amines, which react to form products with a sulphonamide linkage. This library comprised 204 sulphonyl chlorides and 105 amines extracted from the ACD (*32*).

As before, both monomer pools were prepared by adding the link group and terminating with a methyl group. In this case a sulphonamide bond linkage was used rather that an amide linkage. As before both sets of prepared monomers were docked into lipase to generate 2 or 3 GOLD solutions per monomer. Following this, all sulphonyl chloride solutions were checked for overlaying sulphonamide linkages with all amine monomers. Duplicate products of low fitness and products with high van der Waals energy were rejected.

The results are summarized in Table VII. From the 150000 possible superpositions of GOLD solutions the binding modes of 190 products with reasonable

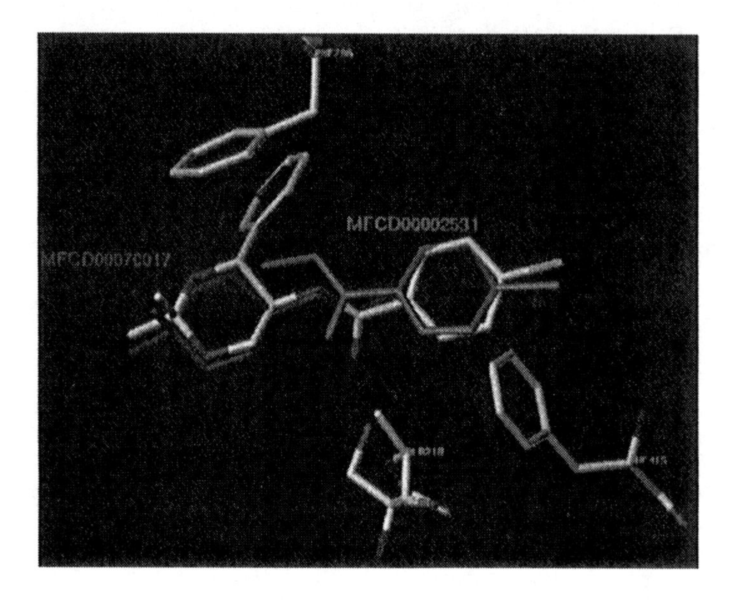

Figure 5. The predictions obtained by GOLD for acid 2531, amine 70017 and product (also available in the ACD, no. 248910) in lipase.

energy were determined. Again the library docking methodology has resulted in considerable data reduction with 190 products selected from the 21420 available.

Table VII. Summary of sulponyl chloride and amine library experiments. All CPU times are for an R4400 processor.

No. of Sulphonyl Chlorides	204
No. of Amines	105
No. of available products	21420
No. of GOLD solutions for sulphonyl chlorides	520
No. of GOLD solutions for amines	280
Total no. of possible superpositions	151200
No. of product superpositions found	369
No. of products following energy screen	384
No. of unique products	190
No. of sulphonyl chlorides in products	66
No. of amines in product	24
CPU time to dock sulphonyl chlorides	48.5h
CPU time to dock amines	39.5
CPU time for superposition and energy screen	1.0h
Total CPU time	89h

As in the case of acids and amides a verification procedure was applied. Good geometries were generated for the products and GOLD was used to predict their binding modes (using the same, two-stage docking procedure as previously). The comparison of binding modes in Table VIII shows the RMSD of heavy atoms between the binding modes predicted by docking the product and by docking the component monomers. The verification stage took 224 CPU hours on an R4400 SGI CPU.

Table VIII. Sulphonyl chloride and amine library. RMSD of direct predictions for product binding mode against product binding mode predictions from docked monomers.

RMSD	*Total*	*Percent*
Under 1.0Å	28	15
Under 2.0Å	97	51
Under 3.0Å	145	76

Reasonable agreement is seen between the binding modes obtained using the two approaches, with half the products having an RMSD of less than 2.0Å and 76% having an RMSD of less than 3.0Å. Approximately one quarter of the products show RMSDs in excess of 3.0Å, indicating a significant difference in the two binding modes. These results, while not too bad in themselves, show significant deterioration from those obtained using the library of acids and amines. The difference is probably due to the increased size and flexibility of the sulphonamide linkage used here, compared with the rigid amide linkage used previously. Furthermore, the Tripos torsional potential used within GOLD is not parameterised for sulphonamide torsions (*33*).

An example of the results obtained is shown in Figure 6, which shows the predicted binding modes of sulphonyl chloride 7450, amine 174223 and product (also present in the ACD, no. 203426).

Comments on Library Docking Experiments. It would appear that docking library monomers is an effective alternative to enumerating a library and then docking in all products. It is a particularly useful technique if selection of a small number of products is desired.

This technique is especially appropriate when dealing with site-directed targeted libraries. In this case each monomer pool is targeted to a different part of the active site. By restricting docking to the appropriate part of the active site GOLD can ensure that each monomer binds as expected. The method is easily extensible to libraries containing more than two monomer pools. However, it is expected that a large number of monomers would need to be docked before three or more monomers with superimposed linkage groups were found. Finally, this technique is not restricted to GOLD, but could be used with any docking program.

Conclusions

This paper documents the continuing improvements to GOLD. Measures have been taken to correct some problems experienced in earlier testing of the software and improved results have been observed on the test suite of 100 PDB complexes. The new algorithm has shown some ability in predicting activity. Finally, a time-efficient mechanism has been adopted for the screening of combinatorial libraries. While the methodology is attractive and initial results are promising, the feasibility of such an approach can only be truly verified by experimental assay and crystallographic study.

References

1. Blaney, J. M.; Dixon J. S. *Perspect. Drug Discov. Res.* **1995,** *1,* 301.
2. Jones, G; Willett, P. *Curr. Opin. Biotechnology,* **1995,** *6,* 652.
3. Miller, M. D.; Kearsley, S. K.; Underwood, D. J.; Sheridan, R. P. *J. Comput.-Aided Mol. Design* **1994,** *8,* 153.
4. Rarey, M.; Kramer, B.; Lengauer, T. *J. Comput.-Aided Mol. Design* **1997,** *11,* 369.
5. Goodsell, D. S.; Olson A. J. *Proteins: Struct. Func. Genet.* **1990,** *8,* 195.
6. Jones, G.; Willett, P.; Glen, R. C. *J. Mol. Biol.* **1995,** *254,* 43.
7. Judson, R. S.; Jaeger, E. P.; Treasurywala, A. M. *J. Mol. Struct.* **1994,** *308,* 191.
8. Gehlhaar, D. K.; Verkhivker, G. M.; Rejto, P. A.; Sherman C. J.; Fogel, D. B.; Fogel, L. J.; Freer, S. T. *Chem. Biol.* **1995,** *2,* 317.

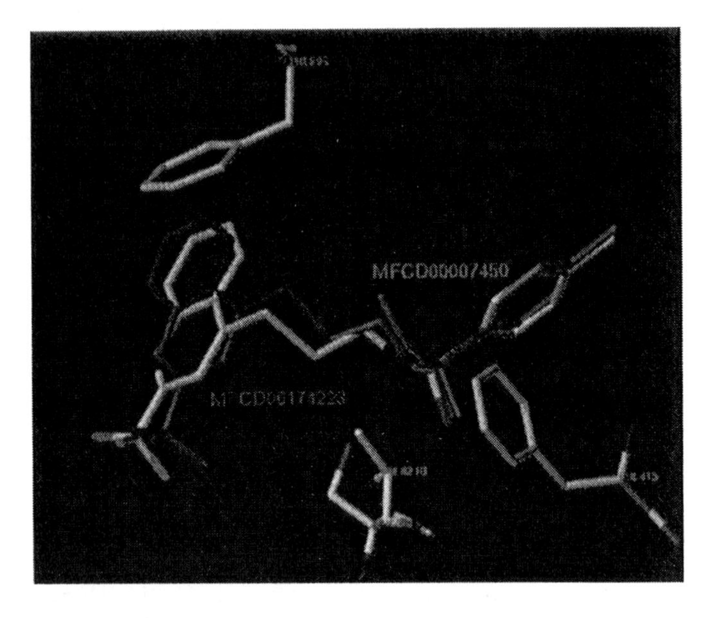

Figure 6. The prediction obtained for sulphonyl chloride 7450, amine 174223 and product (also available in the ACD, no. 203426) in lipase.

9. Davis, L. D. *Handbook of Genetic Algorithms;* Van Nostrand Reinhold: New York, 1991.
10. Holland, J. H.; *Adaptation in Natural and Artificial Systems;* MIT Press: Cambridge, MA, 1992.
11. Goldberg, D. E. *Genetic Algorithms in Search Optimization and Machine Learning;* Addison-Wesley: Reading, MA, 1989.
12. Clark, D E.; Westhead, D. R. *J. Comput.-Aided Mol. Design* **1996,** *10,* 337.
13. Jones, G.; Willett, P.; Glen, R. C.; Leach A. R.; Taylor R. *J. Mol. Biol.* **1997,** *267,* 727.
14. Bernstein, F. C.; Koetzle, T. F.; Williams, G. J. B.; Meyer, F.; Bryce, M. D.; Rogers, M. D.; Rogers, J. R.; Kennard, O.; Shikanouchi, T.; Tusumi, M. *J. Mol. Biol.* **1977,** *112,* 535.
15. Hasel, W.; Hendrickson, T. F.; Still, W. C. *Tetrahedron Comput. Methodol.* **1988,** *1,* 103.
16. Allen, F. H.; Davis, J. E.; Galloy, J. J.; Johnson, O.; Kennard, O.; Macrae, C. F.; Mitchell, E. M., Mitchell, G. F., Smith, J. M.; Watson, D. G. *J. Chem. Inform. Cumput. Sci.* **1991,** *31,* 187.
17. Klebe, G.; Meitzner, T. *J. Comput.-Aided Mol. Design* **1994,** *8,* 583.
18. Jedrzejas, M. J.; Singh, S.; Brouilette, W. J.; Laver, W. G.; Air, G. M.; Lou, M. *Biochemistry* **1995,** *34,* 3144.
19. Bossart-Whitaker, P.; Carson, M.; Babu, Y. S.; Smith C. D.; Laver, W. G.; Air, G. M. *J. Mol. Biol.* **1993,** *232,* 1069.
20. Smith, P. W.; Sollis, S. L.; Howes, P. D.; Cherry, P. C.; Cobley, K. N.; Taylor, H; Whittington, A. R.; Scicinski, J.; Bethell, R. C.; Taylor, N.; Skarzynski, T.; Cleasby, A.; Singh, O.; Wonacott, A.; Varghese, J.; Colman, P. *Bioorg. Med Chem. Lett.* **1996,** *6,* 2931.
21. Sollis, S. L.; Smith, P. W.; Howes, P. D.; Cherry, P. C.; Bethell, R. C. *Bioorg. Med Chem. Lett.* **1996,** *6,* 1805.
22. Varghese J. N.; McKimm-Breschkin, J.; Caldwell, J. B.; Kortt, A. A.; Colman, P. M. *Proteins: Struct. Func. Genet.* **1992,** *14,* 327.
23. White, C. L.; Janakiraman, M. N.; Laver, W. G.; Philippon, C.; Vasella, A.; Air, G. M.; Lou, M. *J. Mol. Biol.* **1995,** *245,* 623.
24. Williams, M.; Bischofberger, N.; Swaminathan, S.; Kim, C. U. *Bioorg. Med. Chem. Lett.* **1995,** *5,* 2251.
25. Stewart, K. D.; Bentley J. A.; Corey, M. *Tetrahedron Comput. Methodol.* **1990,** *3,* 713.
26. Tulinsky, A.; Blevins, R. A. *J. Biol Chem.* **1987,** *262,* 7737.
27. Ajay; Murcko, M. A. *J. Med. Chem.* **1995,** *38,* 4953.
28. Gallop, M. A.; Barrett, R. W.; Dower, W. J.; Fodor, S. P. A.; Gordon, E. M. *J. Med. Chem.* **1994,** *37,* 1385.
29. Bolin, J. T.; Filman, D. J.; Matthews, D. A.; Hamlin, R. C.; Kraut, J. *J. Biol. Chem.* **1982,** *257,* 13650.
30. Hajduk, P. J,; Dinges, J.; Miknis, G. F.; Merlock, M.; Middleton, T.; Kempf, D. J.; Egan, D. A.; Walter, K. A.; Robins, T. S.; Shuker, S. B.; Holzman, T. F.; Fesik, S. W. *J. Med. Chem.* **1997,** *40,* 3144.
31. DesJarlais, R. L.; Sheridan, R. P.; Dixon, J. S.; Kuntz, I. D.; Venkatarghavan, R. *J. Med. Chem.* **1986,** *29,* 2149.
32. Available Chemicals Directory; MDL Information Systems Inc.: San Leandro CA 94577, USA.
33. SYBYL version 6.3; Tripos Inc.: St Louis, MO 63144, USA.
34. Perlman, R. S. *Chem Des. Automat. News* **1987,** *2,* 1.
35. Grochulsky, P.; Li, Y.; Schrag, J. D.; Bouthillier, F.; Smith, P.; Harisson, D.; Rubin, M.; Cygler, M. *J. Biol Chem.* **1993,** *268,* 12843.

Chapter 19

Reduced Dimensionality in Ligand–Protein Structure Prediction: Covalent Inhibitors of Serine Proteases and Design of Site-Directed Combinatorial Libraries

Daniel K. Gehlhaar, Djamal Bouzida, and Paul A. Rejto

Agouron Pharmaceuticals, Inc., 3301 North Torrey Pines Court, La Jolla, CA 92037

Structure prediction of ligand-protein complexes is greatly facilitated when the location of ligand functional groups relative to the protein is known. This situation arises in two applications of practical interest: when a covalent bond is formed between the ligand and the protein, and when a fragment of the ligand dominates the molecular recognition with the protein. In both of these cases, it is shown that the predicted structure corresponds to the experimentally observed structure with increased probability. Using this approach, a library of compounds is screened for potential inhibitors of dihydrofolate reductase and porcine pancreatic elastase; known inhibitors were ranked favorably in both cases.

The computational prediction of binding geometries of compounds in a protein active site, termed the docking problem, is an important component in computer-aided structure-based drug design (1). The problem can be fruitfully approached with an energy function that is sufficiently accurate to discriminate between correct and incorrect binding modes and that is amenable to a conformational search method that can determine the lowest energy state of the complex, which defines the predicted binding mode. Early methods treated both the ligand and the protein as rigid bodies and the search was limited to optimizing the position and orientation of the ligand within the active site, while more recent methods incorporate flexibility of the ligand and limited protein flexibility (2-7). It remains a significant challenge because of the large number of degrees of freedom and the need to calculate interaction energies, but is eased significantly when the number of degrees of freedom is reduced. This can be achieved, for example, by fixing the conformation of the ligand in its bound conformation, but in general this knowledge is not available prior to experimental determination of the structure of the complex.

An alternate method to reduce the number of degrees of freedom is to fix part of the ligand within the active site of the receptor, a simplification that is exploited in this paper for two special cases. In the first case, the ligand forms a covalent complex with the protein, and the location of the ligand is constrained by this chemical bond. In the second case, the portion of the ligand that is primarily responsible for molecular recognition, the anchor, is fixed within the active site. Assuming that the binding mode of the anchor fragment is not affected, this approach, termed fixed anchor docking, allows the efficient evaluation of various substituents. In both cases, the search space is significantly reduced compared to fully flexible docking simulations because the center of mass degrees of freedom have been eliminated and the number of binding modes are limited dramatically. We discuss the methodology employed for the docking simulations and discuss validation studies. In conjunction with a method for estimating the binding affinity of the resulting complexes, we use this approach to search libraries of compounds for covalent inhibitors of porcine pancreatic elastase and to build a virtual site-directed library for dihydrofolate reductase. We demonstrate that known inhibitors of porcine pancreatic elastase and a potent inhibitor of dihydrofolate reductase, methotrexate, are identified using this method.

Serine Proteases. The active site of many viral serine proteases, such as cytomegalovius protease and hepatitis C protease, are solvent-exposed and comparatively shallow (*8-10*). These proteases differ from enzymes such as HIV protease, which has a highly constrained binding site, and has been a successful target for structure-based design (*11-13*). In order to achieve high potency, noncovalent inhibitors of viral serine proteases might need to be relatively large, consistent with the long recognition sequence of these proteases (*14*). Due to the binding energy provided by the covalent bond, inhibitors that form a chemical bond with the protein can be smaller than those that do not, while maintaining the desired potency. Such comparatively small compounds are desirable both for increased oral bioavailability and for ease of synthesis, although the presence of a reactive functional group may lead to non-specific reactions with other serine proteases and reactive oxygens, as well as complicate the delivery of these compounds.

The activated serine oxygen can bond covalently with an electrophilic group via a tetrahedral intermediate (*11*), but the software commonly used for predicting the structure of ligand-protein complexes does not take such interactions into account. While some docking software does allow the modeling of covalent interactions (*2*), the user is required to place the ligand in the correct tetrahedral geometry relative to the protein, a limitation that makes the program unsuitable for searching large chemical databases. We describe a method for the fully automated and rapid flexible docking of inhibitors covalently bound to serine proteases, combining an energy function specifically tuned for molecular docking with an efficient search engine based on evolutionary programming.

Site-Directed Combinatorial Libraries. Another application of structure prediction methods in reduced dimensionality arises in the design of combinatorial chemistry libraries, a field that is beginning to have significant impact on drug discovery (*15,16*). Combinatorial approaches can accelerate the speed of research and increase the amount of structure-activity data typically available by orders of magnitude. Because of the combinatorial explosion, however, the number of compounds that in principle can be synthesized typically far exceeds the number that can be made, let alone the number that can be assayed in a reasonable timeframe. While considerable progress has been made in automatic synthesis and screening methods, smaller, targeted libraries solve many of the problems associated with large generic libraries. In this approach, the library is biased towards features that promote molecular recognition. These features can be obtained from the sequence of known substrates (*17*), transition state mimics (*18*), or the three-dimensional structure of the receptor (*19,20*).

Combinatorial approaches and structure-based drug design approaches complement each other. The large number of compounds inherent in combinatorial methods reduces the need for precise estimates of binding affinity, so critical in traditional single compound design. At the same time, constraints imposed by the structure of the receptor can reduce significantly the size of library that needs to be considered. Due to the combinatorial nature of the synthesis, eliminating even modest numbers of monomers at each stage can have significant results. A library that consists of a central core fragment with three attachment sites and 100 possible monomers at each site contains 10^6 compounds, at the limits of current synthesis and technology. If some fraction of the monomers f can be eliminated at each site, the size of the library scales down as f^n, where n is the number of attachment sites. Hence, eliminating 75% of the monomers in this example with three attachment sites results in a library approximately 1% of the original size, well within the capability of current experimental methodologies for synthesis and screening.

Methodology

Energy Function. The energy function used to predict the structure of ligand-protein complexes contains an intermolecular term for the interaction between the ligand and the protein binding site, and an intramolecular term for the conformation of the ligand. The intramolecular term consists of the van der Waals and torsional strain terms of the DREIDING force field (*21*), and is used to differentiate between low- and high-energy internal geometries of the ligand. The intermolecular potential is a pairwise sum of piecewise-linear potentials (Figures 1a and 1b) between ligand and protein heavy (non-hydrogen) atoms (*7*), with parameters depending on the type of interaction and the size of the atom. Ligand and protein heavy atoms are classified as hydrogen bond donors, acceptors, donors and acceptors (e.g. a hydroxyl oxygen), or nonpolar. Small (fluorine and metal ions), medium (carbon, oxygen, and nitrogen), and large (sulfur, phosphorus, chlorine, and bromine) atoms are assigned atomic radii of 1.4, 1.8, and 2.2 Å, respectively. These parameters were derived from interatomic distances observed in high-quality crystal structures, and then optimized.

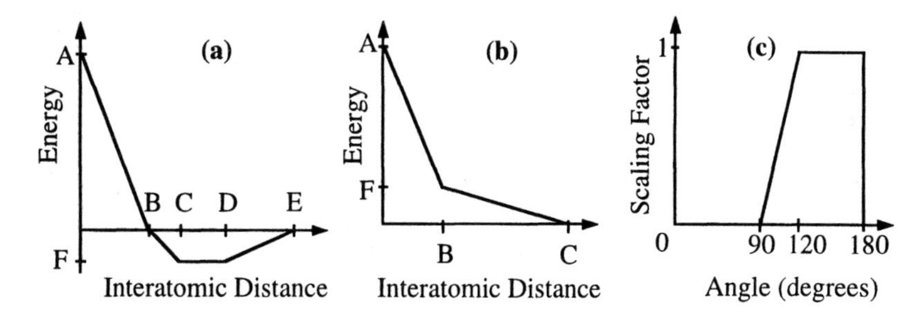

Figure 1. (a) Functional form of the atomic pairwise potential used for the hydrogen bonding (A=15.0, B=2.3, C=2.6, D=3.1, E=3.4, F=-4.0) and dispersion terms (B=0.93σ, C= σ, D=1.25 σ, E=1.5 σ, F=-0.4, where σ is the sum of the atomic radii of the protein and ligand atoms) of the intermolecular potential. (b) Functional form of the atomic pairwise potential used for donor-donor and acceptor-acceptor repulsive interactions. A=15.0, B=3.2, C=6.0, F=1.5. (c) Scaling factor used to modulate the hydrogen bonding and repulsive terms based on the relative orientation of protein and ligand atoms. All units are Å and kcal/mole.

Each pair of interacting atoms is assigned one of three interaction types: a hydrogen bond interaction between donors and acceptors, a repulsive interaction for donor-donor and acceptor-acceptor contacts, and a generic dispersion term for other contacts. Both the hydrogen bond and repulsive terms are modulated by a scaling factor that imparts a crude angular dependence (Figure 1c). The hydrogen bond angle for ligand acceptors is defined by the ligand atom, the protein donor atom, and its associated polar hydrogen, while for ligand donors, it is defined by the ligand atom, the protein acceptor and its neighbor. In both cases, the attractive region of the potential is scaled. The angle for the repulsive term is defined by the ligand atom, the protein donor atom and its polar hydrogen for ligand donors, or, for ligand acceptors, the ligand atom, the protein acceptor and its neighbor. In this case, the long-range region of the potential (between parameters B and C in Figure 1b) is scaled.

Search Engine. Evolutionary programming (*22*), modeled after natural selection where a population of solutions competes for survival, has been effective in a variety of optimization problems. In this study, the population encodes the dihedral angles for all single-bonded atoms, not including resonant bonds. These angles are initialized to a random value and allowed to vary during the optimization.

The search process consists of a fixed number of generation cycles, where in each cycle, the members of the population are scored using the energy function described above. A subset of the population is selected to be "parents" for the next generation, with the remainder discarded. Parents for the production of “offspring” are selected in a stochastic tournament, wherein the energy of each member of the population is compared with a fixed number of randomly selected opponents. A win is assigned to the member with the lowest energy, and the number of competitions a member wins is used to determine whether it survives into the next generation. Sufficient offspring are generated from the parent vectors through the addition of Gaussian mutations, restoring the population to its original size and completing the generation cycle. Although it is possible to calculate the ideal standard deviation size for simple energy functions (*22*), they are not known for more complex response surfaces. Consequently, a self-adaptive strategy was used where the mutation sizes are allowed to vary *(7)*, with selection pressure determining the ideal values as the simulation progresses. The best-scoring individual in the final generation is minimized using a conjugate gradient search and defines the predicted structure.

Over the course of the search, the repulsive component of each pairwise potential in the intermolecular scoring function is scaled linearly, starting at a low value and rising to its final value at the end of the run. This allows the ligand to freely interpenetrate the protein early in the search, promoting population diversity and ensuring a more complete sampling of the search space. For a ligand with ten rotatable bonds, a typical optimization involves approximately 70,000 energy evaluations and uses approximately 15 seconds of CPU time on a Silicon Graphics R10000 processor.

Covalent Docking. While many chemical groups can serve as the site of covalent attack, one may be interested in a particular group or may want to specifically disallow attack at certain sites, such as protecting groups. To be of use for large database searches, the specification of ligand attack sites must be fully automated.

A convenient representation of the covalent attack sites is afforded by the CIF (Crystallographic Information File) format (*23*). A data definition was created that allows an atom to be specified in terms of its properties, such as element type, hybridization state, number of neighbors, and aromaticity, as well as the properties of its neighbors. The description is recursive, with each neighbor description having its own neighbor descriptions, so that any chemical group, regardless of size, may be described. Each group description can be associated with a label string. For example, the attack site specification for a triflouromethyl ketone is given by "C,HeavyAtomNeighbors=3, Neighbor(O, BondOrder=2), Neighbor(C, Neighbor(F), Neighbor(F), Neighbor(F)), Neighbor(C)". This defines a carbon atom with three heavy atom neighbors; one neighbor is a double-bonded oxygen, another is a carbon which itself has three fluorine neighbors, and the third neighbor may be any carbon atom.

Accurate modeling of the tetrahedral state of the covalent inhibitors is necessary for correct prediction of the bound structure. Many inhibitors undergo substantial conformation changes upon formation of the covalent interaction, as the electrophilic atom changes from planar to tetrahedral geometry. These conformational changes can affect ligand atoms far away from the covalent interaction, especially if the covalent attack occurs in a predominantly rigid region of the ligand such as a ring. A summary of the method used to generate the geometry about the tetrahedral center for reactive carbonyls is shown in Figure 2. Starting from a rough approximation of the tetrahedral state, a proton is added to the carbon perpendicular to the plane defined by the carbonyl carbon and oxygen and another adjacent atom, reducing the carbonyl to an alcohol. A short minimization with the BatchMin program (*24*) using the all-atom Amber force field (*25*) and a distance-dependent dielectric is used to correct the geometry. The tetrahedral model is then aligned with the catalytic serine, fixing the distance between the serine oxygen and ligand carbon at 1.45 Å, the bond angle about the serine oxygen at 109.5°, but allowing the serine Cα-Cβ and Cβ-Oγ bonds to rotate freely during the search. The process is then repeated for the opposite enantiomer as it is impossible to predict *a priori* which is most active.

Library Generation. The protocol for directed library design comprises four major steps: database querying, virtual library building, fixed anchor docking, and binding affinity estimation and ranking. It results in predicted binding modes and affinities for each compound in the library. Each substitution site in the central anchor is assigned its own reaction. In the first step, the database is screened for compounds that have at least one moiety that can be linked covalently to the central anchor fragment following some pre-specified chemical reactions. For very large databases, additional requirements based on simple molecular properties such as molecular weight can also be imposed. All fragments that satisfy the above constraints are linked to the central anchor motif, thus building a virtual library of compounds. The third step represents the primary filter, whereby the binding mode of all virtual molecules within the active site is predicted, subject to the constraint that the anchor motif is fixed in its initial

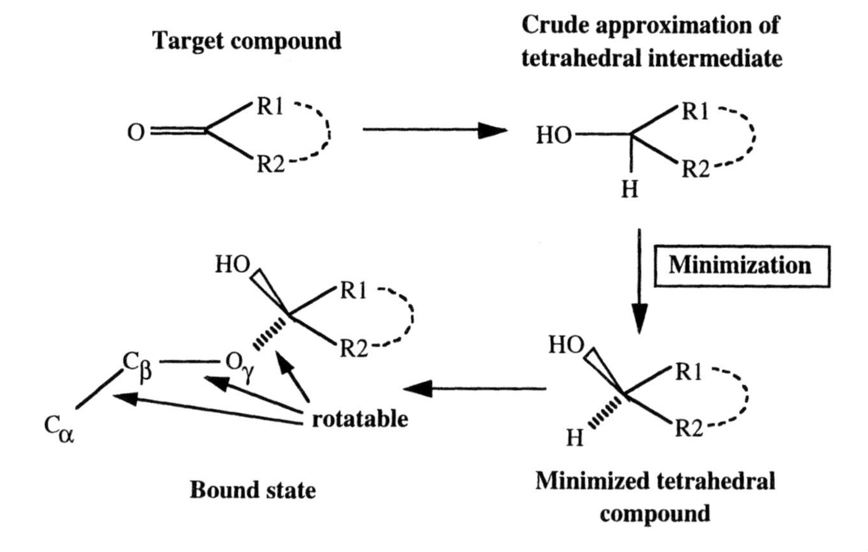

Figure 2. Generation of the tetrahedral intermediate structure from a carbonyl-containing ligand.

position and orientation. After the structure of each compound is predicted, the entire ligand is relaxed within the binding site using a conjugate-gradient minimization, and the final step is binding affinity estimation and ranking of the compounds.

Validation Studies

Covalent Docking. To be useful for database searches, a structure prediction method for covalently bound inhibitors must discriminate not only between alternate binding modes, but also between the active and inactive stereoisomers. It must be relatively insensitive to small changes in the internal geometry of the ligand, so the same structure should result when a ligand is obtained from a high-resolution crystal structure or generated from a molecular modeling program.

Five ligand-protein complexes containing covalently bound ligands were selected from the Brookhaven Protein Data Bank (PDB) (*26*): wheat serine carboxypeptidase (1bcr), human leucocyte elastase (1eas and 1eau), and porcine pancreatic elastase (4est and 8est). Although many structures in the PDB are reported to contain a covalently bound ligand, a surprisingly small number had tetrahedral geometry at the site of covalent attachment; many ligands were in a planar geometry. The lack of a significant number of structures with the ligand in a tetrahedral geometry made validation of our method of generating the tetrahedral state difficult. We therefore chose a somewhat generic parameterization of the geometry at the attack site, with a bond distance of 1.45 Å and bond angle of 109.5°.

The five crystallographic complexes were chosen based on the quality of the structures and the size of the inhibitors, which vary in size from 18 to 38 heavy (non-hydrogen) atoms, and have between six and 13 rotatable bonds. Two of the structures (1bcr and 8est) undergo covalent attack at an aldehyde moiety, while the others have reactive diflouro- or triflouro-methyl ketone moieties. For each system, hydrogen atoms were added to the receptor and non-buried water molecules were removed. A total of three sets of 100 docking simulations were performed for each complex, one using the tetrahedral geometry given in the crystal structure, and one for each of two stereoisomers of a computationally generated ligand conformation. For the latter, the ligands were constructed in a molecular modeling program, minimized to convergence in the Amber force field (*14*), and the tetrahedral geometry was generated using the method described above. No specific knowledge about the crystal structures, such as torsion angles about nonrotatable bonds, was used to generate the modeled ligands.
A structure prediction was defined to be successful when the root mean square deviation (RMS) of the heavy atoms in the predicted structure was within 2.0 Å of the crystallographic structure. For all five complexes, the structure of the bound inhibitor was predicted correctly; the crystallographic binding mode of the ligand had the lowest energy (Table I). Furthermore, correct and incorrect stereoisomers could be distinguished, even when the ligand was obtained by molecular modeling. For the 4est complex, an alternate binding mode was essentially isoenergetic with the correct binding mode, but where the phenyl moiety occupied a pocket that had been filled by a crystallographic water in the original protein structure. When the water was replaced, the success rate increased to 41%.

Table I: Comparison of Docking Simulations Using Three Different Tetrahedral Geometries at the Covalent Attack Site.

	Crystal	*Model Stereoisomer 1*[b]		*Model Stereoisomer 2*	
PDB Entry	*Success Rate*[a]	*Success Rate*[a]	*Energy*[c]	*Success Rate*[a]	*Energy*[c]
1bcr	78%	46%	-63.5	12%	-56.0
1eas	79%	49%	-75.2	41%	-55.5
1eau	61%	58%	-74.7	30%	-50.2
4est	19% (41%)[d]	68%	-86.2	31%	-56.3
8est	99%	88%	-75.8	90%	-66.9

[a]Percent of predicted structures within 2.0 Å RMS of the crystallographic structure.
[b]Stereoisomer corresponding to the crystal structure.
[c]Energy of the best-scoring docked ligand.
[d]Success rate when a crystallographic water was replaced.

Fixed Anchor Docking. We validated the fixed anchor docking method by performing multiple docking simulations of five ligand-protein complexes with known crystal structures. The non-covalent complexes were methotrexate bound to dihydrofolate reductase (DHFR) (3dfr), Viracept bound to HIV-1 protease (*27*), biotin bound to streptavidin (1stp), FK506 bound to FKBP (*28*), and a tripeptide inhibitor bound to thermolysin (4tmn).

For these simulations, a docking simulation is defined to be successful using the more stringent requirement that the predicted structure is within 1.5 Å RMS of the crystal structure. Two sets of docking simulations, each comprising 100 trials of each of the above ligands, were performed. In the first set, a fully flexible docking simulation was performed, while in the second an anchor fragment of the ligands was fixed in its crystallographic position (Figure 3). For the flexible docking simulations, the success rate is correlated with the number of degrees of freedom of the ligand, while for fixed anchor docking, the X-ray binding mode is consistently reproduced (Table II). In the case of FK506, the conformation of the macrocycle is fixed in its crystallographic form, rationalizing its anomalously high success rate. As expected, fixing the binding mode of the anchor fragment substantially increases the rate of successful prediction.

Table II: The number of bonds allowed to rotate during the conformational search and the success rate for flexible and fixed anchor docking for five ligand-protein complexes.

Protein-ligand complex	*Rotatable bonds*	*Flexible docking*	*Fixed anchor docking*
Biotin/Streptavidin	5	99%	100%
Methotrexate/DHFR	7	75%	100%
FK506/FKBP	7	93%	100%
Viracept/HIV-1 Protease	8	57%	100%
4tmn/Thermolysin	13	44%	100%

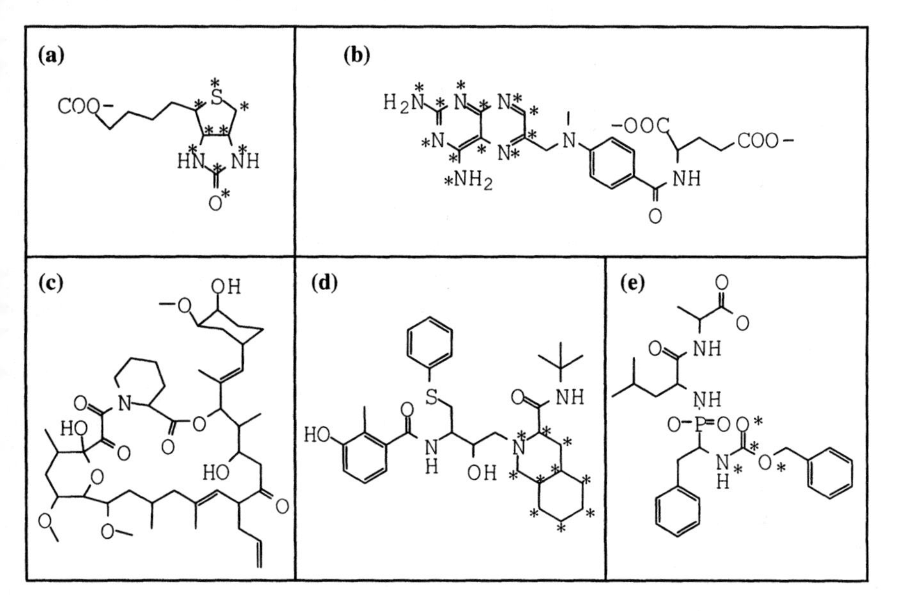

Figure 3. Compounds used for fixed anchor validation are (a) biotin, (b) methotrexate, (c) FK506, (d) viracept, and (e) 4tmn. Atoms fixed during the docking simulation are indicated by stars; for FK506, the entire macrocycle is fixed.

Database Screening

In principle, the goal of a computational database search is to accurately rank the entire set of compounds in order of affinity and thereby eliminate the need for experimental screening. To achieve this goal, the relative binding free energy of a large number of disparate compounds must be computed, a task that is beyond the capability of current methodologies. As a result, computational database screenings typically focus on the more limited goal of increasing the likelihood of finding an active compound in an experimental screen of a subset of the database, as compared to random. A seeded database experiment, where a known inhibitor is added to an otherwise arbitrary database, can be used to assess the effectiveness of the computational procedure by considering the relative ranking of the known inhibitor.

Covalent Inhibitors. To test the effectiveness of finding active covalent inhibitors from a database, the algorithm was used to screen a subset of the Available Chemicals Directory (ACD; MDL Information Systems, San Leandro, CA) for inhibitors of porcine pancreatic elastase (4est). This system was chosen because of the high resolution of its crystallographic structure (1.6 Å), the well-defined binding site, and the existence of a large inhibitor. Approximately 25,000 ACD compounds were screened, each with between zero and twelve rotatable bonds, twelve and forty heavy atoms, and at least one aromatic ring. Covalent attack was allowed for non-aromatic ketones and esters, with the exception of protecting groups, such as carbobenzoxy. While not all of these ketones are sufficiently activated to form a covalent complex, they can be modified to a more active form, and hence these compounds represent a diverse set of potentially reactive groups. The known inhibitor was also included as part of the search.

After the structure of the bound complex was predicted, a number of screens were imposed to eliminate compounds with characteristics that are unlikely to be found in ligand-protein complexes (Figure 4). These include large movement of the serine oxygen upon binding (over 0.5 Å), gross close contacts with the binding site (at least one non-hydrogen bonding pair of heavy atoms with a 30% or greater van der Waals overlap), and a high desolvation penalty. The structures that passed these screens were clustered according to their predicted binding free energies, resulting in 5,858 unique structures with a favorable binding energy (Figure 5). The desolvation penalty and the predicted binding free energies were computed using a modification and extension (*29*) of the LUDI scoring function (*30*) that includes short-ranged repulsive interactions and a term to describe hydrogen bonding geometry (*31*). The known inhibitor was ranked in the top one percent of all compounds that satisfied the screening criteria, and in the top one quarter of one percent of all compounds in the initial database. In addition, the correct stereoisomer and binding mode for this compound were selected. A number of compounds chemically unrelated to the known inhibitor were found, some of which form favorable hydrogen bonds in the active site (Figure 6). None of these compounds have been tested for activity against the enzyme.

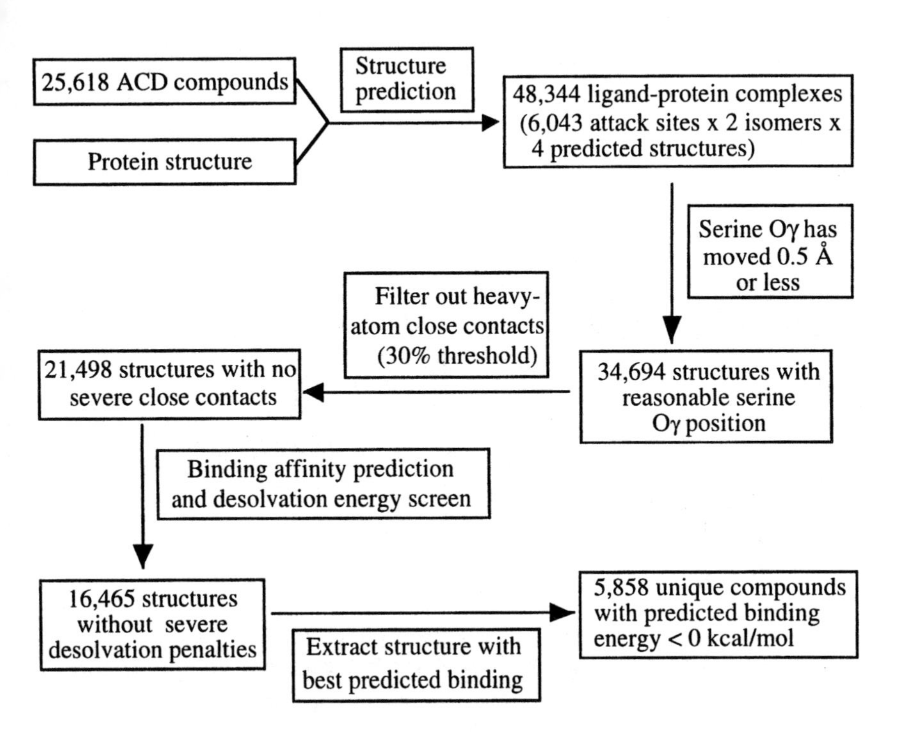

Figure 4. Summary of covalent inhibitor database search processing and results.

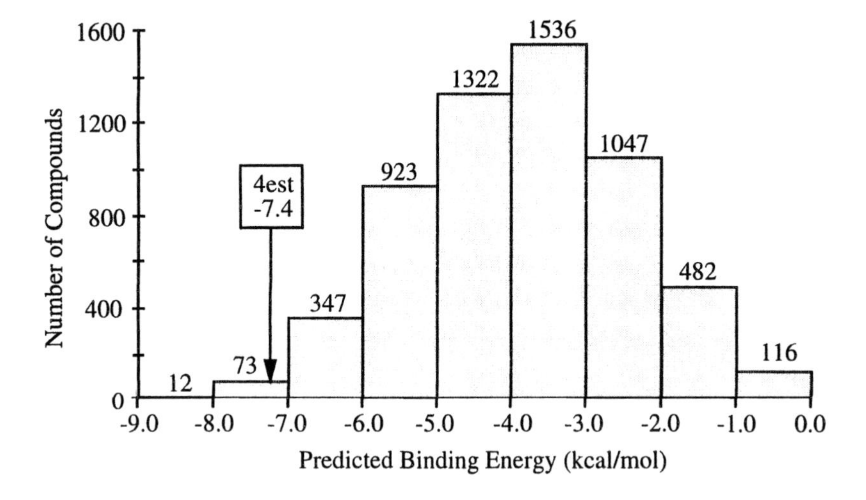

Figure 5. Distribution of predicted binding affinities for the covalent inhibitors. The relative position of the tripeptide inhibitor 4est is indicated.

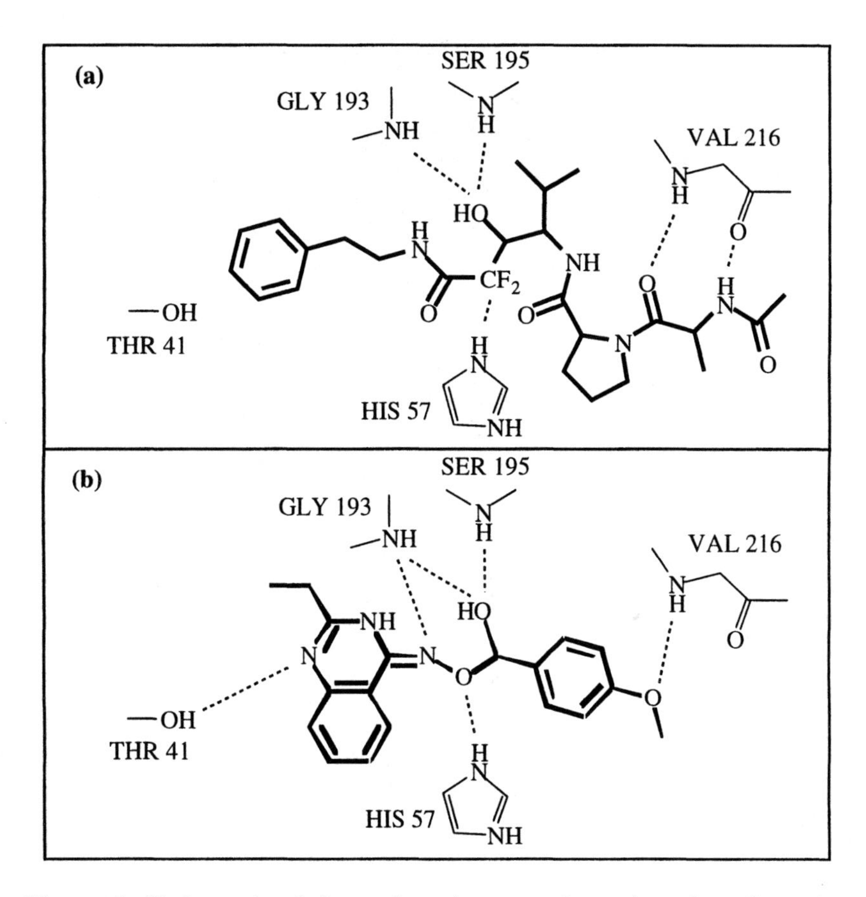

Figure 6. Hydrogen-bond interactions between the active site of porcine pancreatic elastase and (a) inhibitor 4est, (b) a compound identified from the database search, (c) a second compound identified from the database search. The range of the hydrogen bonds, indicated by dashed lines, is between 2.5 and 3.4 Å.

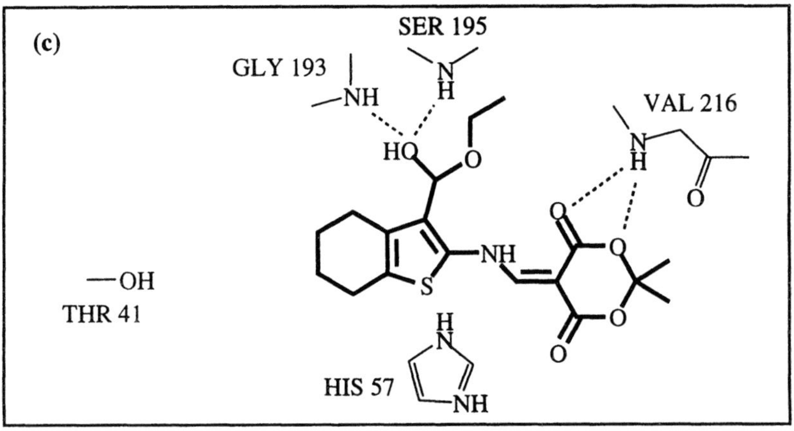

(c)

SER 195

GLY 193

NH

HO

O

VAL 216

NH

O

S

—OH

THR 41

HIS 57

NH

Figure 6. Hydrogen-bond interactions between the active site of porcine pancreatic elastase and (a) inhibitor 4est, (b) a compound identified from the database search, (c) a second compound identified from the database search. The range of the hydrogen bonds, indicated by dashed lines, is between 2.5 and 3.4 Å.

Site-Directed Combinatorial Library Design. The protocol was used to design a site-directed combinatorial library for dihydrofolate reductase (DHFR) (*32*), a popular test case for computational methods for structure prediction and binding affinity estimation. Methotrexate is an unusually high-affinity inhibitor of DHFR and as such is a good inhibitor for validating new methods. While identification of methotrexate is a necessary condition for a useful computational screening protocol, it does not guarantee success in all cases, since lead compounds are not always highly potent.

Methotrexate contains a pteridine ring whose structure is predicted consistently within 0.5 Å of its crystallographic binding mode. This high consensus suggests that the pteridine is essential for molecular recognition and can be considered an anchor motif for DHFR. Given this anchor fragment, direct alkylation was chosen for the attachments of substituents to the pteridine ring. The ACD was restricted to compounds with molecular weight less than 250 and which contain an amine group with at least one hydrogen and one heavy atom neighbor that is part of an aromatic group. From the 7,256 compounds obtained, a virtual library of 7,677 compounds was generated, all of which contain the pteridine motif. The number of compounds in the library is larger than the number of compounds retrieved from the database because some compounds had multiple sites where the reaction could take place.

A series of ten docking simulations was performed for each compound, keeping the pteridine ring fixed within the DHFR active site. All structures with 25% or greater heavy atom overlaps with the active site atoms were discarded, and the binding affinity of the remaining molecules was estimated. All the compounds with a large predicted desolvation penalty were removed. Of the 7,677 compounds in the virtual library, only 516 satisfied all the screening criteria, 7% of the original library. As anticipated, methotrexate is predicted to have the best binding energy, near -12 kcal/mol (Figure 7). In addition to methotrexate, a number of other compounds were generated that form good hydrogen bond interactions within the active site (Figure 8).

Conclusion

Reducing the dimensionality of the conformational space in ligand-protein structure prediction reduces the computational complexity of the problem and can lead to significantly improved prediction. Often, restricted mobility can only be achieved by making strong approximations regarding the nature of the binding process. There are however, at least two special cases where the dimensionality is reduced without need for such assumptions. In ligand-protein complexes with a covalent bond, the restriction is imposed by the requirement that the chemical bond be formed between specific ligand and protein atoms. We have demonstrated that for such systems, the crystal structure of ligand-protein complexes can be reproduced with high fidelity, and that a known inhibitor of the porcine pancreatic serine protease elastase can be identified from a database of random compounds that contain reactive functional groups. Furthermore, methotrexate is identified as a potent inhibitor of DHFR when seeded in a database of amines that can react with the pteridine central core. This method can reduce the size of combinatorial libraries, and, in conjunction with more sophisticated selection of reactant molecules, can aid in the generation of site-directed libraries (*33*).

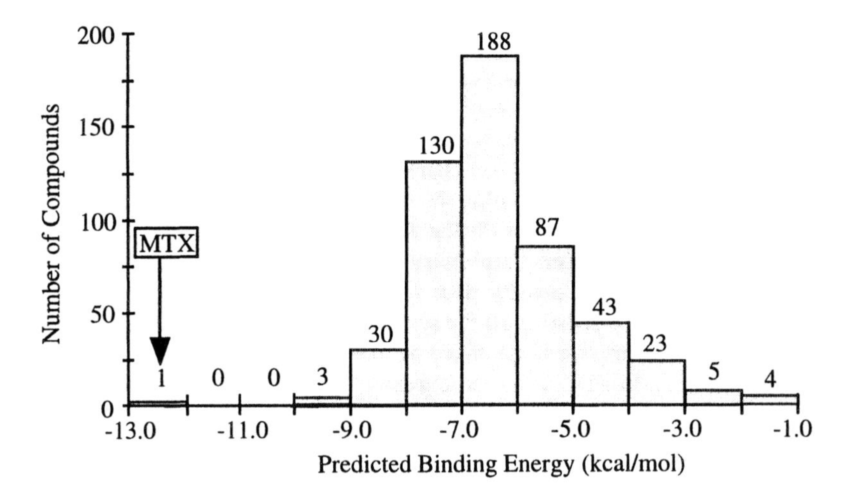

Figure 7. The binding affinity distribution for the virtual library of compounds generated using the pteridine core in DHFR.

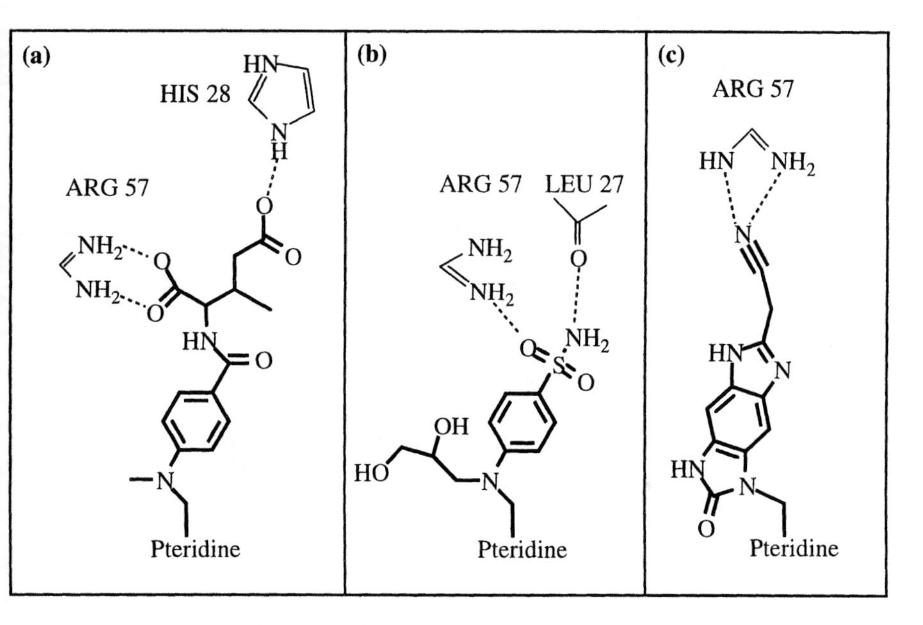

Figure 8. Hydrogen-bond interactions between the active site of DHFR and (a) methotrexate, (b) a compound identified from the database search, (c) a second compound identified from the database search. The range of the hydrogen bonds, indicated by dashed lines, is between 2.5 and 3.4 Å.

Acknowledgments

We thank our colleagues S. Arthurs, A. Colson, Dr. S. Freer, V. Larson, and L. Schaffer for significant contributions to the software described in this work, Dr. Peter Rose for providing the empirical binding affinity method, and Dr. Gennady Verkhivker for useful discussions about molecular recognition.

Literature Cited

1. Kuntz, I. D.; Meng, E. C; Shoichet, B. K. *Accts. Chem. Res.* **1994**, *27*, 117-123.
2. Jones, G.; Willett, P., Glen, R. C.; Leach, A. R.; Taylor, R. *J. Mol. Biol.* **1997**, *267*, 727-748.
3. Kuntz, I. D.; Makino, S. *J. Comp. Chem*, **1997**, *18*, 1812-1825.
4. Desmet, J.; Wilson, I. A.; Joniau, M.; DeMaeyer, M.; Lasters, I. *FASEB J.* **1997**, *11*, 164-172.
5. Welch, W.; Ruppert, J.; Jain, A. N. *Chem. Biol.* **1996**, *3*, 449-463.
6. Rarey, M.; Kramer, B.; Lengauer, T.; Klebe, G. *J. Mol. Biol.* **1996**, *261*, 470-489.
7. Gehlhaar, D. K.; Verkhivker, G. M.; Rejto, P. A.; Sherman, C. J.; Fogel, D. B.; Fogel, L. J.; Freer, S. T. *Chem. Biol.* **1995**, *2*, 317-324.
8. Chen, P.; Tsuge, H.; Almassy, R. J.; Gribskov, C. L.; Katoh, S.; Vanderpool, D. L.; Margosiak, S. A.; Pinko, C.; Matthews, D. A.; Kan, C.-C. *Cell* **1996**, *86*, 835-843.
9. Love, R. A.; Parge, H. E.; Wickersham, J. A.; Hostomsky, Z.; Habuka, N.; Moomaw, E. W.; Adachi, T.; Hostomska, Z. *Cell* **1996**, *86*, 331-342.
10. Kim, J. L. Morgenstern, K. A.; Lin, C.; Fox, T.; Dwyer, M. D.; Landro, J. A.; Chambers, S. P.; Markland, W.; Lepre, C. A.; O'Malley, E. T.; Harbeson, S. 1.; Rice, C. M.; Murcko, M. A.; Caron, P. R.; Thomson, J. A. *Cell* **1996**, *87*, 343-355.
11. Babine, R. E.; Bender, S. L. *Chem. Rev.* **1997**, *97*, 1359-1472.
12. Wlodawer, A.; Erickson, J. W. *Annu. Rev. Biochem.* **1993**, *62*, 543-585.
13. Appelt, K.; Bacquet, R. J.; Bartlett, C. A.; Booth, C. L.; Freer, S. T.; Fuhry, M. A. M.; Gehring, M. R.; Herrmann, S. M.; Howland, E. F.; Janson, C. A.; Jones, T. R.; Kan, C.; Kathardekar, V.; Lewis, K. K.; Marzoni, G. P.; Matthews, D. A.; Mohr, C.; Moomaw, E. W.; Morse, C. A.; Oatley, S. J.; Ogden, R. C.; Reddy, M. R.; Reich, S. H.; Schoettlin, W. S.; Smith, W. W.; Varney, M. D.; Villafranca, J. E.; Ward, R. W.; Webber, S.; Webber, S. E.; Welsh, K. M.; White, J. *J. Med. Chem.* **1991**, *34*, 1925-1934.
14. LaFemina, R. L.; Bakshi, K.; Long, W. J.; Pramanik, B.; Veloski, C. A.; Wolanski, B. S.; Marcy, A.; Hazuda, D. J. *J. Virology* **1996**, *70*, 4819-4824.
15. Thompson, L.A.; Ellman, J.A. *Chem. Rev.* **1996**, *96*, 555-600.
16. Gordon, E. M.; Barrett, R. W.; Dower, W. J.; Fodor, S. P. A.; Gollop, M. A. *J. Med. Chem. 1994*, *37*, 1385-1401.
17. Yu, H.; Chen, J. K.; Feng, S.; Dalgarno, D. C.; Brauer, A. W.; Schreiber, S. L. *Cell* **1994**, *76*, 933-945.
18. Wiley, R. A.; Rich, D. H. *Med. Res. Rev.* **1993**, *13*, 327-384.

19. Kick, E. K.; Roe, D. C.; Skillman, A. G.; Liu, G.; Ewing, T. J.; Sun, Y.; Kuntz, I.D.; Ellman, J .A. *Chem. Biol.* **1997**, *4*, 297-307.
20. Salemme, F. R.; Spurlino, J.; Bone, R. *Structure* **1997**, *4*, 319-324.
21. Mayo, S. L.; Olafson, B. D.; Goddard, W. A. *J. Phys. Chem.* **1990**, *94*, 8897-8909.
22. Fogel, D.B. *Evolutionary Computation: Towards a New Philosophy of Machine Intelligence*. IEEE Press: New York, NY, 1995; 75-155.
23. Hall, S. R.; Allen, F. H.; Brown, I. D. Acta Cryst. **1991**, *A47*, 655-685.
24. Mohamadi, F.; Richards, N. G. J.; Guida, W. C.; Kiskamp, M.; Caufield, C.; Chang, G.; Hendrickson, T.; Still, W. C. J. Comput. Chem. **1990**, *11*, 440-467.
25. Weiner, S. J.; Kollman, P. A.; Case, D. A.; Singh, U. C.; Ghio, C.; Algona, G.; Profeta, S. Jr.; Weiner, P. J. J. Am. Chem. Soc. **1984**, *106*, 765-784.
26. Bernstein, F. C.; Koetzle, T. F.; Williams, G. J.; Meyer, E. F. Jr.; Brice, M. D.; Rodgers, J. R.; Kennard, O.; Shimanouchi, T.; Tasumi, M. *J. Mol. Biol.* **1977**, *112*, 535-542.
27. Appelt, K.; *Perspect. Drug Discovery Des.* **1993**, *1*, 23-48.
28. Van Duyne, G. D.; Standaert, R. F.; Karplus, P. A; Schreiber, S. L.; Clardy, J. *J. Mol. Biol.* **1993**, *229*, 105-124.
29. Rose, P. W. In *Computer-Assisted Molecular Design Course*, UCSF; San Francisco, CA, **1997**; Chapter 12.
30. Bohm, H. J. *J. Comput. Aided Mol. Des.* **1994**, *8*, 243-256.
31. McDonald, I. K.; Thornton, J. M. *J. Mol. Biol.* **1994**, *238*, 777-793.
32. Bolin, J. T.; Filman, D. J.; Matthews, D. A.; Hamlin, R. C.; Kraut, J. *J. Biol. Chem.* **1982**, *257*, 13650-13662.
33. Polinsky, A.; Feinstein, R. D.; Kuki, A. In *Molecular Diversity and Combinatorial Chemistry: Libraries and Drug Discovery*; Chaiken, I. M., Janda, K. D., Eds.; Conference Proceedings Series; American Chemical Society: Washington, D.C. **1996**; Chapter 20, 219-232.

Chapter 20

Development and Validation of the EVA Descriptor for QSAR Studies

David B. Turner [1], Peter Willett [1], Allan M. Ferguson [2], and Trevor W. Heritage [2]

[1] Krebs Institute for Biomolecular Research and Department of Information Studies, University of Sheffield, Western Bank, Sheffield S10 2TN, United Kingdom
[2] Tripos Inc., 1699 South Hanley Road, St. Louis, MO 63144

QSAR models are of great importance in the rationalisation and prediction of the relative bioactivities of sets of compounds. Recently, 3D-QSAR techniques, such as CoMFA, have proved to be an effective means of correlating shape-related features with bioactivity, provided that a suitable relative alignment of the structures concerned can be found. Here we describe the EVA QSAR method. EVA, which is based upon IR-range vibrational frequencies, provides an alignment-free methodology and is shown to produce statistically robust QSARs comparable in most cases to results obtained with CoMFA. The method has been extensively validated on eleven different data sets. A brief discussion of conformational sensitivity is given together with an evaluation of the possibilities for model interpretation. We also report ongoing work centred upon using a genetic algorithm to provide models with enhanced predictivity over "classical" EVA QSAR.

Since the advent of classical QSAR techniques, exemplified by Hansch *(1)*, there has been considerable progress in the development of molecular descriptors and chemometric techniques for use in such studies. The development of 3D-QSAR techniques *(2)* that attempt to correlate biological activity with the values of various types of molecular field, for example steric, electrostatic or hydrophobic, has been of particular interest *(3-5)*. The original, and most well-known of the 3D-QSAR techniques is Comparative Molecular Field Analysis *(3)* (CoMFA) which uses steric and electrostatic field values calculated at the intersections of a three-dimensional grid surrounding the structures in the data set. A major limitation of CoMFA, and most other 3D-QSAR techniques, is the dependency upon the relative orientation of the molecules in the data set *(6,7)*. Despite efforts to improve the efficiency of the alignment process *(7-10)* the selection of the molecular alignment is regarded as the major variable in the analysis. These problems are further exacerbated when the conformational flexibility of the molecules in the data set is considered.

There is, therefore, considerable interest in the development of new descriptions of molecular structure that do not require the alignment of molecules, but that retain the 3D and molecular property information encoded within molecular fields. Alternative descriptions of molecular fields than those used in CoMFA or molecular surface properties, for example methods based on autocorrelation vectors *(11)*, molecular moments *(12)*, or MS-WHIM descriptors *(13)*, may provide effective orientation-independent descriptions of molecular structure. In this chapter we review a novel descriptor of molecular structure, known as EVA (EigenVAlue descriptor), that is derived from calculated infra-red (IR) range vibrational frequencies. As discussed later in this chapter, EVA has been found to yield robust QSAR models that are, for the most part, statistically comparable to those derived using CoMFA, with the advantage that EVA does not require structural alignment.

Rationale

During the late 1980's workers at Shell Research Limited *(14)* reasoned that a significant amount of information pertaining to molecular properties, in particular biological activity, might be contained within the molecular vibrational wavefunction, of which the vibrational spectrum is a fingerprint. The EVA descriptor is thus derived from normal co-ordinate **E**igen**VA**lues (*i.e.* the vibrational frequencies) that are either calculated theoretically or extracted from experimental IR spectra. Typically, a classical normal co-ordinate analysis *(15)* (NCA) is performed on an energy minimized structure, and the resulting eigenvalues represent the normal mode frequencies from which the EVA descriptor is derived. The associated normal coordinate eigenvectors (*i.e.*, the vibrational motions) are not used within the EVA descriptor. The force constants upon which a normal co-ordinate analysis is dependent may be determined using a molecular mechanics, semi-empirical, or *ab initio* quantum mechanical method. The accuracy of the calculated vibrational eigenvalues is, therefore, determined entirely by the quality of the force constants applied or derived.

Determination of the EVA Descriptor

Using the standard Cartesian co-ordinate system as a basis for describing the displacement of an atom from its equilibrium position in a vibrating molecule requires $3N$ coordinates for a molecule containing N atoms. Three of these coordinates describe rigid-body translational motion, and a further three describe rigid-body rotations. Thus, in the general case for a molecule of N atoms there are $3N-6$ vibrational degrees of freedom, or $3N-5$ for a linear molecule such as acetylene (only two coordinates are required to fix the orientation). The number of vibrational degrees of freedom is equivalent to the number of fundamental vibrational frequencies (normal modes of vibration) of the molecule. Whilst each of these fundamental vibrations can be calculated, they may or may not appear in an experimental IR absorption spectrum due to symmetry considerations, *i.e.* they may have zero (or close to zero) intensity *(15)*.

Thus, in order to derive the EVA descriptor, each structure is initially characterized by $3N-6$ (or $3N-5$) vibrational modes. In all but the special case where the molecules in the data set contain the same number of atoms it is not possible to compare the vibrational frequencies directly. This so-called dimensionality problem does not arise during a CoMFA analysis because the molecular fields arising from each molecule are calculated across a fixed set of lattice points; this would be an issue if, for example, one wanted to compare directly the atomic point charges from which the electrostatic fields are derived. Furthermore, even in cases in which it is desired to compare molecules that do contain the same number of atoms, and hence the same number of vibrational modes, it is difficult to establish which vibrations are directly comparable between molecules; this problem arises from inherent and effectively indeterminate contributions made by individual atoms to a given vibrational mode.

In EVA, the dimensionality of the descriptor is unified across the entire data set by a three-step procedure that involves transformation of the sets of vibrational frequencies onto a scale where they are directly comparable (*i.e.*, a fixed-dimensional scale). In the initial step of this standardization process the frequency values are projected onto a bounded frequency scale (BFS) with individual vibrations represented by points on this axis. The bounds chosen for the BFS of 1 and 4,000 cm^{-1} encompass the frequencies of all fundamental molecular vibrations and match the range observed for experimentally-derived IR spectra. The second step in the standardization process requires that each calculated frequency is characterized in terms of a kernel of fixed height, width, and shape. Each of the calculated vibrations is weighted equally during this process. The resulting value associated with each of the calculated vibrations permits the proportion of overlap of vibrations to be determined, and may be considered analogous to, but in no way representative of, peak intensity. Infra-red intensity information is not used in the generation of the EVA descriptor and, as explained below, the technique is not intended to simulate experimental IR spectra.

In practice, in the second step a Gaussian kernel of fixed standard deviation (σ) is placed over each vibrational frequency value for each structure, resulting in a series of $3N$-6 (or $3N$-5) identical, and overlapping Gaussians (Figure 1). The value of the EVA descriptor, EVA_x, at any chosen sampling point, x, on the bounded frequency scale is then determined by summing the amplitudes of each and every one of the $3N$-6 (or $3N$-5) overlaid Gaussian kernels at that point:

$$EVA_x = \sum_{i=1}^{3N-6} \frac{1}{\sigma\sqrt{2\pi}} e^{-(x-f_i)^2/2\sigma^2}$$

where f_i is the ith frequency for the structure concerned.

It is important at this stage to reiterate that the purpose of the above EVA smoothing procedure is not an attempt to simulate the infra-red spectrum of the molecule of interest, since the transition dipole data is ignored, but rather it is to provide a basis upon which vibrations occurring at slightly different frequencies may be compared to one another. The Gaussian function applied to define peak shapes adds a probabilistic element in that the peak maxima are centered at each of the calculated frequency values (f_i) and thus these points are taken to be the most probable values of the respective frequencies. An EVA descriptor sampled at a point for which $x \neq f_i$ is thus considered to be a less probable value of the ith frequency. In such cases, the corresponding contribution of f_i to the final value of the EVA descriptor at point x

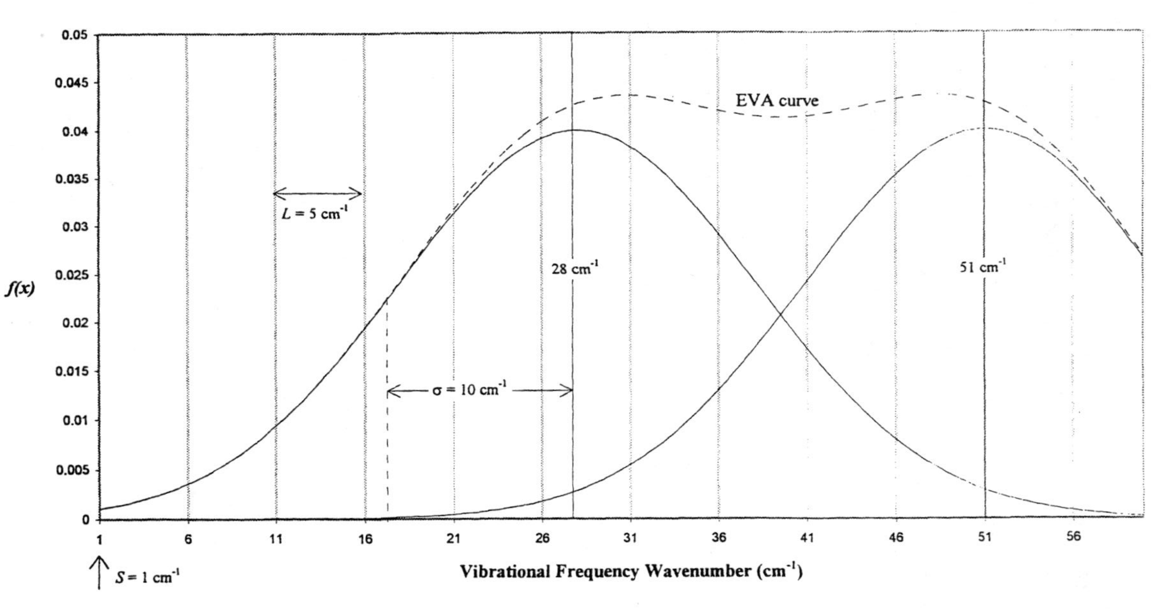

Figure 1. Profile of the "EVA curve" for three arbitrary vibrational frequencies expanded using a σ of 10 cm^{-1}. The "EVA curve" is determined by summing the estimated "intensities" (amplitudes) of the vibrations centred at 28 cm^{-1} and 51 cm^{-1}, respectively. The EVA descriptor is extracted by sampling the frequency scale at fixed intervals of L cm^{-1}. (Reproduced from Ref. 21 with kind permission from Kluwer Academic Publishers).

(EVA_x) will be less than the maximum possible contribution. To a certain extent, this behavior of the EVA descriptor reduces the dependency of the final QSAR model on the accuracy of the original calculated vibrational frequencies, which are sensitive to the molecule geometry optimization criteria and to the theoretical approximations or empirically based parameters of the chosen modeling paradigm (whether quantum, semi-empirical or molecular mechanics). Furthermore, and as discussed in detail below, this behavior has implications for the sensitivity of the descriptor to molecular conformation, in that small changes in vibrational frequencies arising from conformational changes may have insignificant effects on the resulting EVA descriptor values.

In the third, and final, step of the data standardization process the EVA function is sampled at fixed increments of L cm^{-1} along the BFS, giving the 4,000/L values comprising the EVA descriptor. Typically, a descriptor set is derived using a Gaussian standard deviation (σ) term of 10 cm^{-1} and a sampling increment (L) of 5 cm^{-1}, giving 800 descriptor variables. As is the case with CoMFA, the dimensionality of the EVA descriptor is much larger than the number of compounds in a typical QSAR data set and thus data reduction methods, such as Partial least squares to Latent Structures *(16)* (PLS) or Principal Components Regression (PCR), are applied to search for robust correlations with biological activity data. For most molecules of interest to a medicinal chemist geometry optimisation and normal mode calculation is the time-limiting step. However, if molecular mechanics methods are used this is only liable to take about a minute per structure depending mainly upon N and the hardware available. Therefore, whilst slower than CoMFA field calculations, the time needed for frequency calculations need not be prohibitive for QSAR datasets of typical size.

Applications of the EVA Descriptor in QSAR/QSPR Studies

One of the first demonstrations in the public domain of the regressive modeling capability of the EVA descriptor was obtained in a QSPR study *(17)* using Cramer's BCDEF data set *(18)*. The data set consists of measured log$P$ values for a highly heterogeneous set of 135 small organic chemicals, ranging from polycyclic aromatics, such as the highly lipophilic phenanthrene (log$P$ = 4.46), to small hydrophilic moieties, including methanol (log$P$ = 0.64). The EVA descriptors were derived using a 10 cm^{-1} Gaussian σ term and a sampling increment (L) of 5 cm^{-1} based on normal coordinate frequencies calculated using the AM1 *(19)* Hamiltonian in the MOPAC *(20)* semi-empirical MO program. These parameters gave an EVA descriptor consisting of 800 variables per structure, which were regressed against the log$P$ values using PLS. A regression equation based on only five PLS latent variables, that explained 96% of the variance in the log$P$ values, was obtained in this way. Full leave-one-out cross-validation of this data set yielded a crossvalidated-r^2 (*i.e.* q^2) of 0.68. This model was then used to predict the log$P$ value for a test set of 76 "unseen" chemicals, resulting in a predictive r^2 of 0.65. This study demonstrates the value of EVA as both an explanatory and a predictive tool, and, in addition, highlights one of its key advantages over 3D QSAR techniques such as CoMFA. In cases, such as this, where no intuitive alignment of the data set structures exists, it is very difficult or even impossible to apply CoMFA in a meaningful way, but with EVA no such complexity exists. Furthermore, bulk properties such as log*P* have no orientation dependence and thus any attempt to introduce such a dependency for QSAR purposes is entirely arbitrary. The diversity of structures exemplified in this data set also suggests that EVA may be applied to the analysis of diverse sets of compounds rather than just to congeneric series, which is a limitation for most alternative descriptors.

In subsequent studies *(21*,22) the general applicability of the EVA descriptor in QSAR studies has been investigated in detail using eleven data sets exhibiting a range of biological end points (Table I).

Using EVA descriptors derived from AM1 modes, good PLS models (as judged by q^2 scores) can be obtained for nine of the eleven data sets. The exceptions to this are the oxadiazole *(23)* and biphenyl *(24)* data sets for which, at best, only poor models can be obtained. It is important to remind the reader that although the EVA QSAR models presented in Table I are satisfactory, they are based solely upon the default EVA descriptor parameters (σ = 10 cm^{-1} and L = 5 cm^{-1}). Additional studies *(21*,*22*) have been performed in which the effect of changes to these parameters on the quality of the final QSAR models has been investigated and for nearly all of the data sets listed there exist values of σ that give rise to superior PLS models. A range of σ values should therefore be investigated prior to settling on a final model. A protocol recommended by Turner *et al. (21)* suggests that a σ value of 10 cm^{-1} is a reasonable starting point for a QSAR study and thereafter, if satisfactory results are not achieved, to supplement this with analyses based on σ terms of 5, 25, and 50 cm^{-1}. In addition, Table I lists analogous results for these datasets for which AMBER *(25)* molecular mechanics was used for geometry optimisation and to calculate normal mode frequencies. For the most part the AMBER-based EVA QSAR models give poorer results than those obtained with MOPAC AM1.

A useful benchmark for determining the effectiveness of the EVA descriptor in QSAR studies is to compare its statistical performance and model characteristics with analogous CoMFA model for the same data sets. A key limitation in all such comparative studies *(21*,*22*) is that the data sets have been selected *because* a good, published CoMFA model exists. This, therefore, may lead to significant bias in favor of the CoMFA technique, but none-the-less the results do provide interesting insights into the nature and scope of the EVA descriptor.

Examination of Table I shows that, at least in terms of the q^2 scores, the EVA descriptors provide roughly equivalent correlations as CoMFA for the cocaine *(26)*, dibenzofuran *(24)*, dibenzo-*p*-dioxin *(24)*, piperidine *(21)*, sulphonamide *(21)* and steroid data sets *(3)*. Although not as high as CoMFA, good predictive correlations are also obtained using EVA for the β-carboline *(27)*, muscarinic *(28)* and nitroenamine *(21)* data sets. The two cases where EVA performs poorly, the oxadiazole *(23)* and biphenyl *(24)* data sets, also yield the poorest CoMFA results, although statistically significant correlations ($q^2 \approx 0.5$) are obtained with CoMFA.

The robustness of EVA PLS models has been extensively tested by Turner *et al. (21*,29), both in terms of randomization permutation testing *(23)* and the ability of the models to make reliable predictions for test chemicals. Using the standard steroid data set from the original CoMFA study *(3)*, but with structures corrected according to Wagener *et al. (10)*, a predictive-r^2 value of 0.69 is obtained for the ten previously

TABLE I. Summary of EVA QSAR analyses for different sources of normal co-ordinate frequencies, using default EVA descriptor generation parameters ($\sigma = 10$ cm^{-1}, $L = 5$ cm^{-1}) and comparison with CoMFA.[a]

	AMBER		MOPAC AM1		CoMFA (both fields)	
Dataset[b]	q^2	r^2	q^2	r^2	q^2	r^2
β-Carbolines[c] (41)	0.29 (6)	0.97	0.50 (6)	0.97	0.68 (4)	0.89
Biphenyls (14)	0.16 (1)	0.72	0.28 (2)	0.90	0.49 (3)	0.87
Cocaines (13)	0.57 (2)	0.91	0.49 (2)	0.95	0.59 (4)	0.88
Dibenzo-*p*-dioxins (25)	0.48 (2)	0.85	0.68 (2)	0.88	0.66 (1)	0.80
Dibenzofurans (39)	0.61 (1)	0.74	0.78 (4)	0.97	0.72 (6)	0.85
Muscarinics (39)	0.42 (3)	0.88	0.53 (4)	0.95	0.59 (4)	0.84
Nitroenamines (17)	0.47 (2)	0.86	0.49 (3)	0.93	0.84 (3)	0.96
Oxadiazoles (23)	≤ 0	-	≤ 0	-	0.51 (2)	0.85
Piperidines (137)	0.71 (5)	0.84	0.76 (4)	0.84	0.73 (3)	0.80
Steroids (TBG[d]**)** (21)	0.42 (5)	0.99	0.70 (4)	0.98	0.62 (3)	0.92
Steroids (CBG[e]**)** (21)	0.79 (2)	0.90	0.70 (2)	0.87	0.75 (2)	0.91
Sulphonamides[f] (100)	-	-	0.54 (6)	0.80	0.65 (5)	0.82

[a]The leave-one-out crossvalidated- r^2 (q^2) values are reported together with the optimal number of LVs in brackets. Any q^2 values of ≤ 0 are indicated as ≤ 0 and LV_{opt} omitted as meaningless. Models are based on the selection of LV_{opt} by the first minimum in the SE_{cv} score. Full (fitted) models were derived only where $q^2 > 0$.

[b]Bracketed values are the number of compounds in the dataset.

[c]AMBER had the required force-field parameters for only 39 of 41 structures;

[d]Testosterone-Binding Globulin affinity as the target activity;

[e]Corticosterone-Binding Globulin affinity as the target activity;

[f]AMBER force field parameters not available.

unseen test chemicals; the biological endpoint used was the affinity for corticosteroid-binding globulin (CBG) expressed as 1/[log*K*]. This compares to a much lower value for CoMFA combined steric and electrostatic fields of 0.35. The apparently poor CoMFA test set predictive performance is almost entirely due to an extremely poor prediction for the only structure in the test set containing a fluorine atom, omission of which raises the CoMFA predictive-r^2 to 0.84. In contrast, the EVA predictive performance is raised to 0.74 when this compound is excluded, a small but nonetheless significant improvement. Clearly, in terms of the EVA descriptor space this compound can not be considered an extreme outlier, but in terms of CoMFA fields it is too different from the structures in the training set for a reliable prediction to be made.

The main advantage of EVA over CoMFA for QSAR purposes is the fact that orientation and alignment of the structures in the data set is not required. In CoMFA, the superposition of structures is the major variable, providing in some instances different modeling statistics for even quite small changes in the relative positions of the atoms in a pair of structures. However, given the nature of the field-based descriptors used in CoMFA, alignment does facilitate a powerful means of visualizing the important features of a QSAR model in the form of plots of the structural regions that are most highly correlated (either positively or negatively) with the biological property of interest. Despite the undoubted utility of these CoMFA plots, they do not indicate precisely which atoms are responsible for the modeled correlations since the electrostatic and steric fields are composed of contributions from each and every atom in the molecule - the PAC (Predicted Activity Contributions) *(30)* method has been reported to help deal with this problem. A further point to note is that it is not possible to predict the effects that structural changes may have on the resultant CoMFA fields. In contrast to CoMFA, there exist no obvious means of back-tracking from those components of the EVA descriptor which are highly correlated with changes in biological activity to the corresponding molecular structural features; a discussion of the ways of achieving this is presented at the end of this chapter.

EVA Descriptor Generation Parameters

The judicious selection of parameters is a prerequisite to the success of any QSAR method, and EVA is no exception. The most fundamental parameter involved in the derivation of the EVA descriptor is the Gaussian standard deviation (σ). The sampling increment (L), as explained in more detail below, need only be sufficiently small such that descriptor sampling errors are minimised.

Gaussian Standard Deviation (σ) The effect of varying the σ term of the EVA descriptor is illustrated in Figure 2 in which, as σ is increased, the features of the descriptor profile are progressively smoothed. The effect of the application of a Gaussian function during the EVA descriptor standardization process is to "smear out" a particular vibrational frequency such that vibrations occurring at similar frequencies in other structures overlap to a lesser or greater extent. It is this overlap that provides the variable-variance upon which regression modeling is dependent. By definition, each and every Gaussian must overlap, but for the most part this occurs at small

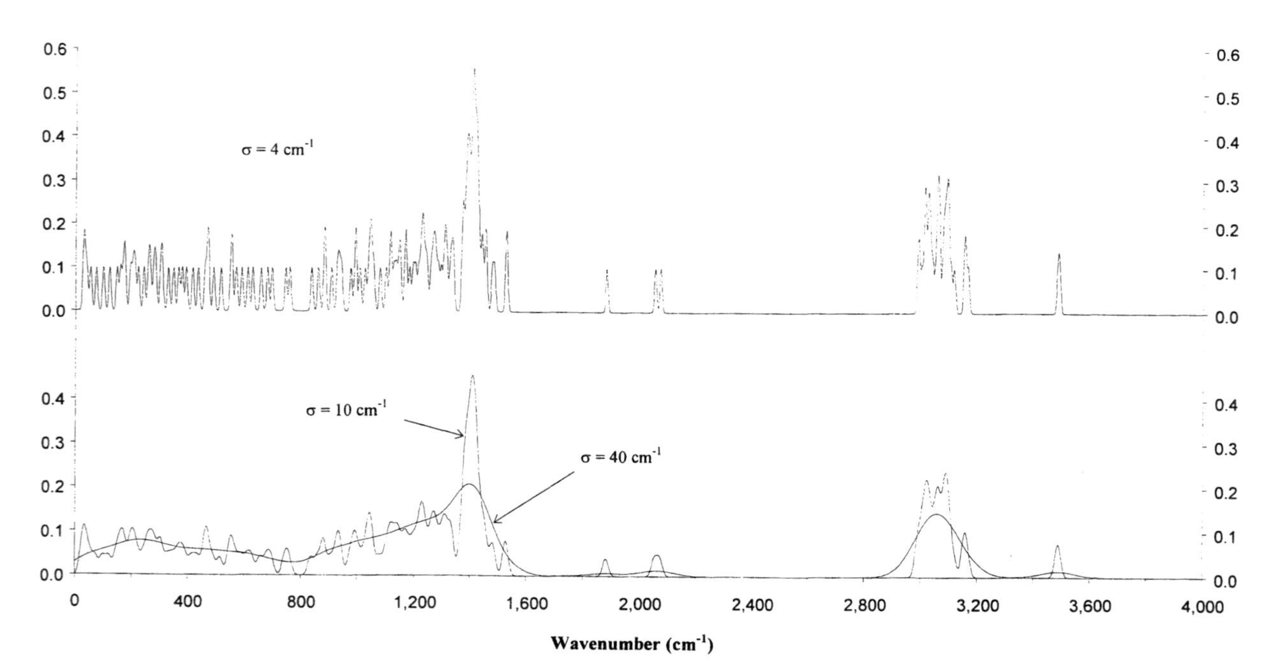

Figure 2. Illustration of the effect of the use of different σ terms on the EVA descriptor profile for an example molecule (deoxycortisol). (Reproduced from Ref. 21 with kind permission from Kluwer Academic Publishers).

(negligible) values and consequently the contribution to variance is typically insignificant. Only where the frequency values are sufficiently close to one another relative to the value of σ is it likely that inter-structural overlap of Gaussians will occur at values of significant magnitude. The selection of the Gaussian standard deviation, therefore, determines the number of, and extent to which, vibrations of a particular frequency in one structure can be statistically related to those in the other structures in the data set.

In addition to inter-structural overlap of Gaussians, the σ term also governs the extent to which vibrations within the same structure may overlap at non-negligible values. Intra-structural Gaussian overlap of this type, which is also dependent on the 'density' (*i.e.* proximity) of vibrations at various regions of the spectrum, causes EVA variables to consist of significant contributions from more than one vibrational frequency. The mixing of information contributed by individual normal co-ordinate frequencies must be considered to be undesirable in terms of ease of model interpretation. However, in order to provide sufficient inter-structural Gaussian overlap, it is inevitable that a certain degree of intra-structural kernel overlap occur.

Thus, small values of σ give rise to minimal intra-structural Gaussian overlap, whilst at larger values of σ significant overlap arises. In the former case there will be a reduction in inter-structural overlap, perhaps to such an extent that there exists no overlap of the Gaussians at significant values. In this instance the descriptor takes on the characteristics of a binary indicator, showing only the presence or absence of specific vibrational frequencies, thereby rendering the descriptor useless for regression analysis, but perhaps still of utility in classification analysis. In cases where larger σ values are used increased mixing of the information encoded by one frequency with that encoded by other frequencies arises.

Sampling Increment (*L*) Detailed investigation into the effect of various combinations of the σ and L parameters on the resulting q^2 value has been carried out by Turner *(21)*. Turner's results indicate that, for the most part, the final q^2 value is insensitive to small changes in either of these parameters, *i.e.* the information content of the EVA descriptor remains consistent. The most significant variations in q^2 are seen as σ is reduced (giving a more spiky spectral profile) and the sampling increment (L) is increased; this is analogous to lowering the spectral resolution. This result is intuitively reasonable since one would anticipate that, as L becomes very large relative to σ, some of the Gaussian peaks (or information encoded within them) will be omitted from the descriptor. In some cases, the information omitted will be predominantly noise resulting in a superior QSAR model, but in other cases signal may be accidentally omitted, resulting in degradation of the QSAR model. This sampling phenomenon can be referred to as *blind variable selection* since variables are selected or excluded from the descriptor on a completely arbitrary basis, which is of course undesirable. The L value at which blind variable selection begins to occur is σ-dependent; the larger the σ term the higher the permissible value of L. Thus, to avoid blind variable selection one might wish to minimize the value of L, but this must be balanced against the additional computational requirements associated with such a practice. Conversely, therefore, the value of L should be maximized in order to reduce the computational overhead, and this leads to the concept of critical L values (L_{crit}) which are σ-

specific and which, if exceeded, may result in a sampling error. In a systematic study the L_{crit} values for various σ have been established *(21)* with the conclusion that a simple rule-of-thumb is that L should be no larger than 2σ. These findings confirm that the intuitively reasonable, and default, selection of an L of 5 cm^{-1} with a σ term of 10 cm^{-1} should result in no significant sampling error and that, in fact, the value of L may be increased to 20 cm^{-1} with no apparent information loss (change in q^2).

The existence of these L_{crit} values is important not least because one of the problems with CoMFA at present is that the coarse grid-point spacing (typically 2Å) that is generally used is such that there is incomplete sampling of the molecular fields, which may result in information loss. The consequence of this is that reorientation of an aligned set of molecules as a rigid body within the defining CoMFA 3D region often results in substantial changes to the resulting QSAR model *(31)* (as evidenced in the q^2 values). EVA, on the other hand, does not suffer from such sampling errors provided that the σ-specific L_{crit} values are not exceeded.

Characteristics of the EVA Descriptor

Although the EVA descriptor is not intended to simulate the infra-red spectrum of a molecule (intensity information is discarded), it is useful to visualize the EVA descriptor in the form of a "spectrum". This permits the examination of the distribution of vibrations in a molecule or in a set of molecules. Figure 3 shows plots of the EVA descriptor for deoxycortisol (one of the most active CBG-binding compounds in the steroid data set of Cramer *(3)*) and estradiol (one of the inactive structures) over the spectral range 1 to 4,000 cm^{-1}. Also shown in Figure 3 is the univariate standard deviation (STDEV) of the descriptor over the entire steroid training set of 21 structures. The density of peaks in the fingerprint region (1 to 1,500 cm^{-1}) indicates that there is considerably more vibrational information in this region than in the functional group region (1,500 to 4,000 cm^{-1}) of the spectrum, as is typical of experimental infra-red spectra. The EVA descriptor values and STDEV are largest at frequencies centered around 1,400 and 3,100 cm^{-1}, corresponding to C-H bending and stretching vibrations. Figure 3 also highlights the errors associated with the calculation of normal co-ordinate frequencies (in this case using MOPAC), since a carbonyl stretching frequency is expected (from experiment) to appear at around 1,700 cm^{-1}, but is represented here by peaks at ≈2,060 cm^{-1}. This feature of the EVA descriptor again indicates that there is no attempt to simulate an experimental IR spectrum, but this does not detract from the usefulness of the descriptor for QSAR purposes, since consistency rather than accuracy across the data set is critical. Furthermore, for QSAR purposes relative, rather than absolute, differences in vibrational frequency across the data set are important. One might expect that this would become more of an issue should heterogeneous data sets be used since the consistency with which errors associated with the reproduction of equivalent vibrational frequencies may be more erratic. In practice, however, reasonable QSAR results have been obtained using a variety of heterogeneous data sets in conjunction with MOPAC *(17,21)*. There is the expectation that better quality normal mode calculations (such as, but not exclusively, *ab initio* quantum-mechanics) are likely to give better QSAR results with heterogeneous data-

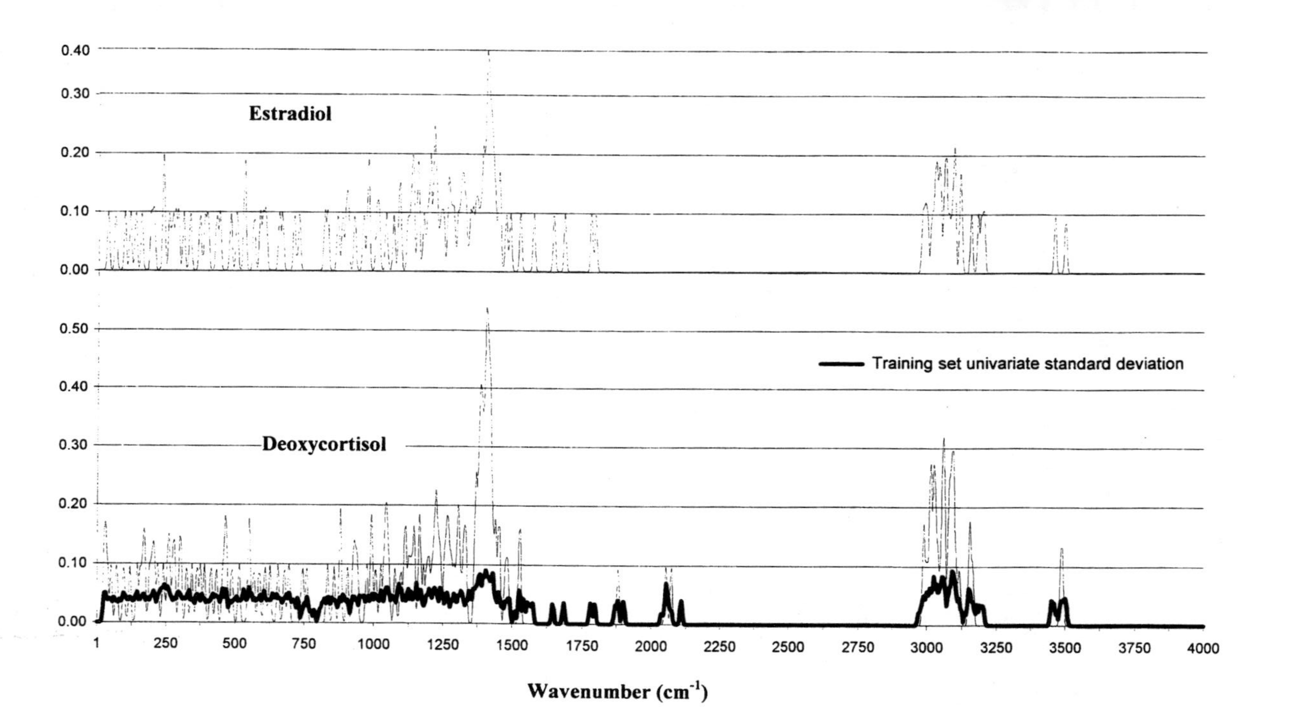

Figure 3. EVA descriptor profile for the steroids estradiol and deoxycortisol together with the univariate standard deviation of the descriptor over the 21 trainign set structures ($\sigma = 4$ cm^{-1} and $L = 5$ cm^{-1}). (Reproduced from Ref. 29 with kind permission from Kluwer Academic Publishers).

sets. However, there is no evidence to support this assertion since, to the authors knowledge, such studies have not been carried out.

Conformational Sensitivity of the EVA Descriptor

The sensitivity of CoMFA to the molecular orientation and alignment, and therefore, to the molecular conformation is well established *(6,7,32)* but, whilst EVA is completely independent of molecular orientation and alignment, the impact of the molecular conformation on EVA QSAR performance has not been discussed thus far. Intuitively, it is obvious that a change in conformation will result in changes in the force constants between atoms and, therefore, in the normal co-ordinate frequencies and displacements. The question is "to what extent are these changes evident within the EVA descriptor?" and "how much of this can be accounted for through the appropriate choice of Gaussian σ?". Some limited studies of these conformational effects have been performed *(29,34)*. In one such study *(34)* performed by Shell Research Limited, five classes of chemical, known to act at the same biological target (encompassing pyrazoles, thiazoles, piperidines, quinolines and thiochromans, totaling more than 250 structures) were clustered, using a nearest neighbor algorithm, based on the EVA descriptor. The conformations of each molecule were repeatedly randomized, new EVA descriptors generated, and the clustering process repeated. The conclusion were that, whilst the nearest neighbor relationships between compounds change, the overall cluster membership is approximately constant. This suggests that, in most cases, a conformational change does not lead to a sufficiently large change in the resulting EVA descriptor to cause a change in the underlying statistical model.

In a more recent study *(29)*, EVA descriptors for test chemicals were generated for conformations which matched those used in the training set, and also for non-matching conformations. At low σ values, the predictions made based on the non-matching conformations are considerably poorer than those made for the matched conformations. This difference gradually decreases until convergence is achieved at $\sigma = 12\ cm^{-1}$; thereafter, the predictions for the two conformations are roughly equivalent. In general, the conformational sensitivity of the EVA descriptor decreases as σ is increased. This is intuitively reasonable since a larger σ permits the significant overlap of more distant vibrational frequencies. As expected, the CoMFA predictions for non-matching conformations are much poorer than any of those obtained using EVA, thereby highlighting the differences in conformational sensitivity of the two methods.

QSAR Model Interpretation

In CoMFA, 3D-isocontour plots are used to visualize those regions of space indicated by the PLS model to be most highly positively or negatively correlated with biological activity. Whilst no such 3D visualization is possible with EVA, a variety of 2D plots have been suggested *(29,33)* that indicate the relative importance of regions of the spectrum in correlating biological activity. Figure 4 shows two such plots based on a two-component PLS model for the steroid data set *(29)*, that in some ways facilitate interpretation of an EVA QSAR model in analogous fashion to the interpretation of an

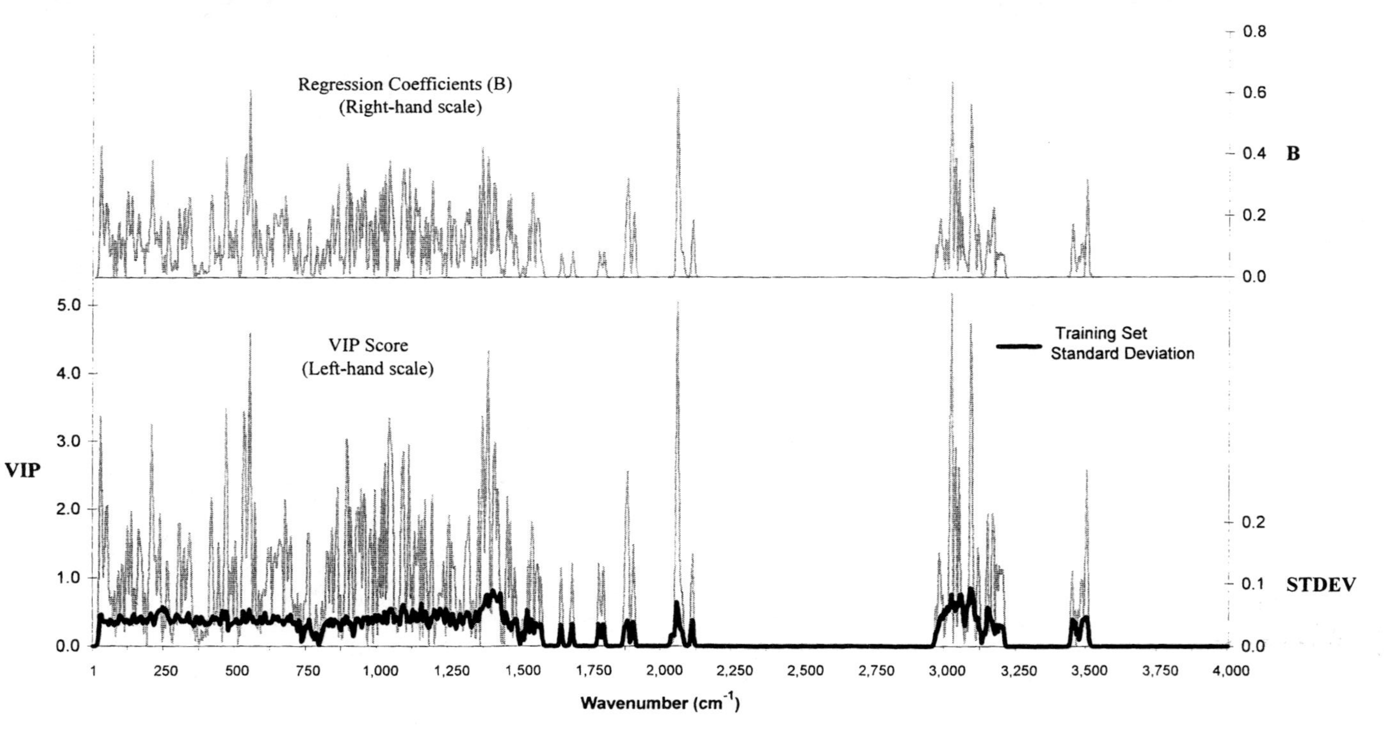

Figure 4. Plot of the most significant variables in an EVA PLS analysis as determined by the absolute magnitude of the regression coefficients (B) and alternatively by the VIP scores. (Reproduced from Ref. 29 with kind permission from Kluwer Academic Publishers).

experimental IR spectrum. The two measures shown in the figure are the magnitudes of the regression coefficients (B), and the Variable Influence on Projection *(33)* (VIP).

To backtrack to the important structural features indicated by the PLS model it is necessary to identify the variables most highly correlated with activity, decompose those variables into the contributing vibrational frequencies, and then to interpret and visualize the underlying normal mode vibrations. Two simple approaches have been proposed for identifying the most important variables in the PLS analysis *(33)*. The first approach suggests that important variables will have regression coefficients in excess of half of the largest coefficient. The second method, based upon the VIP score, states that important variables will have a VIP score greater than 1.0, whilst unimportant variables will have a VIP score less than 0.8 *(33)*. Analysis of the EVA descriptor (σ = 4 cm^{-1}) for the steroid data set by Turner *et al.* *(29)* results in the selection of a large number of EVA variables at a threshold of VIP $\geq$ 1.0 (183 variables), while a threshold of VIP $\geq$ 3.0 yields a more manageable number (17 variables). It is reasonable to use such a high VIP threshold since these are the variables most heavily weighted by PLS, and thus may be used to get some feel for the main structural features used to discriminate between the training set structures.

The decomposition of the selected (important) EVA variables into their contributory normal mode frequencies is most straightforward, and certainly less ambiguous, if each EVA variable is composed of one and only one normal co-ordinate frequency. For this reason, it is important that the smallest σ value is used during the analysis as possible, since, as discussed earlier, σ directly affects the degree of intra-structural Gaussian overlap. Examination of the underlying frequencies for EVA variables with VIP $\geq$ 3.0 is not straightforward. However, for the steroid data set, PLS appears to discriminate between high, medium, and low active structures based on the presence or absence of specific frequencies that are characteristic of the functionalities considered important for binding affinity. For example, the variable with the second highest VIP score at 2,056 cm^{-1} relates to the position-3 carbonyl group stretching mode. This group is one of the features deduced by Mickelson *et al.* *(35)* to be critical for CBG-binding and is present in all of the high and medium activity compounds, as well as the most active of the low activity compounds.

The initial attempts at interpreting QSAR models based upon the EVA descriptor, discussed herein, are encouraging in that the classifications between structures can, to some extent, be rationalized in terms of the features postulated to be necessary for activity. Nonetheless, EVA QSAR models can not, to date, be interpreted to the same extent as CoMFA models in which the correlations may be related to probe interaction energies.

Recent Developments

The promising results presented thus far may lead one to believe that development of the EVA methodology has been completed, but this is not the case. One of the most promising developments has centred upon the selection of localized σ values across the spectrum. The motivation for this is two-fold. First, there are two datasets previously noted for which EVA has not provided good QSAR models. Second, "classical"

EVA, which uses fixed σ values for all vibrations, is a specific example of the more general case where σ may have optimal localised values; *i.e.*, different parts of the spectrum may be better treated with different width Gaussians. In order to direct the search for appropriate localised σ values a genetic algorithm (GA) is used in conjunction with the PLS crossvalidation q^2 score as the objective function to be optimised. Details of this approach, tentatively referred to as EVA_GA, have yet to be published. However, the results obtained with this procedure thus far are extremely promising. Much improved QSAR models have been obtained in every case (seven datasets) attempted to date - including the oxadiazole and biphenyl datasets for which poor models were obtained previously *(21)*. In addition, EVA_GA has provided QSAR statistics better than or equivalent to CoMFA for all these datasets. The most significant aspect of this work is that predictions made upon previously unseen test compounds can be enhanced in comparison to classical EVA, indicating that there is not a tendency to overfit to the training set compounds. Work is continuing both to validate the technique thoroughly and to optimise the GA parameters concerned.

Summary

One of the main problems encountered with field-based QSAR techniques such as CoMFA is the need to align the structures concerned. The selection of such alignments, in terms of the molecular orientation and conformation, is essentially arbitrary, but can have profound effects on the quality of the derived QSAR model. For this reason, a number of groups have attempted to develop new 3D-QSAR techniques that extend beyond this limitation, with varying degrees of success. This chapter has reviewed the progress made with one such methodology, that based upon molecular vibrational eigenvalues, and known as EVA.

EVA provides an entirely theoretically-based descriptor derived from calculated, fundamental molecular vibrations. Molecular structure and conformation are implicit in the descriptor since the vibrations depend on the masses of the atoms involved and the forces between them. The significant advantage that EVA offers relative to CoMFA and related 3D-QSAR techniques is that molecular vibrational properties are orientation independent, thereby eliminating ambiguity associated with the well known molecular alignment problem.

The discussion of the QSAR modeling performance of EVA herein illustrates the general applicability of the descriptor and the robustness of the resultant QSAR models in terms of crossvalidation statistics. In addition, extensive randomization testing of some of the PLS models discussed herein *(29)* shows that the probability of obtaining similar correlations by chance to those actually obtained using the EVA descriptor is essentially zero. Randomization and related statistical tests have played a crucial role in conclusively demonstrating that EVA can be used to correlate biological activity or other properties and generate statistically valid QSAR models. In most, but not all, cases examined EVA compares favorably with CoMFA, both in terms of the ability to build statistically robust QSAR models from training set structures, and to use those models to predict reliably the activity of "unseen" test chemicals. Furthermore, EVA has yielded predictively useful QSAR models for quite heterogeneous data sets, where the application of CoMFA is either difficult or impossible.

In addition to the development of the use of localised σ values in the generation of the EVA descriptor briefly described above (EVA_GA) there are a number of other areas in which EVA could be developed. There is considerable interest in exploring several aspects of the descriptor, including the correlation with specific types of effects, such as hydrophobic, steric, or electrostatic. In addition, despite the example provided herein of taking significance-of-variable plots coupled with techniques for selecting these variables as a means to interpreting an EVA QSAR model, there is need for more sophisticated techniques for the decomposition of EVA variables into the underlying normal mode vibration(s) and thereby to the groups of atoms that are characteristic of those vibrations. A further area that requires investigation is the sensitivity of EVA to the molecular conformation used, and to what extent this governs the choice of σ parameter.

As the EVA methodology matures, other applications, besides 3D-QSAR, will begin to emerge that take advantage of the strengths of the technique. One such example *(36)* centers on the use of EVA for similarity searching in chemical databases, in which the overall conclusions are that EVA is equally effective for this purpose as the more traditional 2D fingerprint method, although, depending on the similarity measure applied, the hits returned by EVA and 2D similarity measures may be structurally quite different. A consequence of this finding is that EVA-based similarity searching may provide an alternative source of inspiration to a chemist browsing a database. The applicability of EVA to similarity searching greatly contrasts with the complexities inherent to field-based similarity searching *(10)* in chemical databases.

Finally, the technique described herein that yields the standardized EVA descriptor from the calculated vibrational frequencies is not limited to that purpose and may, in principle, be applied in any circumstance where the property or descriptor is non-standard. For example, the standardization procedure could be applied to interatomic distance information, either for a single conformation, or as a means of summarizing conformational flexibility. Furthermore, the same procedure could be applied to other descriptions of molecular structure that are dependent on the number of atoms, such as electron populations, partial charges, or vibrational properties other than normal co-ordinate eigenvalues (EVA), including transition dipole moments (intensity) or eigenvector data (directionality of the vibrations). The EVA standardization methodology, therefore, provides a novel and potentially widely applicable means of transforming non-standard data.

Literature Cited

1. Hansch, C.; Fujita, T. *J. Am. Chem. Soc.* **1964**, *86*, 1616.
2. *3D QSAR in Drug Design: Theory, Methods and Applications*; Kubinyi, H., Ed.; ESCOM: Leiden, 1993, Vol. 1.
3. Cramer, R.D.; Patterson, D.E.; Bunce, J.D. *J. Am. Chem. Soc.* **1988**, *110*, 5959.
4. Kim, K.H.; Martin, Y.C. *J. Org. Chem.* **1991**, *56*, 2723.
5. Kellogg, G.E.; Semus, S.F.; Abraham, D.J. *J. Comput-Aided Mol. Design* **1991**, 5, 545.
6. Klebe, G.; Abraham, U.; Mietzner, T. *J. Med. Chem.* **1994**, *37*, 4130.

7. Kroemer, R.T.; Hecht, P. *J. Comput-Aided Mol. Design* **1995**, 9, 396.
8. Good, A.C. *J. Mol. Graph.* **1992**, *10*, 144.
9. Good, A.C.; Hodgkin, E.E.; Richards, W.G. *J. Chem. Inf. Comput. Sci.* **1992**, *32*, 188.
10. Thorner, D.A.; Wild, D.J.; Willett, P.; Wright, P.M. *J. Chem. Inf. Comput. Sci.* **1996**, *36*, 900.
11. Wagener, M, Sadowski, J.; Gasteiger, J. *J. Am. Chem. Soc.* **1995**, *117*, 7769.
12. Silverman, B.D.; Platt, D.E. *J. Med. Chem.* **1996**, *39*, 2129.
13. Bravi, G.; Gancia, E.; Mascagni, P.; Pegna, M.; Todeschini, R.; Zaliani, A. *J. Comput-Aided Mol. Design* **1997**, *11*, 79.
14. Ferguson, A.M.; Heritage, T.W. (1990) Shell Research Limited Internal Reports (not publicly available).
15. Herzberg, G. *Molecular Spectra and Molecular Structure. II. Infrared and Raman Spectra of Polyatomic Molecules, Eighth Edition;* D. Van Nostrand Company Inc.: New York, 1945.
16. Lindberg, W.; Persson, J.-A.; Wold, S. *Anal. Chem.* **1983**, *55*, 643.
17. Ferguson, A.M.; Heritage, T.W.; Pack, S.E.; Phillips, L.; Rogan, J.; Snaith, P.J. *J. Comput-Aided Mol. Design* **1997**, *11*, 143.
18. Cramer III, R.D. *J. Am. Chem. Soc.* **1980**, *102*, 1837.
19. Stewart, J.J.P. *J. Comp. Chem.* **1989**, *10*, 221.
20. Stewart, J.J.P. *J. Comput.-Aided Mol. Design* **1990**, *4*, 1.
21. Turner, D.B.; Willett, P.; Ferguson, A.M.; Heritage, T.W. *J. Comput.-Aided Mol. Design* **1997**, *11*, 409.
22. Turner, D.B. *An Evaluation of a Novel Molecular Descriptor (EVA) for QSAR Studies and the Similarity Searching of Chemical Structure Databases*, Ph.D. Thesis, University of Sheffield, 1996.
23. Jonathan, P.; McCarthy, W.V.; Roberts, A.M.I. *J. Chemometrics* **1996**, *10*, 189.
24. Waller, C.L.; McKinney, J.D. *J. Med. Chem.* **1992**, *35*, 2660.
25. Weiner, S.J.; Kollman, P.A.; Case, D.A.; Singh, U.C.; Ghio, C.; Alagona, G.; Profeta Jr., S.; Weiner, P. *J. Am. Chem. Soc.* **1984**, *106*, 765.
26. Carroll, F.I.; Gao, Y.G.; Rahman, M.A.; Abraham, P.; Parham, K.; Lewin, A.H.; Boja, J.W.; Kuhar, M.J. *J. Med. Chem.* **1991**, *34*, 2719.
27. Allen, M.S.; Laloggia, A.J.; Dorn, L.J.; Martin, M.J.; Costantino, G.; Hagen, T.J.; Koehler, K.F.; Skolnick, P.; Cook, J.M. *J. Med. Chem.* **1992**, *35*, 4001.
28. Greco, G.; Novellino, E.; Silipo, C.; Vittoria, A. *Quant. Struct.-Act. Relat.* **1991**, *10*, 289.
29. Turner, D.B.; Willett, P.; Ferguson, A.M.; Heritage, T.W. *J. Comput.-Aided Mol. Design*, Accepted for publication.
30. Waszkowycz, B.; Clark, D.E.; Frenkel, D.; Li, J.; Murray, C.W.; Robson, B.; Westhead, D.R. *J. Med. Chem.* **1994**, *37*, 3994.
31. Cho, S.; Tropsha, A. *J. Med. Chem.* **1995**, *38*, 1060.
32. Itai, A.; Tomioka, N.; Yamada, M.; Inoue, A.; Kato, Y. In *3D QSAR in Drug Design: Theory, Methods and Applications*; Kubinyi, H., Ed.; ESCOM: Leiden, 1993, Vol. 1; pp. 200-225.
33. Wold, S.; Johansson, E.; Cocchi, M. In *3D QSAR in Drug Design: Theory, Methods and Applications*; Kubinyi, H., Ed.; ESCOM: Leiden, 1993, Vol. 1; pp. 523-550.
34. Heritage, P.W. Shell Research Limited Internal Reports (not publicly available).
35. Mickelson, K.E.; Forsthoefel, J.; Westphal, U. *Biochemistry*, **1981**, *20*, 6211.
36. Ginn, C.M.R.; Turner, D.B.; Willett, P.; Ferguson, A.M.; Heritage, T.W. *J. Chem. Inf. Comput. Sci.* **1997**, *37*, 23.

Chapter 21

PRO_ANALOG: Automated Analog Building According to Principles of Medicinal Chemistry

Richard G. A. Bone, Michael A. Firth [1], Richard A. Sykes, Christopher W. Murray, and Jin Li

Proteus Molecular Design, Ltd., Lyme Green Business Park, Macclesfield, Cheshire SK11 OJL, United Kingdom

PRO_ANALOG is a suite of utilities which addresses the matter of expanding the analog space around a lead molecule. It has been implemented within the Prometheus suite of "computer-aided molecular design" programs. A lead molecule will have a large number of possible derivatives which should be investigated for pharmaceutical activity. It is likely that a medicinal chemist would only consider an incomplete set of these. We have implemented functionalities which allow a medicinal chemist or molecular modeller to select categories of possible transformations and derivatisation sites and, in a combinatorial sense, produce a library of all the analog molecules that would arise from such modifications. Molecular transformations are encoded with a new SMILES-string matching technology. Use of predefined lists of familiar functional groups from the medicinal chemistry literature ensures that the derived molecular structures are likely to be synthetically accessible. A range of transformation classes enables almost any molecular structure to be constructed. Statistical techniques used in molecular diversity can be applied to cluster the structures and ranking can be achieved by passing them to ligand-receptor docking software. The members of the libraries may then be assessed for their likely efficacy prior to embarking on laboratory synthesis. The method is illustrated here with a set of thrombin inhibitors.

In the development of a drug candidate from a "lead" molecule, a significant proportion of the effort lies in the choice and exploration of medicinal chemistry around that lead. Hand in hand with deducing "structure activity relationships", many

[1] Current address: International Research IS, Zeneca Pharmaceuticals, Mereside, Alderley Park, Macclesfield, Cheshire, United Kingdom, SK10 4TG.

series of derivatives must be synthesised and tested. Especially in the early stages of such a task, there can be a daunting array of possibilities. The chemist is faced with two problems: how to prioritise amongst the possible derivatisation routes; and how to combine systematically the effects of successive derivatisations to explore the analog space fully. We describe here the functionality and application of a new software package which addresses these issues.

PRO_ANALOG is the latest addition to the Prometheus suite of computer-aided molecular design (CAMD) programs (*1*). It enables explicit specification of derivatisations via advanced SMILES-string (*2*) matching and thence the generation of virtual libraries of such structures. Effecting molecular transformations with editing operations on SMILES strings or other line-notations is not a new idea. For example, Martin and van Drie (3) demonstrated how the ALADDIN control language could be used to modify SMILES strings in the context of substructure searching. Ho and Marshall (*4*) have utilised properties of SMILES strings when generating databases of new molecular structures. Daylight's "Reaction Toolkit" (*5*) is based on line-notation representations of chemical transformations. Nevertheless, our implementation has broad applicability and differs from previous ideas principally in its systematisation.

In essence, our approach involves the user heavily in the chemistry-dependent decisions; the computer itself facilitates the more mundane tasks of structure manipulation and library building. Our scheme is shown in figure 1. The first stage of systematic generation of analogs requires identification of several derivatisation sites by the chemist and specification of lists of functional groups or isosteres to place at each one. This level of interactivity ensures that the molecular modeller or medicinal chemist retains control over the nature of the chemistry to be explored. Representation of synthetic transformations is achieved using utility functions written in the programming language GLOBAL2 (*6*) and expressed as SMILES-string manipulations. In this way, chemical transformations are encoded as rules which may even be saved for future use. Further functions take lists of derivatisations and apply them to the lead molecule structure and its growing lists of analogs. Combinatorial enumeration enables all possible combinations of analogs resulting from derivatisation at each site to be produced. The merits of one set of analogs over another may be assessed by computing docking scores for each set, using the program PRO_LEADS (*7*). Sets which consistently produce better docking scores can then be considered for synthesis.

In this way, we are able to both achieve a labour saving in the automated generation of derivatives (as well as systematisation in that effort) and provide the possibility to "score" a given library, thus offering a means to assess the benefit of a set of transformations.

In this paper we describe the software tools comprising PRO_ANALOG and give an example of its application. The core of the utility is SUPER-SMILES, an enhanced implementation of SMILES which allows us to express derivatisations as molecular transformation rules. We outline the features of SUPER-SMILES which are critical to the development and application of PRO_ANALOG. We go on to describe the types of transformations which are built into the package and which enable us to derivatise molecules in highly specific ways. Then we show how combinatorial libraries may be built and judged for their probable efficacy. In practice,

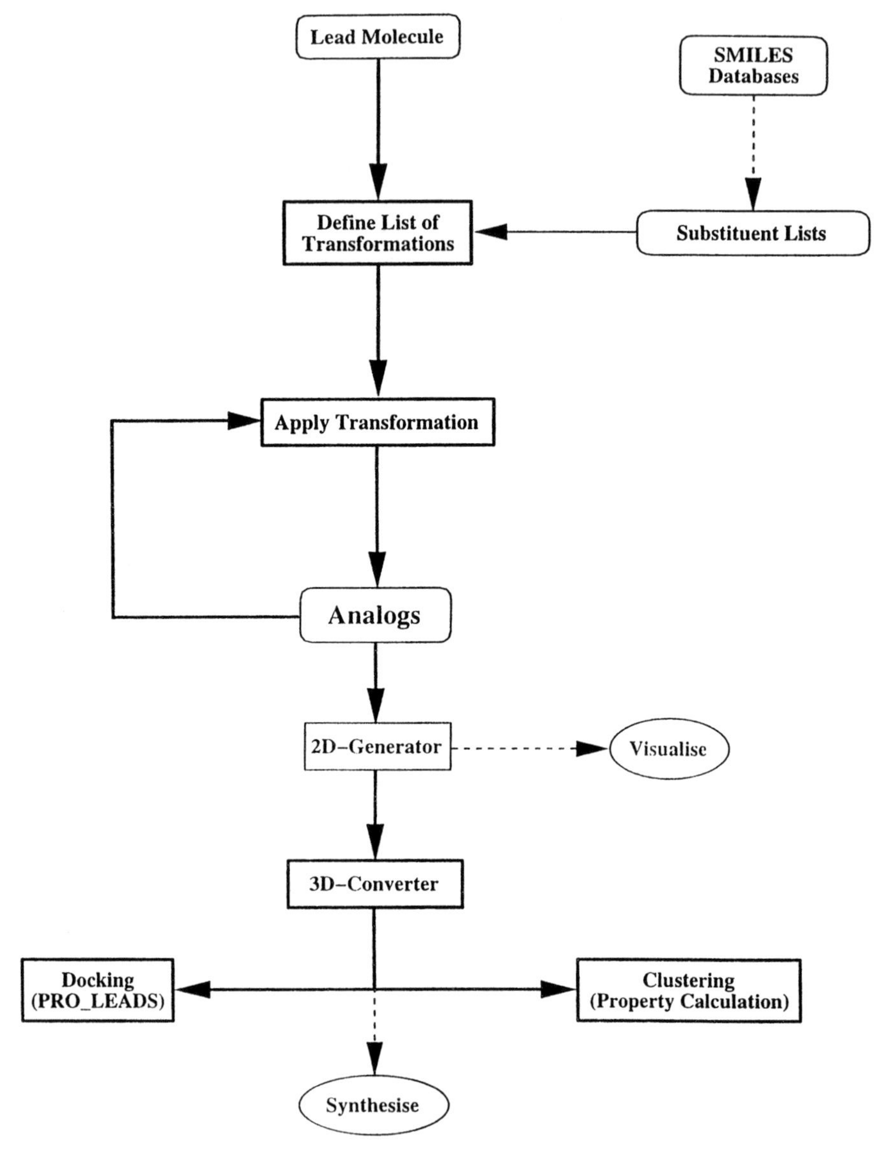

Figure 1. Typical use and application of PRO_ANALOG in drug design process.

all of this functionality is becoming accessible via a graphical user interface currently under development.

GLOBAL2

Underpinning the Prometheus software environment is the programming language GLOBAL2 which is based on the original implementation of GLOBAL (*6*). The language may be executed within an interactive environment or from pre-written scripts and finds its main application as the construction of driver routines for molecular design procedures. In this context it behaves as an interpreted language. But at root, it is a fully functional computer language and has many of the features intrinsic to other more widely used languages. It has a blocked structure and function declaration similar to that of Pascal; typed variables, loop structures, logical constructs and formatted I/O as are found in FORTRAN and C; aspects of object-orientation found in C++ such as class structures and inheritance; and general-purpose file and string-manipulation facilities as commonly associated with Perl and Tcl. Additionally it is possible to call FORTRAN and C subroutines and functions from GLOBAL2 as well as access the UNIX system interface, e.g., for performing file operations.

With time, many desirable utilities to facilitate molecular structure manipulation have been created with GLOBAL2. By now, most operations are performed with molecule data structures, calling upon a large library of chemistry-specific functions. The work in this paper relies upon a small subset of these. Principally we use the built in capabilities for manipulating character strings. We show how it is possible to construct with ease, some quite complicated molecular structures using just string-concatenation and splicing.

SMILES and SUPER-SMILES

SUPER-SMILES is a set of extensions to Daylight's SMILES (*8*). It is effectively a super-set of SMILES: almost all SMILES strings can be interpreted by it and lead to a structural representation in the usual way. The enhanced functionality has been designed to facilitate substructure matching and reaction building within the framework of a well-established line-notation for chemical structure. In outline, SUPER-SMILES allows complete structure specification including all bonds and hydrogens but does not support lower-case definitions of aromatic rings. It enables substructure identification with a powerful pattern matching syntax, has facilities for atom-label and macro specification and allows transformations to be defined with insertion/deletion facilities. The definition of the SMILES language (with SUPER-SMILES extensions) has been implemented using a parser-generator.

Molecule Building and Pattern Matching. The conversion of SUPER-SMILES strings to molecular structures (*i.e.*, a bond connection table) is achieved with the functions `SmilesMatch` and `SmilesUniqueMatch` which have been implemented using a finite state machine. This underlying structure enables us both to build molecular structures and match SUPER-SMILES patterns. `SmilesMatch` returns the successive matches of a SUPER-SMILES pattern against a reference molecule. A pattern may match several times against the reference, and topological symmetry can often mean that certain matches are equivalent. `SmilesUniqueMatch` guarantees uniqueness of the results of pattern-matching. A structure may be built by supplying a "null" molecular object as reference; `SmilesMatch` then itself returns a structure corresponding to each molecule represented by a pattern. Another function, `SmilesCanon`, adopts the Weininger algorithm (*8*) for converting a molecular structure into a "canonical" SMILES string representation. These functions form the core of our ability to carry out molecular transformations with SMILES-based syntax.

As part of our suite, we have devised a method of deriving 2D-coordinates from SMILES-strings (or a connection table). This feature facilitates testing of SMILES-string writing and also visualisation of the results of analog building and library formation; it will be described elsewhere (*9*).

Databases of Functional Groups. The requirement for hydrogens, bond-orders and aromatic rings to be fully specified leads us to set up a fragment database of commonly used moieties, largely for convenience. The ability of SUPER-SMILES to recognise macros (pre-defined fragment or pattern names) makes this a particularly useful feature. Over 250 SUPER-SMILES strings have been encoded into a file of categorised functionalities. (Examples of the categories present include `Alkyls`, `Cycloalkyls`, `Carbonyls`, `Halocarbons`, `HeteroAromatics`, `Alkoxys` and `AminoAcids`.) Each fragment-name can be "cut and pasted" when forming transformation rules or used as a macro within SUPER-SMILES or as a (string) variable name within GLOBAL2. For example, the `Alkyls` package includes the definitions:

- `Methine = "-C-H";`
- `Methylene = "-[C-H2]";`
- `Methyl = "-[C-H3]";`
- `Ethyl = "-[C-H2]-[C-H3]";`

Our default form for fragments is that they add "on the left", *i.e.*, the leading bond (on the lefthand end of the string) is always present. (Therefore when constructing an entire molecule, a "capping group" must be placed initially.)

The string-syntax of GLOBAL2 allows build up of larger functional groups (and molecules) from smaller patterns. In which case, a propyl chain may be built as follows:

- `Propyl = Methylene + Ethyl;`

String concatenation sets the variable `Propyl` to `"-[C-H2]-[C-H2]-[C-H3]"`.

The user is at liberty to introduce any other functional group definitions. Recent conversion of the NCI database (*10*) to a SMILES format will facilitate access to SMILES representations of many more molecules of even greater complexity.

SUPER-SMILES Utility Functions. A problem with a linear notation is that polyatomic functional groups (especially ring systems) may have more than one possible attachment point. It would not be desirable to store multiple representations of the same functional group to cover all possible environments. Therefore we have two utility functions, `SmilesShifter` and `SmilesSplicer` which exploit embedded labels in certain of the SUPER-SMILES strings in our database. Labels begin with an underscore and take a number which corresponds to IUPAC conventions.

Pyridine has been encoded (as a macro) in a "canonical" form:

- `N'_1':1:C'_2'(-H):C'_3'(-H):C'_4'(-H):C'_5'(-H):C'_6'(-H):1`

The labels are in quotes (`' '`) and the colons (`:`) are aromatic bond types.

`SmilesShifter` acts on such a labelled SUPER-SMILES string to return a string whose leftmost atom is that specified and with any hydrogen attached to that atom automatically stripped off. For example, to obtain a 3-substituted pyridine:

- `SmilesShifter(Pyridine,3);`

gives us `"-C:1:C(-H):N:C(-H):C(-H):C:1-H"`. (The labels are removed in the shifting process.)

If we wish to attach some other atom or group at a pre-defined position in the macro, then we are effectively specifying simultaneously "right" and "left" attachment points. `SmilesSplicer` acts in the same way as `SmilesShifter` but allows additionally a "substituent" to be attached in a specified place on the string. (We are able to do this with appropriate use of `SmilesMatch` and `SmilesCanon`.) For example, (using the macro definition of `Amino`):

- `SmilesSplicer(Pyridine,3,Amino,2);`

expands to `"-C:1:C(-[N-H2]):N:C(-H):C(-H):C:1-H"`.

Each of these functions can be used within GLOBAL2 to build up more complicated molecules in an intuitive way, for example:

```
•   Histidine = Amine + Methine +
      "(" + SmilesShifter(Imidazole,4) + ")C(=O)-O-H";
```

The other amino acids may be similarly built up. Stereochemistry is handled in the same way as Daylight's set of definitions (see, *e.g.*, (*11*)).

Pattern Matching. The specification of substructures within a molecule is useful not only for verification of SUPER-SMILES representations but for identification of environments at which to carry out derivatisations. In the spirit of both flexibility and precision, many features are supported.

Wild cards. Certain types for atoms (`*`) and bonds (`|`, or, by default, absence of explicit bond-type) allow simple specification of all or multiple parts of a molecule. For example, the ring substructure `"C-1-*-C-C-C-C-1"` could represent either (the carbon skeleton of) cyclohexane or (the heavy-atom skeleton of) piperidine.

Constraints. The directive `Not` allows exclusion of particular atoms from a matching environment. For example, the substructure "`N(-H)(Not -C=O)-C`" specifies a (primary or secondary) amine but excludes amides.

`Once` enforces a single match of the `SmilesMatch` utility in the case of topological symmetry. Therefore, "`{Once N(-C)(-C)-H}`" will match a secondary amine but force a single symmetry-unique match to result.

Logicals. The conjunction `Or` allows more than one match, where possible. For example, in order to match an imine or a carbonyl group (but not an alkene), one can type "`C=(O Or N)`".

The "exclusive Or", (`Xor`) may be used where several alternative substructural representations can be used and it is not necessary to try each member in the list once one match has been found. This is particularly useful when specifying aromatic rings where either Kekulé or valence-bond forms may be present, viz: the pattern "`C:1:C:C:C:C:C:1 Xor C-1=C-C=C-C=C-1`" will find the first match of a phenyl ring starting at any given position.

Other Features. Additionally, there are facilities for skipping the scope of the pattern to disjoint parts of the molecule (`Goto, Start, Revisit`) but these will be described elsewhere (*12*).

Transformations in SUPER-SMILES

SUPER-SMILES transformations, allow amended ("derivatised") molecules to be generated from a reference ("lead") molecule or molecules. The SUPER-SMILES directives `Add` and `Delete` may be applied at any matching point within a pattern. It is necessary to specify a matching environment, corresponding to some portion of the molecule. Insertions and/or deletions are then applied to that context.

A simple example of substitution is therefore:

- `O=C (Delete -Cl) (Add -O-H)`

In this case, the transformation represents nucleophilic substitution at the carbonyl group, specifically hydrolysis of an acid chloride functionality. If there is no "`C(=O)-Cl`" group present, no deletion/addition occurs.

Classes of Transformations. Whilst the simple addition/deletion syntax is relatively straightforward, more complex operations such as insertions, ring formation or substitution of atoms at branch points are quite tricky to specify. For ease of use, underlying SUPER-SMILES transformations for different classes have been encoded by a series of GLOBAL2 functions, each of which returns a SUPER-SMILES string representing the specific transformation. Arguments passed can include functional groups to add/delete and the environment in which they are to be found. The arguments are spliced/embedded into the requisite SUPER-SMILES syntax. The resulting SUPER-SMILES rule may be stored and re-used for future analogs or leads.

We have developed 8 different transformation classes (Substitution, Addition, Insertion, Deletion, Bridge, Replacement, Extraction, AromaticSubs). Consequently, the type of functionalisation that we can effect is quite large and our range of transformation classes is much more powerful than the simple single-point attachments of other analog builders. Other benefits include the fact that we are not limited to actual synthetic chemical methods in our choice of transformations. Consequently, portions of the lead space may be explored which would not necessarily be immedately accessible by straightforward laboratory reactions. Furthermore, multi-step transforms may be swiftly represented by a single SUPER-SMILES transformation rule.

Categories of Transformation Function. Understanding that the scope of a pattern may be quite wide and is context dependent, the classes of transformation each have several distinct instances of applicability. In general, we have:

- <Generic> (apply transformation wherever possible)
- Specific (specify environment/attachment points)
- Single (apply the transformation only once where symmetry is present)
- Smart (takes care of hydrogens at attachment points, when present)

In practice, access to all these functions is via a graphical user interface which permits visual selection of derivatisation sites and menu-driven access to the various functionalities.

Examples. In figure 2, we show examples of the applicability of each of the transformation classes. In each case we give the function call, comprising the function name and its SUPER-SMILES arguments which would be required to effect the given transformation.

Aromatic Substitution (a). In the case of phenyl substitution, when exactly one group is already present, there is no need to specify the reference attachment point in the molecule.

Substitution (b). In this case we show the application of the rule if all hydrogens are fully specified in the reference molecule.

Replacement (c). Substitution of an atom at a branch point demands use of a different set of functions from those which effect straightforward terminal substitution. It is necessary to use a "Specific" function because there is more than one carbon atom in the molecule.

Addition (d). In the case where no hydrogens are present in the structure, we can use "Addition" to extend new groups off a predefined attachment point.

Insertion (e). Chain lengthening can be carried out between bonded atoms.

Deletion (f). Terminal atoms can be removed easily enough.

Extraction (g). Portions of chains can likewise be deleted leaving an intact molecule. This operation only has meaning for a portion of a molecule which has exactly two connections.

Ring Formation (h). We can insert a group between two positions which themselves are not directly bonded.

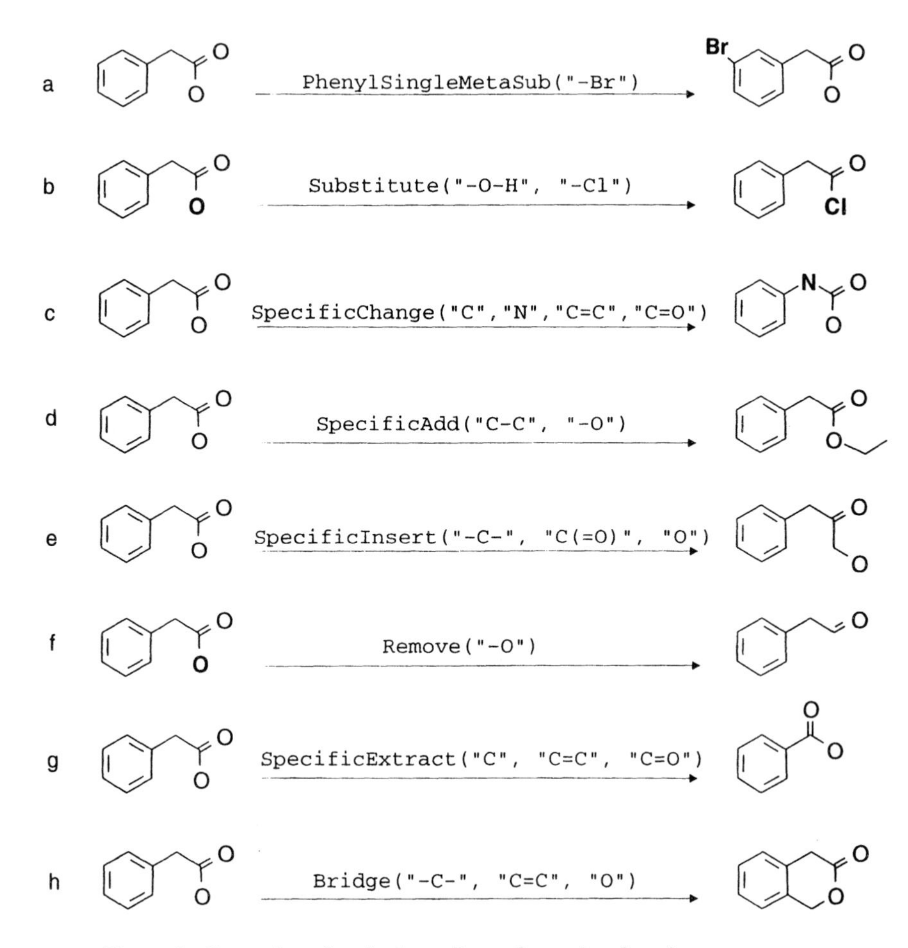

Figure 2. Examples of each class of transformation function.

Combinatorial Library Building

A major area of application for analog building is in library design. The transformation functions that we have developed allow a medicinal chemist or molecular modeller to interactively define SUPER-SMILES rules for analog building and therefore to explore lead optimization. In the "brute force" scenario, where all possible derivatisations need to be considered, a large number of possibilities clearly results. In general, though, the chemist will have identified a small number of sites in the lead molecule, at which, in each case, a list of possibilities needs to be investigated. A "combinatorial library" can be built in which, systematically, derivatisations at each site are combined with derivatisations at each of the other sites in turn. Accordingly, PRO_ANALOG contains the software technology to cope with building lists of (site-specific) transformation rules and the ability to assemble the application of such lists of rules into libraries of analogs.

Transformations Derived from Lists of Substituents. We have "accumulator" functions which take as an argument, a transformation function type, arguments to be passed to it, and a list of substituents. These accumulators return lists of SUPER-SMILES transformation rules.

For example: within GLOBAL2 we set up a list of substituents, each of which is to be applied to a common substitution position in a target molecule:

- `SubsList = ["-Br", Methyl, "-O-H", "-C1CCCCC1"];`

We can pass such a list to an accumulator in order to produce a list of transformation rules.

- ```
  SubsRules =
   SubstituteAccumulate(Substitute,SubsList,"-Cl");
  ```

In this case, each member of `SubsList` in turn is passed to the function `Substitute` along with the argument "`-Cl`" so that a number of chloride replacements can be specified.

Analog Library Construction. Enumeration classes have been constructed in GLOBAL2 which allow recursive application of lists of transformations to a (growing) list of analogs. The function, `CombinatorialLibrary`, takes as arguments, separate lists of transformation rules. The lead molecule (or, even, series of molecules) can be read in from an MDL SD-file *(13)* and the derivative molecules are exported to a separate SD-file. An interesting point of note is that there is no restriction that rules in a given list all refer to the same derivatisation position. For example, different sites on the same benzene ring could be mixed in the same rule-set.

Case Study: Thrombin Inhibitors

In order to demonstrate the potential utility of virtual library building, we mimicked a QSAR case study from the literature of thrombin inhibitor synthesis. Our premise is
  ```

to show that the virtual experiment could, in principle, be used to assess, a priori, the likely efficacy of various members of these derivatives. In this case a receptor structure is known so it makes sense to attempt to characterise a set of analogs by how well they dock, rather than by, say, molecular similarity criteria (*14*) or by a diversity analysis.

Nα-(2-naphthylsulphonylglycyl) 4-amidinophenylalanine piperidide (NAPAP) is a known strong thrombin binder for which a receptor-bound crystal structure is available. In the years before that structure was known, NAPAP itself was identified in a series of Nα-arylsulphonylaminoacylated 4-amidinophenylalanines investigated by Stürzebecher and coworkers (*15*), the core structure of which is shown in figure 3. With the benefit of hindsight, we can show how computational docking studies corroborate the experimental results.

We work with two lists of substituents, `L1` and `L2` containing SUPER-SMILES strings for groups placed at positions R1 and R2 respectively.

The 4 R1-substituents were n-butylamine, pyrollidine, piperidine and morpholine. They give rise to a number of amides with small hydrophobic groups attached to the nitrogen.

The 8 R2-substituents were chosen to explore different ways of orienting and displacing an arylsulphonic group away from the α-amino group of phenylalanine. They are: tosyl; α- and β-naphthylsulphonyl; tosyl-glycyl; α- and β-naphthylsulphonyl glycyl; tosyl β-alanyl and tosyl ε-aminocaproyl. The main effect being explored is a lengthening of the chain between the α-amino group and the arenesulphonyl group. We may build each of the substituents from SUPER-SMILES macros and utilities, predefining `Tosyl` as `"Sulpho + SmilesShifter(Toluene,4)"`:

- `Tosyl`
- `SulphoLinker + SmilesShifter(Naphthalene,1)`
- `SulphoLinker + SmilesShifter(Naphthalene,2)`
- `Carbonyl + Methylene + AmineLinker + Tosyl`
- `Carbonyl + Methylene + AmineLinker + Sulpho + SmilesShifter(Naphthalene,1)`
- `Carbonyl + Methylene + AmineLinker + Sulpho + SmilesShifter(Naphthalene,2)`
- `Carbonyl + EthylLinker + AmineLinker + Tosyl`
- `Carbonyl + PentylLinker + AmineLinker + Tosyl`

We need to assemble transformation rule-sets for each substitution position. We use 4-amidino phenylalanine as the core molecule, in which case R1 is `"-O-H"` and R2 is `"-H"`. For the R1-substituents, we can use `Substitute` functions, for example:

- `Substitute(Hydroxyl,Butyl);`

For the amino-substituents we can use `Add` functions.

- `SmartSpecificSingleAdd(Tosyl, "C1CCCCC1[CH2]C(H)N(H)");`

The `SmartSpecificSingleAdd` function implicitly takes care of the presence of hydrogen attachment to the amine nitrogen, and additionally forces only one symmetry-unique match.

Using `Accumulator` functions, our sets of substituents (held in lists `L1` and `L2`) and the above transformation rules, we can construct rule-sets `T1` and `T2` respectively. Finally, we build a combinatorial library comprising 32 molecules, accounting for each possible pair of substituents using the two rule-sets with a single function call:

- `CombinatorialLibrary("core.sd", "lib.sd", T1, T2);`

In order to "score" each member of the library we proceed using a protocol which includes the following steps: protonate the amidino group; use Converter (*16*) to obtain a 3D-geometry for each molecule; optimize each 3D-structure to a local minimum using molecular mechanics; identify rotatable bonds in each ligand; carry out docking runs on each ligand with the bovine thrombin receptor from which bound waters had been removed, "1ETS" (*17*).

In correspondence with the known bound structure of NAPAP (*17*), we constrained the amidino group to be planar with the benzene ring; but full flexibility of rotation around the aromatic-sulphonyl group was permitted in order to allow the bulky aromatic group to attain optimum fit in the binding site. The docking method (*18*) utilised a fixed receptor geometry, an empirical scoring function (*19*) and a Tabu-search method of optimization, starting at 20 random initial configurations (*7*).

In the published work of Stürzebecher, the pyrollidinyl and piperidinyl substituents consistently performed better than the n-butylamino or morpholinyl substituents for all the R2-substituents (with the exception of the pyrollidinyl analog with α-naphthylsulphonyl). The best substituent sets overall were with α- and β-naphthylsulphonylglycyl as R2-substituent, with the best-performing substituent combination being β-naphthylsulphonylglycyl with piperidinyl, *i.e.*, NAPAP itself.

The principal features of NAPAP binding comprise: hydrogen-bonding contacts by the amidino group in the "S1-pocket"; hydrophobic interactions in the "P" and "D" pockets; and a β-sheet interaction. These are shown schematically in Figure 4. The P-pocket comprises residues His-57, Tyr-60 and Trp-60. The residue Leu-99 flanks both the P and D-pockets. The D-pocket itself is made up of Trp-215, Ile-174 and Asn-98. There is a β-sheet interaction between Gly-216 and the glycyl sulphonamide link.

In table I we give, for each library member, the docking scores and itemize the principal features of the predicted binding mode. In this test case we are looking for parallels with the published activity data, as well as successful replication of known thrombin-binding interactions. Even now, docking calculations come with a significant caveat: it is very hard to both reproduce energies of binding and reasonable docked geometries. Nevertheless, a study of the table shows that the series with naphthylsulphonylglycyl substituents in the R2 position (which includes NAPAP) give lowest binding energies and best correspondence with the binding features of NAPAP itself.

But not all of the substituent combinations are able to recover the key features of NAPAP binding. This is to be expected. Whereas all ligands are able to make the

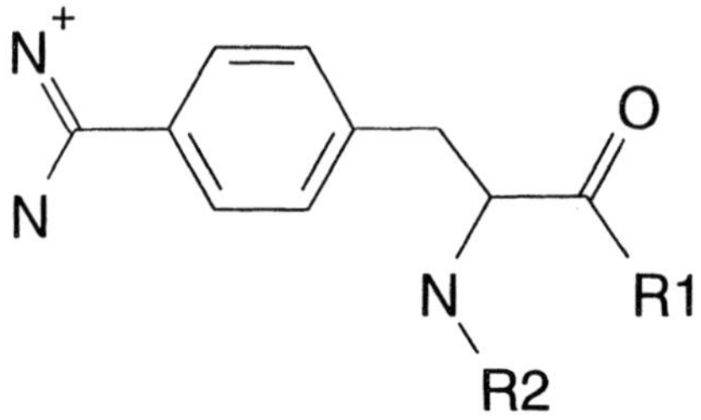

Figure 3. Core structure for thrombin inhibitor design.

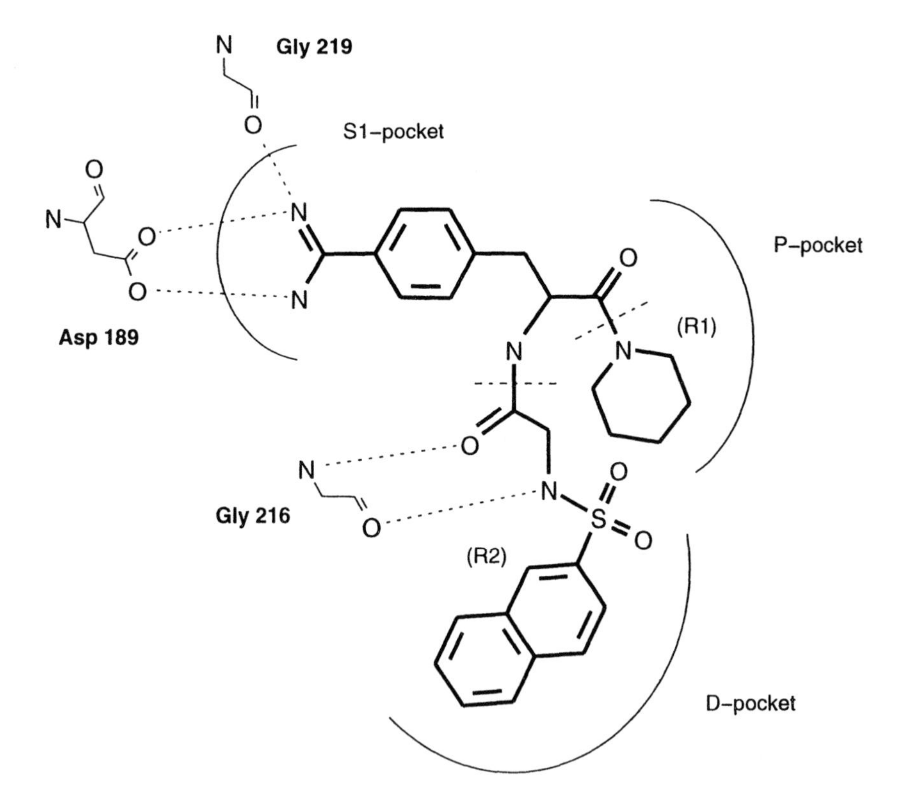

Figure 4. Schematic of NAPAP's interactions in the thrombin binding site.

Table I. Computed binding characteristics of NAPAP Analogs

R1	R2	K_i/μMol.	Best E[a]	ΔE[b]	Binding Features [c]
Morph.	Tos-ε-Acp	—	–42.057	—	S1; P; D
Pip.	Tos-ε-Acp	13	–43.539	—	S1; P; D; Half-β
Pyr.	Tos-ε-Acp	—	–41.827	1.646	S1; P; D
n-But.	Tos-ε-Acp	—	–44.394	2.627	S1; P; D
Morph.	Tos-β-Ala	22	–44.463	—	S1; P; D; β
Pip.	Tos-β-Ala	0.17	–41.352	—	S1; P; D
Pyr.	Tos-β-Ala	2.1	–46.088	—	S1; P; D; Half-β
n-But.	Tos-β-Ala	—	–41.154	2.922	S1; P; D
Morph.	β-Nas-Gly	0.23	–44.340	—	S1; P; D; β
Pip.[d]	β-Nas-Gly[d]	0.006	–46.942	—	S1; P; D; β
Pyr.	β-Nas-Gly	0.013	–45.435	—	S1; P; D
n-But.	β-Nas-Gly	2.3	–50.788	—	S1; P; D; β
Morph.	α-Nas-Gly	0.57	–45.552	—	S1; P; D; Half-β
Pip.	α-Nas-Gly	0.014	–50.110	—	S1; P; D; β
Pyr.	α-Nas-Gly	0.058	–45.041	—	S1; P; D
n-But.	α-Nas-Gly	15	–45.146	—	S1; P; D
Morph.	Tos-Gly	4.1	–45.995	—	S1; P; D; β
Pip.	Tos-Gly	0.048	–46.409	—	S1; P; D
Pyr.	Tos-Gly	0.052	–41.376	—	S1; P; D; Half-β
n-But.	Tos-Gly	14	–43.339	—	S1; P; D
Morph.	β-Nas	3.1	–44.960	4.934	S1; P; D; Half-β
Pip.	β-Nas	0.42	–46.325	1.767	S1; P; D
Pyr.	β-Nas	1.9	–46.642	—	S1; P; D
n-But.	β-Nas	3.6	–44.241	—	S1; P; D; (Ser-214)
Morph.	α-Nas	3.3	–42.048	0.845	S1; P; D
Pip.	α-Nas	2.8	–42.606	∞	
Pyr.	α-Nas	4.9	–44.785	0.512	S1; P; D
n-But.	α-Nas	5.7	–43.062	∞	
Morph.	Tos	5.4	–42.453	—	S1; P; D
Pip.	Tos	2.3	–43.386	2.382	S1; P; D
Pyr.	Tos	5.9	–42.698	—	S1; P; D
n-But.	Tos	23	–42.215	1.643	S1; P; D

[a] Energy of lowest energy solution (kJ/Mol.).
[b] Energy of lowest NAPAP-like solution ("∞" if none) relative to absolute lowest.
[c] Principal ligand-receptor interactions in lowest energy NAPAP-like solution. Features in parenthesis are interactions not exhibited by NAPAP.
[d] NAPAP.

distinctive hydrogen-bonding interactions in the S1-pocket, optimal penetration of the "D" or "P" pockets is harder to obtain and failure to probe the P-pocket is the reason why two of the ligands do not adopt full NAPAP-like binding at all. In the scoring function used, there is no built-in penalty for hydrophobic groups which do not adequately fill cavities. The aryl groups may adopt a number of different positions within the D-pocket, in each case giving rise to a visually plausible fit but one in which the orientation is not always best.

In addition, it appears to be difficult to pick up the β-sheet interaction. For the Tos-β-Acp and Tos-β-Ala series the chain is too long to match the peptidic link with residue Gly-216 effectively. On the other hand, with R2 substituents Tos and α-Nas and β-Nas, there is no peptidic group or the chain is too short. In cases where the chain length should be appropriate, but the β-sheet is not formed, it may be due to inadequate exploration of the internal conformational flexibility of the ligand. In such circumstances, favourable binding must result from good positioning of the D- and P-pocket groups. Therefore, it can be seen that attaining the β-sheet alignment is favourable, but extra stability may also be conferred if D/P-pocket fitting is simultaneously improved.

For a given R2-group the other substituent combinations do not show a strong pattern of consistency. The docking energies for NAPAP-like binding do not seem to differentiate convincingly between the morpholine and piperidine substituents at R1 or between n-butyl and the others. This is understandable because there is very little steric distinction between these groups and the true discriminator of their observed behaviour is likely to be a strong preference for the piperidine group to lie close to the aryl sulphonic group. A longer-range interaction of that nature is hard for a scoring function to pick up. Additionally, we do not show clearly that n-butyl is a poor R1-substituent, but there is a slight tendency in the results to disfavour morpholinyl.

Whilst it would be unreasonable to expect that, in every example, the experimental data could be reproduced comprehensively, general trends in the results of a docking study could be persuasive in highlighting preferred substituents in a library. The thrombin study represents a modestly successful test case and shows that analog building coupled with docking calculations is a potentially useful way of discriminating between series of related ligands.

Although the library considered here is small, it is of a convenient size to illustrate the principles of the approach. Libraries containing several hundred analogs have been created in similar ways during other applications of PRO_ANALOG to drug discovery projects at Proteus. Library generation itself takes merely a few seconds or minutes of computer time on a small desktop workstation, once the SUPER-SMILES rules have been defined. Similarly, conversion of the library members to 3D form and subsequent geometry optimisation is not a rate-limiting step these days and can be carried out in a matter of minutes. The docking exercise is the most intensive part of the study and, for a thorough analysis, can cost an hour per ligand, but at this time it remains a critical method of validation.

References

1. Clark, D. E.; Frenkel, A. D.; Levy, S. A.; Li, J.; Murray, C. W.; Robson, B.; Waszkowycz, B.; Westhead, D. R., *J. Comp.-Aid. Molec. Des.*, **1994**, *9*, 13

2. Weininger, D. *J. Chem. Inf. Comput. Sci.*, **1988**, *28*, 31–38
3. Martin, Y. C.; van Drie, J. H. "Identifying Unique Core Molecules from the Output of a 3-D Database Search", in *Chemical Structures 2*, Ed. W. A. Warr, p. 315, (Springer-Verlag Berlin, 1993)
4. Ho, C. M. W.; Marshall, G. R., *J. Comp.-Aided Mol. Des.*, **1995**, *9*, 65–86
5. Reaction Toolkit, Daylight Chemical Information Systems, Inc., Mission Veijo, CA, 1997
6. Ball, J.; Fishleigh, R. V.; Greaney, P.; Li, J.; Marsden, A.; Platt, E.; Pool, J. L.; Robson, B., in Bawden, D.; Mitchell, E. M., (Eds.) *Chemical Structure Information Systems: Beyond the Structure Diagram*, Ellis Horwood, Chichester, 1990, p. 107
7. Westhead, D. R.; Clark, D. E.; Murray, C. W. *J. Comp.-Aided Mol. Des.*, **1997**, *11*, 209–228
8 Weininger, D.; Weininger, A.; Weininger, J. L. *J. Chem. Info. Comp. Sci.*, **1989**, *29*, 97
9. Bone, R. G. A., *paper in preparation.*
10. Milne, G. W. A.; Nicklaus, M. C.; Driscoll, J.S.; Wang, S.; Zaharevitz, D. National Cancer Institute Drug Information System 3D Database. *J. Chem. Inf. Comp. Sci.,* **1994**, *34*, 1219; http://epnws1.ncifcrf.gov:2345/dis3d/3ddatabase/nci_smil.html
11. http://www.daylight.com/dayhtml/smiles/smiles-isomers.html
12. Bone, R. G. A.; Firth, M. A.; Sykes, R. A. *paper in preparation.*
13. Dalby, A.; Nourse, J. G.; Hounshell, W. D.; Gushurst, A. K. I.; Grier D. L.; Leland, B. A.; Laufer, J. *J. Chem. Inf. Comp. Sci.,* **1989**, *29*, 172
14. Johnson, M. A.; Maggiora, G., *Concepts and Applications of Molecular Similarity;* John Wiley and Sons, New York, 1990
15. Stürzebecher, J.; Markwardt, F.; Voigt, B.; Wagner, G.; Walsmann, P. *Thromb. Res.*, **1983**, *29*, 635–642
16. *Converter* version 2.3, Biosym/MSI, San Diego, CA, 1993
17. Brandstetter, H.; Turk, D.; Hoeffken, H. W.; Grosse, D.; Stürzebecher, J. Martin, P. D.; Edwards, B. F P.; Bode, W., *J. Mol. Biol.*, **1992**, *226*, 1085
18. Baxter, C. A.; Murray, C. W.; Clark, D. E.; Westhead, D. R.; Eldridge, M. D. [submitted], (1998)
19. Eldridge, M. D.; Murray, C. W.; Auton, T. R.; Paolini, G. V.; Mee, R. P. *J. Comp.-Aided Mol. Des.*, **1997**, *11*, 425–445

Chapter 22

Is Rational Design Good for Anything?

Donald B. Boyd

Department of Chemistry, Indiana University—Purdue University at Indianapolis, Indianapolis, IN 46202–3274

An overview is given of some documentable examples where rational approaches have been significantly influential in the drug discovery process. The scope of rational design is defined to include not only computational chemistry and structure determination, but also other nonrandom approaches. In contrast to the enumerated successes, a case study is used to illustrate how quantitative analysis can be used to reveal when a strategy for finding a drug may be untenable. An attempt is made to prognosticate the evolving role of rational design in the new era of combinatorial chemistry.

Prior to the last few decades, the discovery of medicines was mainly an empirical endeavor. Relatively little was known about medically relevant biological targets and three-dimensional molecular structures of the receptors. Compounds were synthesized or obtained from natural sources such as plants or microorganisms found in soil and water samples. The compounds were tested for biological activity on whole organisms such as bacteria or laboratory animals. In those early days, it was not unheard of for a chemist to administer a newly synthesized compound to himself or his associates to learn what biological effects it had.

Starting in the 1960s, increased understanding of how compounds exerted their biological effects made possible the testing of compounds in purified, isolated enzymes and other biochemical systems. The usual strategy was to use in vitro testing for preliminary evaluation of the compounds and then, on the more promising ones, use in vivo (animal) testing, which generally required more material to be synthesized. Most of the compounds being tested were laboriously synthesized one at a time. If sufficient compound was available, it could be broadly screened in assays set up to look for various kinds of biological activity. Regarding microorganisms as a source of drugs, isolation of materials from fermentation broths turned up the same active compounds over and over again, so by the early 1980s it became clear that this strategy was giving diminishing returns.

The sheer labor intensity of the organic chemical and biological effort led gradually to greater interest in so-called rational design approaches such as analyzing quantitative structure-activity relationships (QSAR) or considering the mechanism of an enzymatic reaction. The hope of these new approaches was that the odds of finding a useful compound would be increased.

Pharmaceutical researchers knew from experience that the odds were very low for finding a useful drug by random screening. An often quoted figure, which was at least qualitatively correct, was that 10,000 compounds had to be evaluated to find one that would become a medicine. Random screening was inefficient. Increasing numbers of scientists became convinced that rational approaches would help them focus on those parts of "compound space" (a term not then widely used) that would most likely lead to a new drug.

With advancing computer technology and improved methods for solving structures by X-ray crystallography and nuclear magnetic resonance (NMR) spectroscopy, the 1980s saw a growing interest in rational approaches. By the early 1990s, random screening appeared less and less efficient, and the pendulum in pharmaceutical discovery research was swinging toward modern molecular modeling techniques (*1,2,3*) and iterative structure-based ligand design (SBLD) (*4,5*) as the most promising ways to discover the medicines of the future.

However, about five years ago, a new technology appeared on the scene. Robotic systems run by computers made possible combinatorial chemistry (*6*) and high-throughput screening (HTS). Suddenly, the random approach was economically far superior to designing one compound at a time. Combinatorial chemistry and HTS went hand-in-hand: combinatorial chemistry made possible the generation of huge numbers of new compounds to feed into the screening machines, and HTS could run through the new combinatorial compound libraries at a voracious rate. In addition, old corporate compound libraries consisting of hundreds of thousands of materials could be mined quickly and economically. Often little or no consideration was given to the fact that the old compounds had been setting in bottles on shelves for years, if not decades, and might no longer have the chemical structures marked on the labels. This does not matter in a random approach because the whole enterprise is founded on chance. In any case, hits from these old materials would have to be followed up by identification of the active component(s) and fresh syntheses.

Looking at this brief history, we see that the pendulum was swinging toward rational design until a few years ago and now seems to have reversed course. Will rational design be abandoned in favor of the random approach for finding useful new bioactive compounds? Are rational design methods doomed to join slide rules and IBM punch cards in the technological scrap heap? Clearly, the answer to these questions is no, as evidenced by the work presented at the symposium on which these proceedings are based. Our goal in this brief space is to cite examples where rational approaches have been fruitful. We concentrate on compounds that have made it all the way to the pharmaceutical market. In addition, we give an example of a drug discovery project that had a rationale but lacked rationality, thereby dooming its prospect for success. Computer modeling detected the lack of rationality.

What Exactly Is Rational Design?

When most researchers hear the term "rational drug design" they probably think in terms of using considerations of the three-dimensional structure of the ligands and/or their receptor to discover medically useful compounds. However, this is not the whole story. Rational design is a collection of approaches that do not depend on chance alone to succeed.

Clearly, one facet of rational design includes the computational chemistry (*7*) techniques of (a) molecular modeling, (b) property prediction, such as by using quantum mechanics (*8*) or group additivity (*9*), (c) statistical modeling, such as using QSAR correlations involving molecular descriptors to predict bioactivities of new structures (*10*), and (d) other algorithms, such as docking methods based on energy minimization (*11*) or genetic algorithms (*12*). Rational design need not necessarily involve predicting binding affinities by molecular mechanics or free energy perturbation theory (*13*). Performing molecular dynamics simulations is not synonymous with rational drug design.

A second facet of rational design includes use of experimental and computational techniques for determining the three-dimensional structure of receptors. Thus, X-ray crystallography and multidimensional NMR spectroscopy are used to solve the structures of proteins, nucleic acids, and other biomacromolecules. Computing plays an important role here too because the experimental structures are now usually refined with the molecular dynamics technique of simulated annealing (*14,15*). Distance geometry, another computational chemistry technique (*16*), is used to transform NMR-derived distance constraints into three-dimensional atomic coordinates. When an experimental structure is unavailable, homology modeling is often used to build a tentative three-dimensional structure based on a known protein structure with a similar amino acid sequence.

Yet another important approach of rational design is using compound databases (libraries) to find interesting compounds. The two- or three-dimensional structures in these computer databases can be searched to retrieve, for instance, molecules that meet requirements for a particular pharmacophore, i.e., the minimal features of molecular structure that are essential for evoking an intended biological response. The libraries can also be analyzed by quantitative techniques to map the molecules in compound space based on some set of QSAR descriptors (*17*). By knowing how the compounds are distributed in this space, the diversity or similarity of the structures can be ascertained. Compound libraries, rather than individual compounds, can be "designed."

A long-standing approach to finding an enzyme inhibitor is to consider the biochemical mechanism of action of the enzyme; drugs that are transition state analogs may result (*18*). Another facet of rational design is to try to devise a ligand that will resemble a natural substrate. Thus, a peptidomimetic approach has been quite productive (*19*).

A relatively new facet of rational design is genomics. Here the goal is to find specific associations between gene products and disease states, thereby making it possible to design drugs that will block or stimulate those targets as appropriate. Handling all the genetic sequences has given rise to another computer-based subdiscipline called bioinformatics.

Thus, rational design encompasses many techniques and scientific disciplines including traditional medicinal chemistry (*20*). The main point of the term rational design is to differentiate its methods from blind screening, i.e., testing all compounds available or randomly selected ones. Unfortunately, the term "rational design" is upsetting to some traditional medicinal chemists because they think it implies that the research they have been doing was irrational. This implication is unintended. N.B., many traditional medicinal chemistry studies have been entirely rational; they were based on a sound rationale and were carefully planned to test some hypothesis. Finally, it should not be forgotten that educated guesswork, intuition, trial and error, and plain old-fashioned luck have played a role in both rational and random drug discovery in the past and will continue to do so in the future.

What Have You Done for Us Lately?

Because of the wide use of the term "computer-aided drug design," computational chemists are often challenged about whether their computer-based approaches have actually designed a drug. Shown in Figure 1 are pharmaceutical products for which computing played a vital role in their discovery. The earliest example of such a compound, to the author's knowledge, is the antibacterial agent norfloxacin. Structural modifications that led the chemists at Kyorin Pharmaceutical Company to this compound were guided with the assistance of QSAR (*21*). The compound has been on the market since 1983 under various brand names including Noroxin®.

Spurred by this advance, the 6-fluoro-quinolones became a major class of antibacterial agents.

Among drugs entering the market more recently is dorzolamide hydrochloride, which is sold under the brand name Trusopt® by Merck for treatment of glaucoma (*22*). Iterative SBLD and ab initio molecular orbital calculations helped yield this carbonic anhydrase inhibitor. Molecular modeling, QSAR, molecular shape analysis, and docking played a role in the discovery of donepezil hydrochloride, an acetylcholinesterase inhibitor (*23*). Eisai markets this compound as Aricept® for use in patients with Alzheimer's disease. Yet another example is losartan sodium. This is an angiotensin II receptor antagonist that was discovered at DuPont and has been sold by Merck under the brand name Cozaar® since 1995 for control of hypertension. Molecular modeling of the octapeptide angiotensin II and a lead compound from the patent literature have been repeatedly acknowledged (e.g., *24,25*) for helping set the original direction for the structure-activity relationship (SAR) that was pursued by the medicinal chemists to a successful conclusion. A drug for migraine, zolmitriptan, is a 5-HT_{1D} agonist; it was discovered at Wellcome and is marketed by Zeneca under the brand name Zomig®. Molecular modeling and the active analog approach helped define the pharmacophore (*26*).

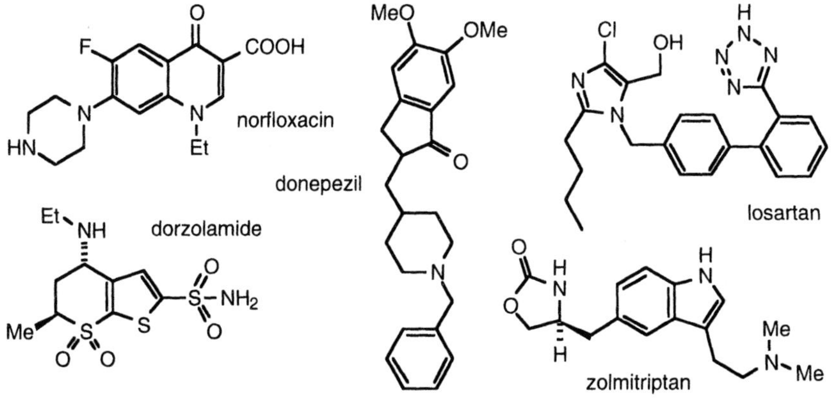

Figure 1. Chemical structures of compounds that reached the pharmaceutical market with the aid of rational approaches.

Iterative SBLD has been quite successful in predicting inhibitors of human immunodeficiency virus (HIV-1) protease. The receptor site of this enzyme is rather deep, and much crystallographic data are available. In Figure 2, some of the marketed HIV-1 protease inhibitors are shown. Indinavir sulfate was designed with the help of X-ray crystallography and molecular mechanics calculations (*27,28*). Merck began marketing of this pharmaceutical in 1996 under the brand name Crixivan®. A second product is nelfinavir mesylate. Discovered at Lilly and Agouron (*29,30*), the compound is being marketed as Viracept® by Agouron.

Another rational design approach yielded the HIV-1 protease inhibitor saquinavir. Considerations of the enzyme mechanism and the transition state of the substrate (*31*) led to a compound marketed by Roche first as Invirase® and more recently in a more potent dosage form as Fortovase®. A fourth inhibitor for treating AIDS is ritonavir, which was designed using a peptidomimetic strategy and is marketed by Abbott as Norvir® (*32,33*).

People outside of the pharmaceutical field sometimes ask if any drug has ever been designed "from the ground up" by computational methods. The answer is not yet, at least not to the author's knowledge. The question itself, while legitimate, may reveal a lack of familiarity with the difficulties of the drug discovery process, which almost always has been – and probably will continue to be – iterative and multidisciplinary. The medicinal chemist makes a compound, and the biologist and other scientists test it. This cycle is repeated many, many times because there are so many properties that a compound must exhibit (*34*) before it can be introduced into medical practice.

Perhaps one of the closest ways computational chemistry could come to designing a molecule from the ground up is with de novo methods (*35,36*). Although these methods for finding or designing ligands are becoming more powerful all the time, it would take serendipity if one of the hypothetical ligands, when synthesized, turned out to possess *all* the properties required to become a pharmaceutical product.

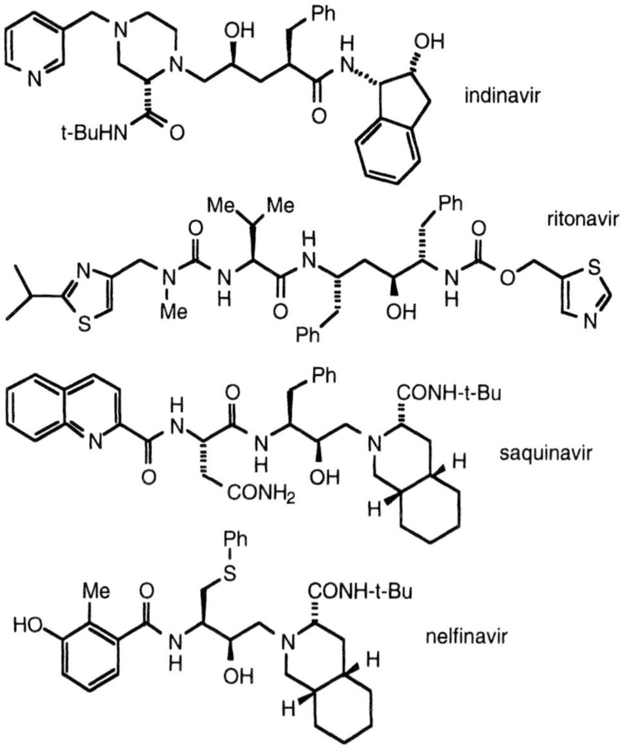

indinavir

ritonavir

saquinavir

nelfinavir

Figure 2. Chemical structures of some marketed HIV-1 protease inhibitors that were discovered with the help of rational approaches.

To keep an objective perspective of rational design, it is worth recalling that having good structural information for a target is no guarantee for success. Numerous investigators at pharmaceutical companies and elsewhere have applied SBLD for years to find new inhibitors that would block dihydrofolate reductase (DHFR), a target related to oncolytic and anti-infective therapies. None of those many rational structural and modeling studies yielded a new medicine (*37*). Drug discovery is so challenging that no method works in every case.

Detecting the Lack of Rationality in a Project

In this section, we present briefly an example of how the quantitative thinking of a computational chemist can detect that a medicinal chemistry project lacks rationality and therefore has little chance of success. This case study involved finding a drug for treatment of a central nervous system (CNS) disorder, whose exact nature is not germane to the principles we wish to illustrate. In some CNS projects, the scientists may have only a hypothesis about what effect a compound will have on a human brain. Will the compound treat depression? Be hallucinogenic? Be an anxiolytic? Improve cognitive scores? Reduce appetite? Bluntly put, the old paradigm was often to find a compound hypothesized to have some CNS effect and meeting safety requirements for clinical testing, to test it in humans, and then to see empirically what, if any, desirable pharmacological activities it may exhibit. After such a CNS compound was approved for marketing, additional uses for it, with luck, might show up when prescribed to a wider patient population.

The case study discussed here was not based on the old CNS paradigm. Rather, the specific disease for which the compound was being sought was known, and a well thought out approach was being followed to find an inhibitor of an enzyme hypothesized to be related to a pathology. Traditional medicinal chemistry was used, i.e., screening was used to find some low-level leads, and then a standard repertoire of chemistry explored the SAR around several leads. One-compound-at-a-time custom synthesis was used. The medicinal chemists decided what new compounds to make based on their many years of combined experience and on the activity (potency) data they were getting back from the biologists.

One of the first things that was checked when a computational chemistry perspective was brought to bear on this project was the biological data. There were two assays operative. A primary one was based on in vitro inhibition of the target enzyme. A secondary assay evaluated compounds in a whole cell system to detect levels of the peptide thought to be related to the enzyme's action. Comparing data from the two assays gave the type of scatter shown in Figure 3. It was immediately apparent that a fundamental flaw existed in the project's screening tactics. There was no salutary correlation between the data from the two assays! If both assays were associated with the same target, then the data points should have shown more of a relationship. The biologists running the assays had convinced themselves, based on their earlier work, that a strong correlation existed, but apparently no one had actually checked this thoroughly. A quantitative approach thus provided fresh insight, albeit unwelcome news.

Another factor besides potency that has to be considered in a drug discovery project is distribution of the compound to the site of action in the body of the patient. For a compound to be CNS-active, it must be able to cross the blood-brain barrier (BBB). There are some general ideas about what physical properties a compound must exhibit in order to be able to cross the BBB, but there are no hard and fast rules. One helpful clue, however, comes from QSAR investigators who showed that many drugs that are able to reach the CNS are lipophilic and have an ionizable group such that the compounds exhibit an octanol-water distribution coefficient ($\log D_{O/W}$) in the range 1.5 - 2.5 (*38*).

The compounds that had been synthesized and tested in this case study were checked to see if they met the $\log D_{O/W}$ criterion. The CLOGP program (*39,40*) was used to compute octanol-water partition coefficients, which for these compounds were the same as distribution coefficients because the compounds had no groups that would be ionized near physiological pH. The results are shown in Figure 4. As is immediately obvious, the vast majority of the compounds lack a sufficiently low distribution coefficient. Less than 5% of the compounds are near the desirable range.

Thus, most of the compounds are so lipophilic that their prospects of crossing the BBB and getting to their intended target is not good.

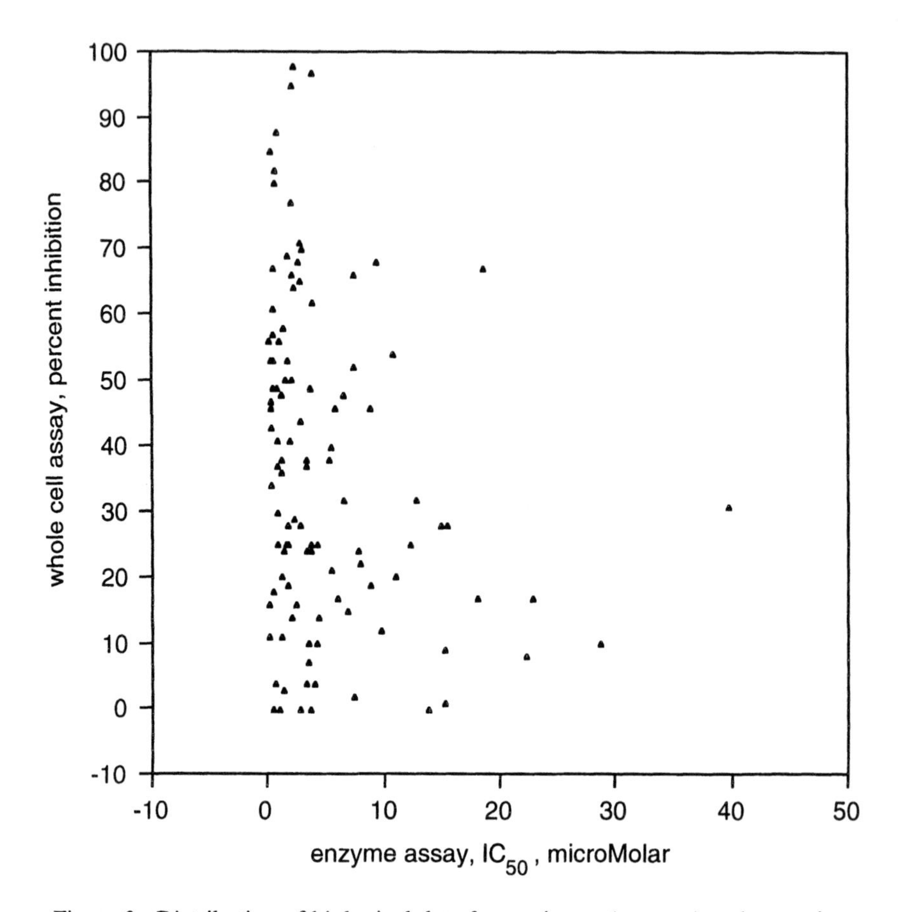

Figure 3. Distribution of biological data from primary (enzyme) and secondary (whole cell) assays used to evaluate compounds being synthesized. Data from the primary screen were reported as percent inhibition. Data from the secondary screen were reported as concentration of compound required for 50% inhibition of production of the supposed product of the enzyme.

This example shows how scientists can be so focused on their routine and their goal (primarily increasing potency) that they neglect to use simple quantitative tools at their disposal to learn if their approach is rational. If it is not rational, then discovery research resources (personnel) can be redeployed in directions with a greater chance of success in finding a drug. The point is that a computational perspective can empower scientists working on a project by supplying important information.

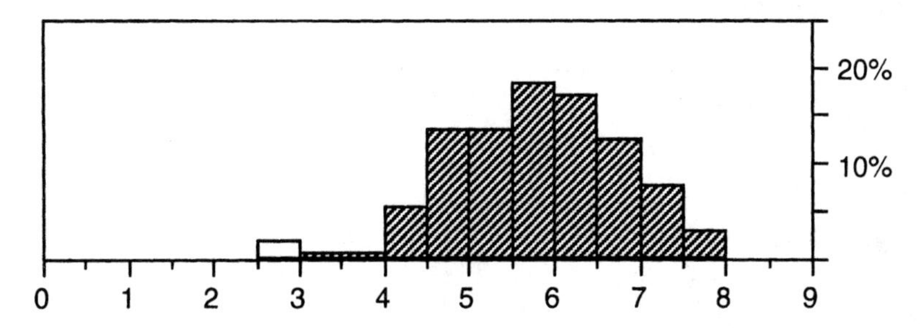

Figure 4. Histogram of the distribution of computed $\log D_{O/W}$ values for compounds being synthesized to explore the SAR. Less than 5% of the compound population had $\log D_{O/W}$ values less than 3 (unshaded bar). At the peak of the histogram, more than 15% of the compound population had $\log D_{O/W}$ between 5.5 and 6.0. Many compounds were even above 6, i.e., they were extremely lipophilic.

Outlook for Rational Design

Predicting the future is treacherous. Several years before the dawning of the age of combinatorial chemistry and HTS, the well-respected former editor of the *Journal of Medicinal Chemistry* wrote (*41*): "The time of random structural variations in molecular modification ... is past." It has not turned out like this. Combinatorial chemistry and HTS have carried the hit-or-miss approach to a technological pinnacle unimaginable a few years ago.

To put the combinatorial chemistry/HTS approach in perspective, it helps to use an analogy: trying to discover a drug is as difficult as looking for a needle in a haystack. With combinatorial chemistry and HTS, the philosophy is to create a bigger "haystack" in hopes that a useful compound will be somewhere in it. In contrast, the philosophy of rational design is that the odds of finding a useful compound can be beat, and the goal is to reduce the size of the haystack!

It is not necessary to ask whether one philosophy is better; clearly both random and rational approaches can and should be used. As mentioned, combinatorial chemistry need not be a completely random enterprise; compound libraries can be rationally designed using QSAR techniques to produce structures in specific volumes of multidimensional molecular descriptor space so as to fill unexplored "voids" in an SAR or in a region around a lead compound. Because of the quantity of data that has to be tracked, stored, retrieved, and analyzed, information management underlies the modern random approach. So with either a random or rational approach, computing is vital. In fact, the term "computer-aided drug design" could take on a much broader meaning.

Rational approaches, many of which are based on modern computer technology, are helping to stimulate ideas and test hypotheses for new ligands. Computational and structural chemists can now point to specific examples where their approaches have proved useful in helping medicines reach patients. This chapter and these proceedings confirm that rational design is contributing not only good science, but also pharmaceutically interesting compounds. The answer to the question posed in the title of this chapter is clearly *yes*, rational design *is* good for finding those elusive needles in the haystack. Looking toward the future, we see an increasing proportion of the newly approved pharmaceuticals being discovered with the aid of rational design approaches.

Acknowledgments

The author thanks Dr. M. Rami Reddy, Dr. Abby L. Parrill, and the sponsors of the ACS symposium for the opportunity to share these results. The author is grateful to Max M. Marsh (Indiana University) for his invaluable advice. Communications to the author can be sent via the Internet (boyd@chem.iupui.edu) or World Wide Web (http://chem.iupui.edu/).

References

1. Balbes, L. M.; Mascarella, S. W.; Boyd, D. B. In *Reviews in Computational Chemistry*, Lipkowitz, K. B.; Boyd, D. B., Eds.; VCH Publishers: New York, NY, 1994, Vol. 5; pp 337-379.
2. Boyd D. B. In *Encyclopedia of Computer Science and Technology*; Kent, A.; Williams, J. G., Eds.; Marcel Dekker: New York, 1995, Vol. 33 (Suppl. 18); pp 41-71.
3. Boyd, D. B. In *Ullmann's Encyclopedia of Industrial Chemistry*, Wiley-VCH, Weinheim, 1998, 6th Edition; in press.
4. Appelt, K.; Bacquet, R. J.; Bartlett, C. A.; Booth, C. L. J.; Freer, S. T.; Fuhry, M. A. M.; Gehring, M. R.; Herrmann, S. M.; Howland, E. F.; Janson, C. A.; Jones, T. R.; Kan, C.-C.; Kathardekar, V.; Lewis, K. K.; Marzoni, G. P.; Matthews, D. A.; Mohr, C.; Moomaw, E. W.; Morse, C. A.; Oatley, S. J.; Ogden, R. C.; Reddy, M. R.; Reich, S. H.; Schoettlin, W. S.; Smith, W. W.; Varney, M. D.; Villafranca, J. E.; Ward, R. W.; Webber, S.; Webber, S. E.; Welsh, K. M.; White, J. *J. Med. Chem.* **1991**, *34*, 1925-1934.
5. Boyd, D. B.; Milosevich, S. A. F. *Perspect. Drug Discovery Des.* **1993**, *1*, 345-358.
6. Gordon, E. M.; Barrett, R. W.; Dower, W. J.; Fodor, S. P. A.; Gallop, M. A. *J. Med. Chem.* **1994**, *37*, 1385-1401.
7. Lipkowitz, K. B.; Boyd, D. B., Eds. *Reviews in Computational Chemistry*; VCH Publishers: New York, 1990, Vol. 1; pp vii-xii.
8. Boyd, D. B. *Drug Inf. J.* **1983**, *17*, 121-131.
9. Carrupt, P.-A.; Testa, B.; Gaillard, P. In *Reviews in Computational Chemistry*; Lipkowitz, K. B.; Boyd, D. B., Eds.; Wiley-VCH: New York, 1997, Vol. 11; pp 241-315.
10. Plummer, E. L. In *Reviews in Computational Chemistry*; Lipkowitz, K. B.; Boyd, D. B., Eds.; VCH Publishers: New York, 1990, Vol. 1; pp 119-168
11. Schlick, T. In *Reviews in Computational Chemistry*; Lipkowitz, K. B.; Boyd, D. B., Eds.; VCH Publishers: New York, 1992, Vol. 3; pp 1-71.
12. Judson, R. In *Reviews in Computational Chemistry*; Lipkowitz, K. B.; Boyd, D. B., Eds.; VCH Publishers: New York, 1997, Vol. 10; pp 1-73.
13. Straatsma, T. P. In *Reviews in Computational Chemistry*; Lipkowitz, K. B.; Boyd, D. B., Eds.; VCH Publishers: New York, 1996, Vol. 9, pp 81-127.
14. Lybrand, T. P. In *Reviews in Computational Chemistry*; Lipkowitz, K. B.; Boyd, D. B., Eds.; VCH Publishers: New York, 1990, Vol. 1; pp 295-320.

15. Torda, A. E.; van Gunsteren, W. F. In *Reviews in Computational Chemistry*; Lipkowitz, K. B.; Boyd, D. B., Eds.; VCH Publishers: New York, 1992, Vol. 3; pp 143-172.
16. Blaney, J. M.; Dixon, J. S. In *Reviews in Computational Chemistry*; Lipkowitz, K. B.; Boyd, D. B., Eds.; VCH Publishers: New York, 1994, Vol. 5, pp 299-335.
17. Martin, E. J.; Spellmeyer, D. C.; Critchlow, R. E., Jr., Blaney, J. M. In *Reviews in Computational Chemistry*; Lipkowitz, K. B.; Boyd, D. B., Eds.; VCH Publishers: New York, 1997, Vol. 10; pp 75-100.
18. Kalman, T. I., Ed. *Drug Action and Design: Mechanism-Based Enzyme Inhibitors*; Elsevier North Holland: New York, 1979.
19. Damewood, J. R., Jr. In *Reviews in Computational Chemistry*; Lipkowitz, K. B.; Boyd, D. B., Eds.; VCH Publishers: New York, 1996, Vol. 9; pp 1-79.
20. Boyd, D. B. In *Rational Molecular Design in Drug Research*, Proceedings of the Alfred Benzon Symposium No. 42; Liljefors, T.; Jørgensen, F. S.; Krogsgaard-Larsen, P., Eds.; Munksgaard: Copenhagen, 1998; pp 15-23.
21. Koga, H.; Itoh, A.; Murayama, S.; Suzue, S.; Irikura, T. *J. Med. Chem.* **1980**, *23*, 1358-1363.
22. Greer, J.; Erickson, J. W.; Baldwin, J. J.; Varney, M. D. *J. Med. Chem.* **1994**, *37*, 1035-1054.
23. Kawakami, Y.; Inoue, A.; Kawai, T.; Wakita, M.; Sugimoto, H.; Hopfinger, A. J. *Bioorg. Med. Chem.* **1996**, *4*, 1429-1446.
24. Duncia, J. V.; Chiu, A. T.; Carini, D. J.; Gregory, G. B.; Johnson, A. L.; Price, W. A.; Wells, G. J.; Wong, P. C.; Calabrese, J. C.; Timmermans, P. B. M. W. M. *J. Med. Chem.* **1990**, *33*, 1312-1329.
25. Duncia, J. V.; Carini, D. J.; Chiu, A. T.; Johnson, A. L.; Price, W. A.; Wong, P. C.; Wexler, R. R.; Timmermans, P. B. M. W. M. *Med. Res. Rev.* **1992**, *12*, 149-191.
26. Glen, R. C.; Martin, G. R.; Hill, A. P.; Hyde, R. M.; Woollard, P. M.; Salmon, J. A.; Buckingham, J.; Robertson, A. D. *J. Med. Chem.* **1995**, *38*, 3566-3580.
27. Dorsey, B. D.; Levin, R. B.; McDaniel, S. L.; Vacca, J. P.; Guare, J. P.; Darke, P. L.; Zugay, J. A.; Emini, E. A.; Schleif, W. A.; Quintero, J. C.; Lin, J. H.; Chen, I. W.; Holloway, M. K.; Fitzgerald, P. M. D.; Axel, M. G.; Ostovic, D.; Anderson, P. S.; Huff, J. R. *J. Med. Chem.* **1994**, *37*, 3443-3451.
28. Holloway, M. K.; Wai, J. M. In *Computer-Aided Molecular Design: Applications in Agrochemicals, Materials, and Pharmaceuticals*; Reynolds, C. H.; Holloway, M. K.; Cox, H. K., Eds.; ACS Symp. Series 589; American Chemical Society: Washington, DC, 1995; pp 36-50.
29. Patick, A. K.; Mo, H.; Markowitz, M.; Appelt, K.; Wu, B.; Musick, L.; Kalish, V.; Kaldor, S.; Reich, S.; Ho, D.; Webber, S. *Antimicrob. Agents Chemother.* **1996**, *40*, 292-297.
30. Kaldor, S. W.; Kalish, V. J.; Davies, J. F., II; Shetty, B. V.; Fritz, J. E.; Appelt, K.; Burgess, J. A.; Campanale, K. M.; Chirgadze, N. Y.; Clawson, D. K.; Dressman, B. A.; Hatch, S. D.; Khalil, D. A.; Kosa, M. B.; Lubbehusen,

P. P.; Muesing, M. A.; Patick, A. K.; Reich, S. H.; Su, K. S.; Tatlock, J. H. *J. Med. Chem.* **1997**, *40*, 3979-3985.

31. Roberts, N. A.; Martin, J. A.; Kinchington, D.; Broadhurst, A. V.; Craig, J. C.; Duncan, I. B.; Galpin, S. A.; Handa, B. K.; Kay, J.; Kröhn, A.; Lambert, R. W.; Merrett, J. H.; Mills, J. S.; Parkes, K. E. B.; Redshaw, S.; Ritchie, A. J.; Taylor, D. L.; Thomas, G. J.; Machin, P. J. *Science* **1990**, *248*, 358-361.
32. Kempf, D. J.; Marsh, K. C.; Denissen, J. F.; McDonald, E.; Vasavanonda, S.; Flentge, C. A.; Green, B. E.; Fino, L.; Park, C. H.; Kong, X. P.; Wideburg, N. E.; Saldivar, A.; Ruiz, L.; Kati, W. M.; Sham, H. L.; Robins, T.; Stewart, K. D.; Hsu, A.; Plattner, J. J.; Leonard, J. M.; Norbeck, D. W. *Proc. Natl. Acad. Sci. U. S. A.* **1995**, *92*, 2484-2488.
33. Kempf, D. J.; Sham, H. L.; Marsh, K. C.; Flentge, C. A.; Betebenner, D.; Green, B. E.; McDonald, E.; Vasavanonda, S.; Saldivar, A.; Wideberg, N. E.; Kati, W. M.; Ruiz, L.; Zhao, C.; Fino, L.; Patterson, J.; Molla, A.; Plattner, J. J.; Norbeck, D. W. *J. Med. Chem.* **1998**, *41*, 602-617.
34. Lipkowitz, K. B.; Boyd, D. B., Eds. *Reviews in Computational Chemistry*; Wiley-VCH: New York, 1997; Vol. 11; pp v-x.
35. Murcko, M. A. In *Reviews in Computational Chemistry*; Lipkowitz, K. B.; Boyd, D. B., Eds.; Wiley-VCH: New York, 1997, Vol. 11; pp 1-66.
36. Clark, D. E.; Murray, C. W.; Li, J. In *Reviews in Computational Chemistry*; Lipkowitz, K. B.; Boyd, D. B., Eds.; Wiley-VCH: New York, 1997, Vol. 11; pp 67-125.
37. Everett, A. J. In *Topics in Medicinal Chemistry*, Leeming, P. R., Ed.; Proc. 4th SCI-RSC Med. Chem. Symp., Cambridge, U.K., Sept. 6-9, 1987; R. Soc. Chem., London, 1988; pp 314-331.
38. Hansch, C.; Bjoerkroth, J. P.; Leo, A. *J. Pharm. Sci.* **1987**, *76*, 663-687.
39. Hansch, C.; Leo, A. *Exploring QSAR: Fundamentals and Applications in Chemistry and Biology*, American Chemical Society: Washington, DC, 1995.
40. DayMenus Software Manual, Version 3.63, Daylight Chemical Information Systems, Inc.: Irvine, CA, 1991.
41. Burger, A. *Progr. Drug Res.* **1991**, *37*, 287-371.

Author Index

Subject Index

C

E

F

H

I

K

L

M

Q

R